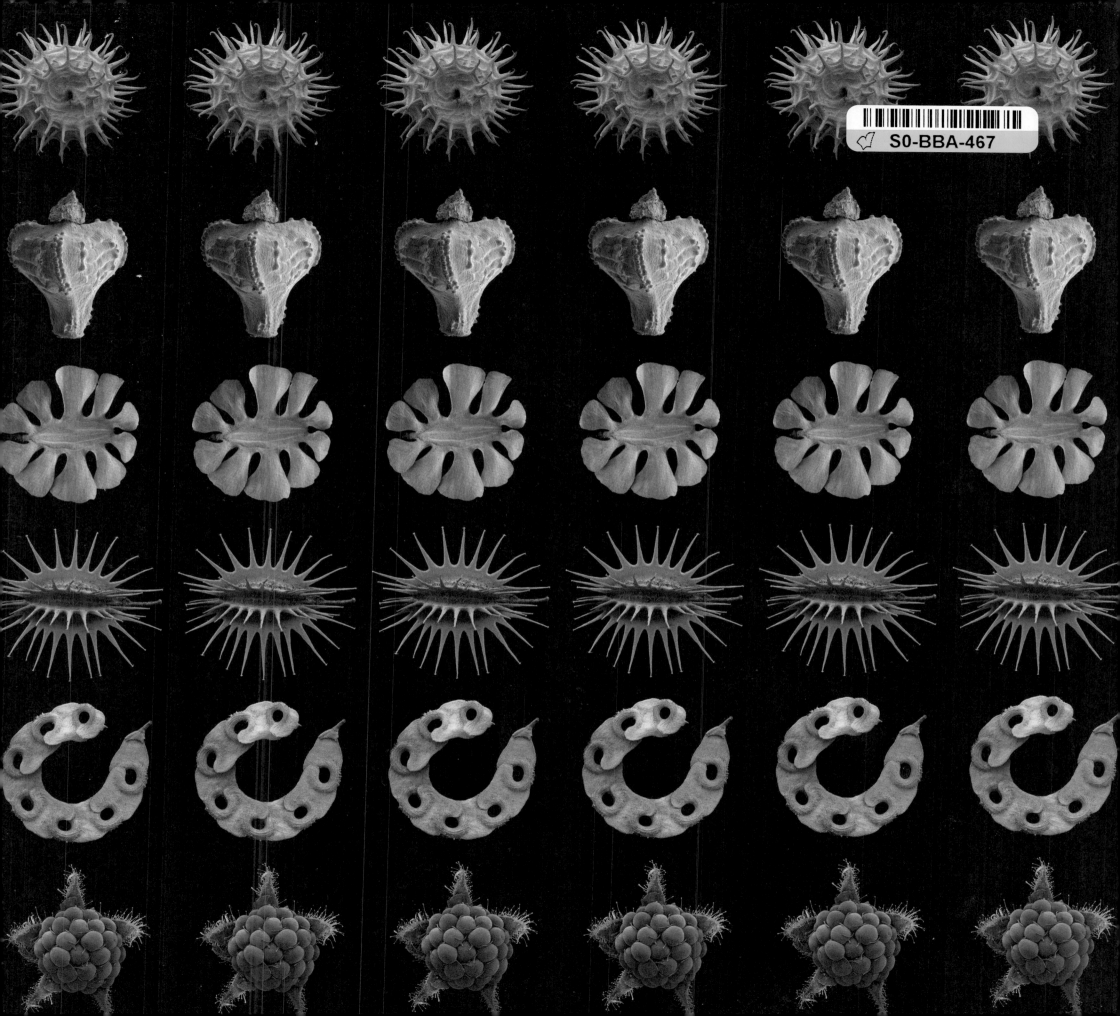

FRUIT

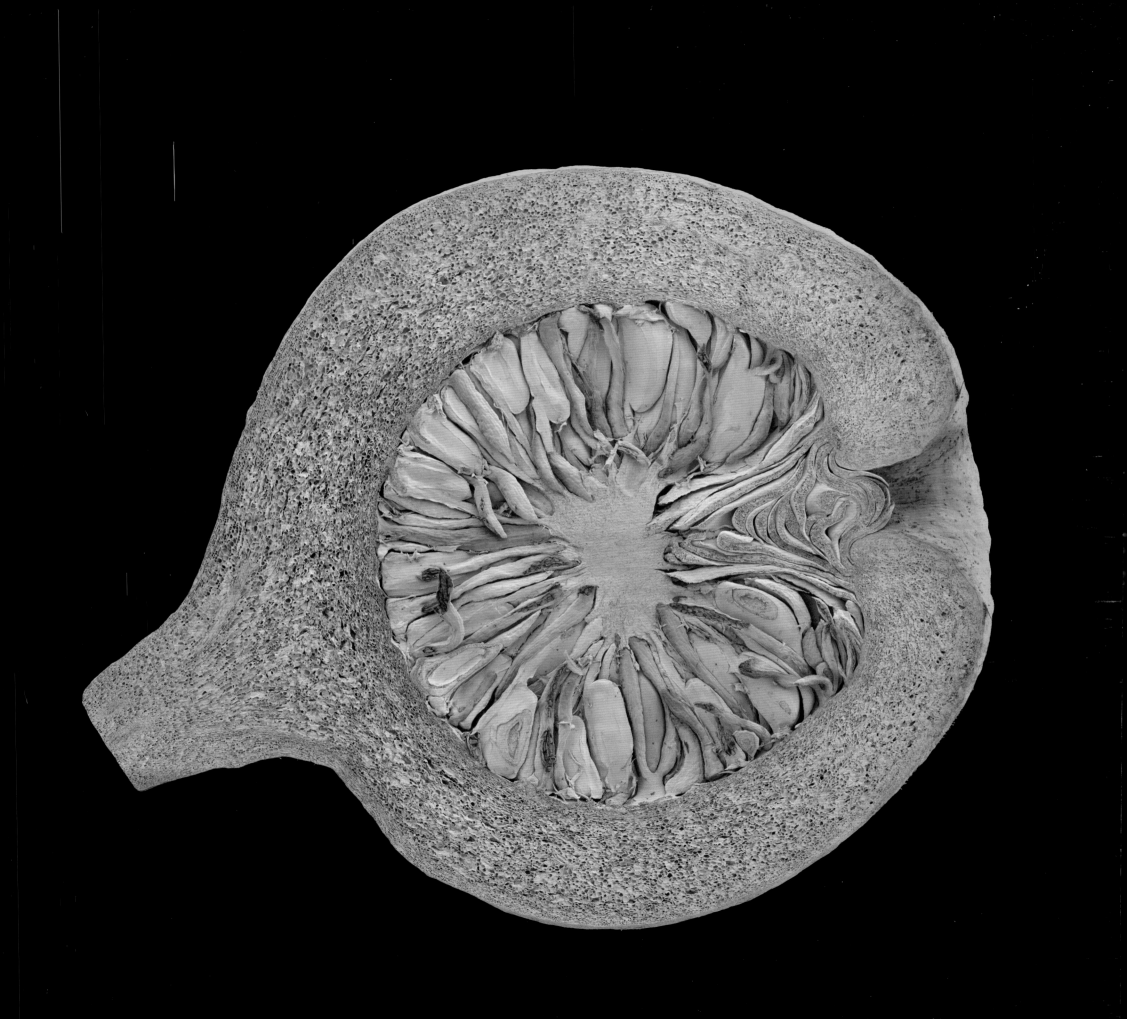

FRUIT

EDIBLE, INEDIBLE, INCREDIBLE

Wolfgang Stuppy & Rob Kesseler

Edited by Alexandra Papadakis

FIREFLY BOOKS

A Firefly Book

Published by Firefly Books Ltd. 2008

First printing

Publisher Cataloging-in-Publication Data (U.S.)

Stuppy, Wolfgang.
    Fruit : edible, inedible, incredible / Wolfgang Stuppy and Rob Kesseler ; edited by Alexandra Papadakis.
[264] p. : 283 col. photos. ; 30 x 28 cm.
Includes bibliographical references and index.
ISBN-13: 978-1-55407-405-1
ISBN-10: 1-55407-405-3
1. Fruit — Pictorial works. 2. Fruit — Seeds — Pictorial works. 3. Fruit. 4. Fruit — Seeds. I. Kesseler, Rob.
II. Papadakis, Alexandra. III. Title.
581.4/640282   dc22   QK660.S887 2008

Library and Archives Canada Cataloguing in Publication

Stuppy, Wolfgang
    Fruit : edible, inedible, incredible / Wolfgang Stuppy and Rob
Kesseler ; edited by Alexandra Papadakis.
Includes bibliographical references.
ISBN 978-1-55407-405-1
1. Fruit — Pictorial works. 2. Fruit — Seeds — Pictorial works. 3. Fruit.
4. Fruit — Seeds. 5. Fruit — Microscopy. I. Papadakis, Alexandra II. Kesseler, Rob III. Title.
QK660.S96 2008   581.4'640282   C2008-901528-2

Published in the United States by
Firefly Books (U.S.) Inc.
P.O. Box 1338, Ellicott Station
Buffalo, New York 14205

Published in Canada by
Firefly Books Ltd.
66 Leek Crescent
Richmond Hill, Ontario L4B 1H1

Developed by
Papadakis Publisher
A member of New Architecture Group Ltd.
Kimber, Winterbourne, Berkshire, RG20 8AN, U.K.
www.papadakis.net

Editorial and design director: Alexandra Papadakis
Editor: Sheila de Vallée
Design Assistant: Hayley Williams

Cover illustrations: (front) *Fragaria* x *ananassa* (Rosaceae) – garden strawberry; only known in cultivation – fruit (glandetum); diameter 1.5cm; (back) *Anagallis arvensis* (Myrsinaceae) – scarlet pimpernel; native to Europe – fruit (pyxidium) with seeds

Published in collaboration with the Royal Botanic Gardens, Kew

PLANTS PEOPLE
POSSIBILITIES

Printed and bound in China

For Emma, with love

Wolfgang Stuppy, Kew & Wakehurst Place, March 2008

In the spirit of *Fruit* I dedicate this book to all my Godchildren: Michael Berlinger, Alfred Chubb, Alice Chubb, Katie Nabbs, Dimitri Roulleau-Gallais, Elenaki Manessi, Laura Stevenson

Rob Kesseler, London, March 2008

Picture Captions

page 1: *Calamus aruensis* (Arecaceae) – rattan palm; native to New Guinea, the Solomon Islands, Aru Islands and the ti of Cape York (Australia) – immature fruits. Unique within the palm family (Arecaceae), the fruits of rattan palms (subfamily Calamoideae) are covered with reflexed over-lapping scales. The scales, whose precise function is unknown are arranged in neat vertical rows that create a pattern resembling reptile skin. One rattan species, *Salacca zalacca* from Malesia, was named 'snake palm' after the surface pattern of its edible fruits; diameter of one fruit 2.6mm

page 2: *Ficus villosa* (Moraceae) – villous fig; native to tropical Asia – longitudinal section of fruit (syconium). The approximately 750 members of the genus *Ficus* (figs) bear their tiny flowers inside a peculiar inflorescence called a *syconium*, which, after pollination, matures into a fruit tha we commonly refer to as a fig. Morphologically a syconium can be compared to a sunflower head curving up its margi to form an urn leaving just a small opening (ostiole) at the top. The entrance of the fig cavity is closed by numerous tightly packed bracts, which, at the time of pollination, giv way to a narrow passage through which the fig's pollinators – tiny fig wasps of the family Agaonidae – can enter the flower-lined cavity. In *Ficus villosa* and other species, the cavity of the syconium is filled with a mucilagi-nous fluid prior to pollination; diameter of fruit 1.2cm

page 3: *Ficus villosa* (Moraceae) – villous fig; native to tropical Asia – cluster of fruits (syconia)

page 5: *Valerianella coronata* (Valerianaceae) – no commo name (literally "crowned lamb's lettuce"); native to the Mediterranean and south-west and central Asia – fruit (pseudosamara); the inferior ovary is formed by three united carpels but only one of them bears a seed, which is why one half of the bottom part of the fruit (the sterile locule pair) is smaller than the other. The enlarged parachute-like calyx with the tips of the six united sepals drawn out into hooked spines assists both wind and animal dispersal; diameter 5.2mm

CONTENTS

8   **Preface** *by Ken Arnold*

12   **Foreword** *by Stephen D. Hopper*

14   **Fruit – Edible, Inedible, Incredible**

18   **What is a Fruit**

21   **What is a fruit and what is a vegetable?**

22   **Angiosperms, Gymnosperms** and those that copulate in secret

25   The naked-seeded ones

30   The non-naked-seeded ones

34   An abominable mystery

36   Angiosperm extremists

38   **No Flower, no Fruit?**

38   Is a pine cone a fruit?

42   **No Carpel, no Fruit?**

46   A shameless display

48   Not quite the ovary of Eve

49   Unwitting couriers

49   Wind, sex and gender separation

50   **What's in a Fruit?**

52   Babylonian confusion

55   Enhanced female performance

57   How to be a carpologist

59   The true meaning of fruits

59   **Simple Fruits**

60   The truth about berries

62   The miraculous miracle berry

65   Golden apples

65   Fragrant citrons

66   Buddha's hand

66   Sizeable pepos

69   Soft shell, hard core *or* how to be a drupe

72   Nuts about nuts

73   Walnuts or waldrupes?

73   Glans quercus

74   Two fruits in one – cashew nut and cashew apple

77   Wheat "grain" and sunflower "seed" – caryopsis and achene

78   Samaras – nuts gone airborne

82   Cypselas – achenes gone airborne

86   Pods and such like

86   Capsules or seven ways to open a fruit

87   Teeth, fissures, cracks and lids

90   Follicle and coccum

91   Pods as in "pea pods"

92   Sweet bean pods

95   The World's largest bean pod

96   Seeds in prison

98   Inside-out drupes

98   To be or not to be a drupe

99   **Multiple Fruits – Several fruitlets from a single flower?**

103 **Schizocarpic Fruits** *or* how to emulate the multiple experience

113 **Anthocarpous Fruits – the carpologists' touchstone**

116 **Compound Fruits – A single fruit from several flowers?**

121   The breadfruit and the Mutiny on the Bounty

123   The largest fruit a tree can bear

124   Figs, gnats and sycophants

131   Angiosperms with cones?

134 **Carpological Troublemakers**

137   Bogus fruits and how to debunk them

138 **So what *is* a Fruit?**

140   The biological function of fruits and seeds

140 **Dispersal – the many ways to get around**

143   Wind dispersal

146   Wings

146   Monoplanes

149   Flying discs

149   Spinning cylinders

149   Shuttlecocks

52 Woolly travellers
52 Love-in-a-puff and other balloon travellers
53 Anemoballism
53 Water dispersal
57 Dispersal by raindrops
57 Plants that do it for themselves
57 Hygroscopic tension
162 Hydraulic pressure
162 **Animal Dispersal**
63 Becoming attached
66 The story of the sadistic *Tribulus*
70 In the claws of the devil
170 How to catch a bird
73 Dispersal by scatter-hoarders
76 Dispersal by ants
80 **Combining Strategies**
84 **Directed Dispersal**
84 **Fleshy Fruits**
87 The evolution of fleshy fruits
91 The good, the bad and the ugly, or why fruits are poisonous
96 Enough is as good as a feast
97 Young and dangerous
99 Climacteric fruits
200 One bad apple spoils the barrel
200 Dispersal syndromes, the sign-language of fruits
202 The bird-dispersal syndrome
205 How to catch the eye of a bird
206 Fleshy seeds
208 Flashy seeds
210 Dangerous beauty
210 Colourful appendages
214 Arillate seeds and the fate of New York
215 Dispersal by mammals
217 The bat dispersal syndrome

221 Monkey fruits – the primate-dispersal syndrome
222 Monkey apple
222 The Queen of Fruits
222 Cacao – food of the gods
224 The baobab
224 Durian – the King of Fruits
226 A big fruit needs a big mouth – the megafaunal dispersal syndrome
228 Africa's large mammals and their fruits
229 Sausages that grow on trees
229 Fruits that only elephants like
230 When the elephants are gone
231 The aardvark and its cucumber
231 *Mallotus nudiflorus* and the Indian rhinoceros
232 The nitre bush and emus
232 Galápagos tomatoes and giant tortoises
233 More inseparable couples
233 Till death do us part
234 The dodo and the tambalocoque – a textbook fairy tale
236 Anachronistic fruits
238 Size no longer matters
239 The largest fruit of America
240 Osage orange
241 How can it be true?
244 Where have all the mammoths gone?
250 **The Millennium Seed Bank Project**
254 **Lusciousness – The crafted image in a digital environment**
256 **Appendices**
258 Glossary
261 Bibliography
262 Index of Plants illustrated
264 Footnotes
264 Picture Credits
264 Acknowledgments

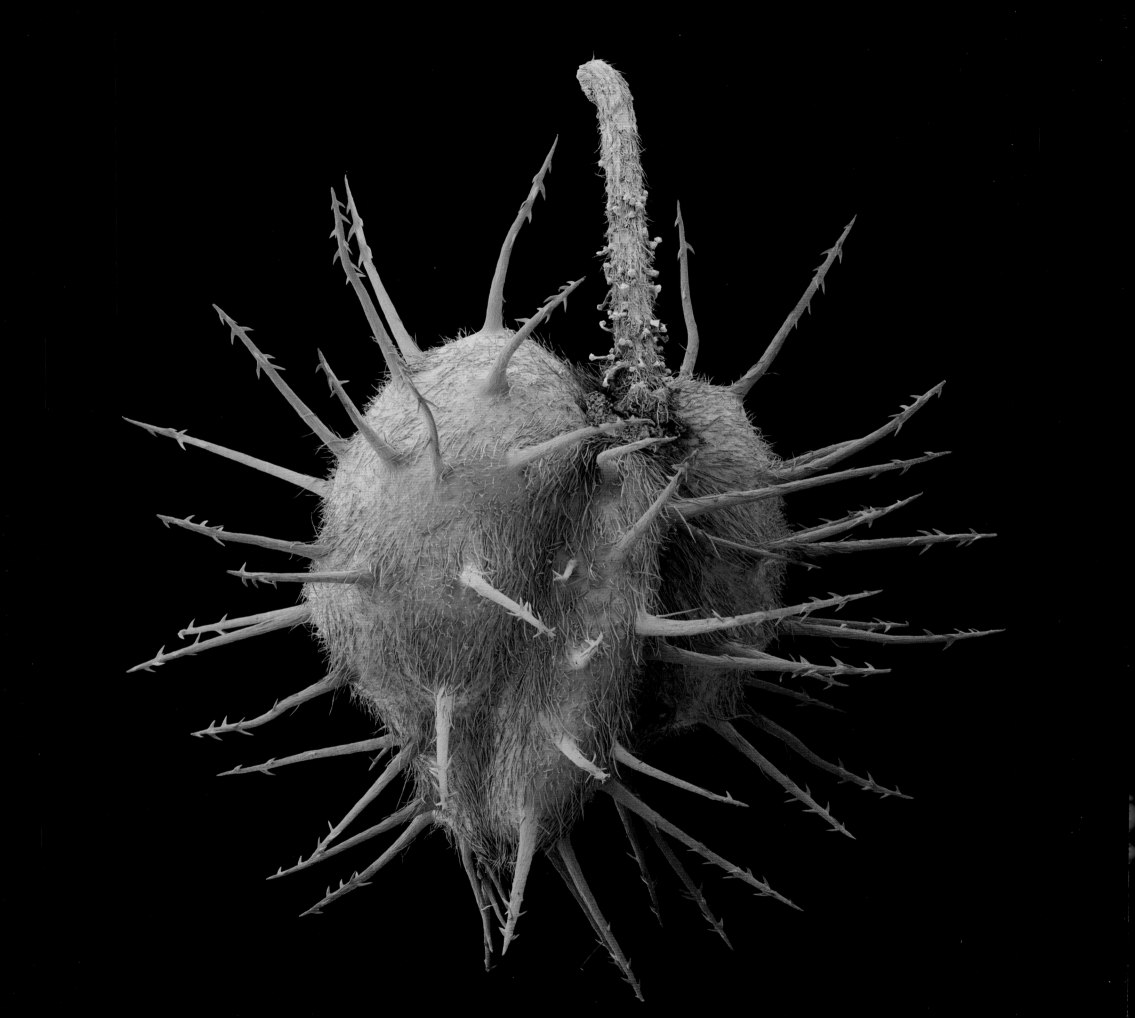

PREFACE

KEN ARNOLD

HEAD OF PUBLIC PROGRAMMES, THE WELLCOME TRUST

Krameria erecta (Krameriaceae) – Pima rhatany; native to the southern USA and northern Mexico – fruit (achene); the barbed spines covering the single-seeded indehiscent fruit of this small shrub are a clear adaptation to achieve dispersal by attachment to the fur of passing animals (epizoochory); the fruit (without spines) is 8mm long

As a foodstuff, a topic of scientific investigation, a metaphor for sin or hard work, and much else, fruits occupy an especially fecund place in our material and psychological responses to the living world. There is a lot to them and they matter much to us. Too much, one might argue, for one perspective alone to satisfy our curiosity. This book, drawing equally on the industry and insight of both science and art, allows us to think through what fruits mean to us in the most magisterial fashion imaginable.

The quest to understand fruits, and indeed plants in general, has involved picturing them time and again; though attempts to do so based on disciplined observations only really began in earnest during the Renaissance. This was when the natural historical project of capturing and contemplating the material world through a visual science took off. The technical quest to survey and understand what was found in the animal, vegetable and mineral kingdoms, and indeed in 'artificial' as well as 'natural' realms, necessitated the invention and then development of a fleet of inquiries based on the art of describing what could be seen.

These investigations were, from the start, hybrid missions that harnessed a variety of practices and techniques, theoretical speculations and forms of analysis. They were also repeatedly influenced by technical innovations. The first major development came in the mid-17th century with the application of single lens and compound microscopes – tools that in many ways heralded the very invention of the idea of extending unaided observation through scientific instruments. Applied with artful audacity by the likes of Robert Hooke and Dutchman Anton van Leeuwenhoek, the microscope revealed whole new worlds of detail to be identified and explained beneath surfaces discernible to the naked eye.

Another major technical spur to enhancing the visual investigation of nature's productions arrived with the invention of photography, which promised a means of capturing what was beheld for both concerted study and celebration. And then, at the end of the 19th century, came a further revolution in looking – the application of x-rays through which the morphology of bodies (particularly live ones) could be investigated without killing and pulling them apart. The convergent integration of these milestones, along with numerous other innovations and breakthroughs, underpins the application of scanning electron microscopy that provided the raw material for the sumptuous images adorning this book.

Eryngium paniculatum (Apiaceae) – cardoncillo; native to Argentina and Chile – fruitlet. The fruits of the carrot family (Apiaceae) develop from an inferior ovary composed of two joined carpels. At maturity the two carpels separate into two individual single-seeded indehiscent fruitlets. As an adaptation to wind dispersal, the fruitlets of the cardoncillo possess a peripheral ring of wing-like outgrowths from the fruit wall (light blue) and carry 2 or 3 persistent sepals (dark blue) at the apex, which also act as wings; fruitlet 4.8mm long

Eryngium leavenworthii (Apiaceae) – Leavenworth's eryngo; native to North America – fruitlet. Unlike the cardoncillo, the fruitlets of Leavenworth's eryngo are adapted to animal dispersal. The persistent sepals are modified to act as claws that help attach the fruitlet to the fur or feathers of a passing animal; fruitlet 9mm long

Running in parallel to the ever more technically sophisticated means of examining and explaining the biology of fruit was another more purely aesthetic approach to the subject, in which nature was depicted, often frozen in still life, in order to provide a point of departure for contemplation and introspection. The work of both troops of investigators frequently crossed paths, with many an artist borrowing the newest, most penetrating scientific equipment for their own ends, re-applying its technical prowess to their concern with the human and moral values inherent in fruit and other natural products. Rob Kesseler and Wolfgang Stuppy's partnership is then by no means the first in which insights from different perspectives have shared pages of the same book, though rarely to such stunning effect.

What their collaborative endeavours uncover are the technical wonders of fruit, created almost atom by atom, pixel by pixel: specimens carefully chosen; angles of incidence determined; lines and surfaces sharpened or refocused; and colours selected and applied, enhanced or toned down, revealing their surface morphologies in high definition. Scientific understanding and technical manipulation were crucial to this process too, the measured practice of coding up the information adding meaning and symbolism to final works that gain their own life and character: by turns curious, attractive, enticing, mysterious, troubling and even ominous.

Say the word 'fruit,' and chances are we will first picture something juicy and filling. People have for millennia consumed literally thousands of their varieties, either cooked, dried, preserved, or just raw. Say it again, and we might instead ponder their biological function: a vehicle for genes packaged into seeds to be passed on to the next generation. Say it a third time, and we might now start thinking in metaphorical terms of our own physical efforts, or the realms of the forbidden – bits of knowledge at once delicious and dangerous.

So, while we ponder technical questions of scale, form, structure, variety, mechanism, and purpose, let's not forget that we are also leafing through a body of artwork that conjures up worlds within worlds, where beauty lurks within imperfection and where perfection can still dramatically reappear out of order.

This then is a book of learning and of sharing, but above all of looking at ravishing fruit for the eyes.

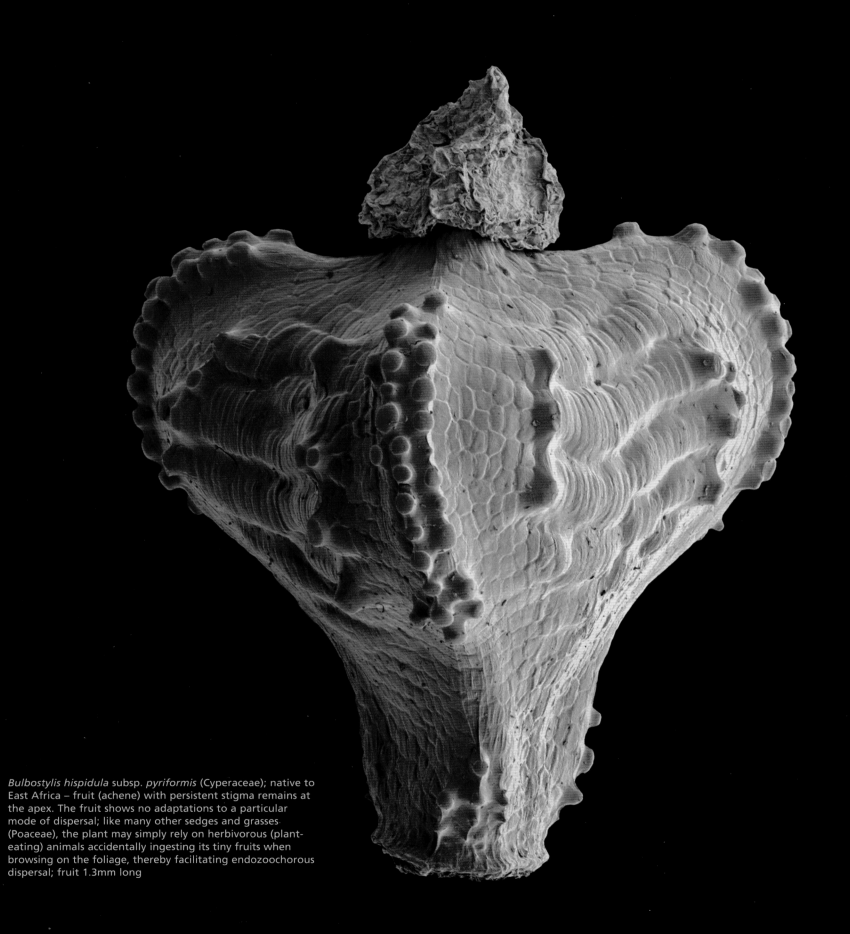

Bulbostylis hispidula subsp. *pyriformis* (Cyperaceae); native to East Africa – fruit (achene) with persistent stigma remains at the apex. The fruit shows no adaptations to a particular mode of dispersal; like many other sedges and grasses (Poaceae), the plant may simply rely on herbivorous (plant-eating) animals accidentally ingesting its tiny fruits when browsing on the foliage, thereby facilitating endozoochorous dispersal; fruit 1.3mm long

FOREWORD

PROFESSOR STEPHEN D. HOPPER FLS

DIRECTOR, THE ROYAL BOTANIC GARDENS, KEW

Like most people, I must confess at the outset to being an incorrigible frugivore. I have enjoyed consuming fruit since my earliest memories, and do so to this day. It is an honour, consequently, to have been invited by the authors of this book to write a few words about their collaborative and brilliant merger of the science and art of fruit. This is the third such book in an award-winning series celebrating the diversity of plant reproductive structures. Its predecessors were *Pollen – The Hidden Sexuality of Flowers*, by Rob Kesseler and Madeline Harley, published in 2004, and *Seeds – Time Capsules of Life* by the present team, published in 2006. I cannot think of a more fitting contribution to this series.

Apart from their obvious nutritional value, fruits offer an enthralling assemblage of insights, inspiration and wonderment. Rob Kesseler's imaginative images have captured such pleasures and made them available to a general readership. Complementarity is assured through Wolfgang Stuppy's lively text – authoritative but accessible. A powerful combination indeed.

We are told that there are more than 150 different technical fruit names coined by botanists over the past two centuries. This is heady stuff. Yet the book takes the reader along a path that unlocks the riches behind such dry nomenclature. I for one enjoyed reading every word, and learnt much more about fruits than I already knew. As the stories of evolution, biology and the use of fruits unfold, the book becomes a compelling read. There is a fertile field here from many points of view. I'm sure that no reader will regard a humble fruit in quite the same way once they have savoured what's in store herein.

This celebration of the beauty and intrinsic interest of fruits contains a significant and deeper message. Fruits are the containers of seeds, of new life on which all animals, including ourselves, are intimately dependent. Without fruits and their dispersal of seeds to safe sites, extinction, the death of birth, is inevitable. We cannot afford to let this happen, if for no other reason than self interest and our very survival. In a time when we travel down unprecedented pathways of climate change, caring for plants, the primary consumers of carbon, was never more important nor urgent. We must stop destroying plant life, and turn to ways that focus on nurturing and supporting green photosynthesizing organisms. We can only do so if plants continue to bear fruit, in all their amazing diversity. This book hopefully will encourage many to go beyond the aesthetic pleasure it so bountifully offers to helping plants and people survive into the future.

The Royal Botanic Gardens, Kew is proud to play its part in inspiring and delivering science-based plant conservation worldwide, enhancing the quality of life. Kew's Millennium Seed Bank, in particular, has engaged a hundred partner institutions in more than fifty countries in helping save plant life. Together, we all can contribute to such a pressing and important cause. It is a real pleasure to say that Kew remains a staunch partner in this publishing venture.

I congratulate the authors, publisher and all involved in this fine production.

Viva fruits – edible, inedible, incredible!

FRUIT

EDIBLE INEDIBLE INCREDIBLE

Calamus longipinna (Arecaceae) – rattan palm; native to New
Guinea and the Solomon Islands – detail of fruit surface.
Typical of rattan palms (subfamily Calamoideae), the fruit is
covered with reflexed overlapping scales that create a
pattern resembling reptile skin. Underneath the thin fruit
wall there is usually a seed covered in a thick, edible, fleshy
seed coat offering a reward for dispersing animals. Among
the natural dispersers of rattan fruits are the palm civet
(*Paradoxurus hermaphroditus*), siamang monkeys (*Hylobates
syndactylus*), the Torresian imperial-pigeon (*Ducula
spilorrhoa*) and cassowaries (*Casuarinus casuarinus*); scales of
the fruit illustrated: 1.6mm long

The word "fruit" conjures up a mouth-watering treat — crunchy apples, aromatic strawberries, juicy oranges. The well travelled among us may also recall the splendid cornucopia of tropical fruits that thrive in the warmer climates of our planet and have become increasingly prominent on supermarket shelves. There are some 2,500 edible tropical fruits worldwide but most are used only locally by indigenous people. Those that have gained commercial importance are some of the finest tasting fruits that nature has to offer, such as mango, durian and mangosteen.

Whether they are tropical, subtropical or temperate, we enjoy fruits in a great many ways, either fresh, dried, cooked or preserved, in yoghurts, ice creams, jams and biscuits, as juices or alcoholic drinks. Some serve as spices in the form of peppercorns, cardamom and chilli peppers. The most valuable of all, the fermented pods ("beans") of the vanilla orchid (*Vanilla planifolia,* Orchidaceae), are traded as a highly priced ingredient in chocolate, ice cream and many other sweet dishes. Others, such as the fruits of the West African oil palm (*Elaeis guineensis,* Arecaceae) and olive (*Olea europaea,* Oleaceae) are pressed for their valuable oils. Countless other fruits are important to humans as a source of natural raw materials such as fibres, dyes and medicine, or simply ornaments.

As wonderful a gift from nature as it may seem, providing us with delicacies and other useful commodities is not the reason that plants produce these glorious fruits. It is therefore legitimate to ask why fruits exist and why we are so attracted to them.

As this book will reveal, fruits are part of an elaborate plot. Their true nature is revealed by what is buried in their core: their seeds. Seeds are the most complex and precious organs plants ever produce, as it is the seeds that carry the next generation. They constitute the only means by which most plants travel, and thus bear the ultimate responsibility for the dispersal and reproductive success of the species. This key role played by fruits and seeds in the survival of a species explains the great variety of dispersal strategies that plants have developed during the course of evolution. The kind of strategy pursued — whether it involves wind, water, animals and humans or the plant's own explosive force — is reflected in a plethora of different colours, sizes and shapes — some of them edible, many of them inedible, and quite a number of them incredible!

Following our earlier book, *Seeds — Time Capsules of Life*, in which we revealed the astounding beauty of seeds, we now embark on an exploration of the natural history of one of nature's most *fruitful* inventions.

Vitis labrusca 'Isabella' (Vitaceae) – Isabella grape; only known in cultivation though locally naturalized in Europe – the Isabella grape is probably a hybrid between the northern fox grape (*Vitis labrusca*) from eastern North America and an unknown cultivar of grape vine (*Vitis vinifera*). Its fruits (berries) are eaten fresh and used to prepare juice and wine. During ripening, they change colour from green via yellow, pink and blue until they are finally dark blue-black with a heavy bloom. The colourful appearance of the illustrated cluster of grapes is the result of the asynchronous ripening of the individual berries

WHAT IS A FRUIT?

Calotis breviradiata (Asteraceae) – short-rayed burr daisy;
native to Australia – fruit (cypsela); the bristly hairs and
feathery pappus rays may primarily facilitate wind
dispersal (anemochory) and animal dispersal
(epizoochory); 2.8mm long

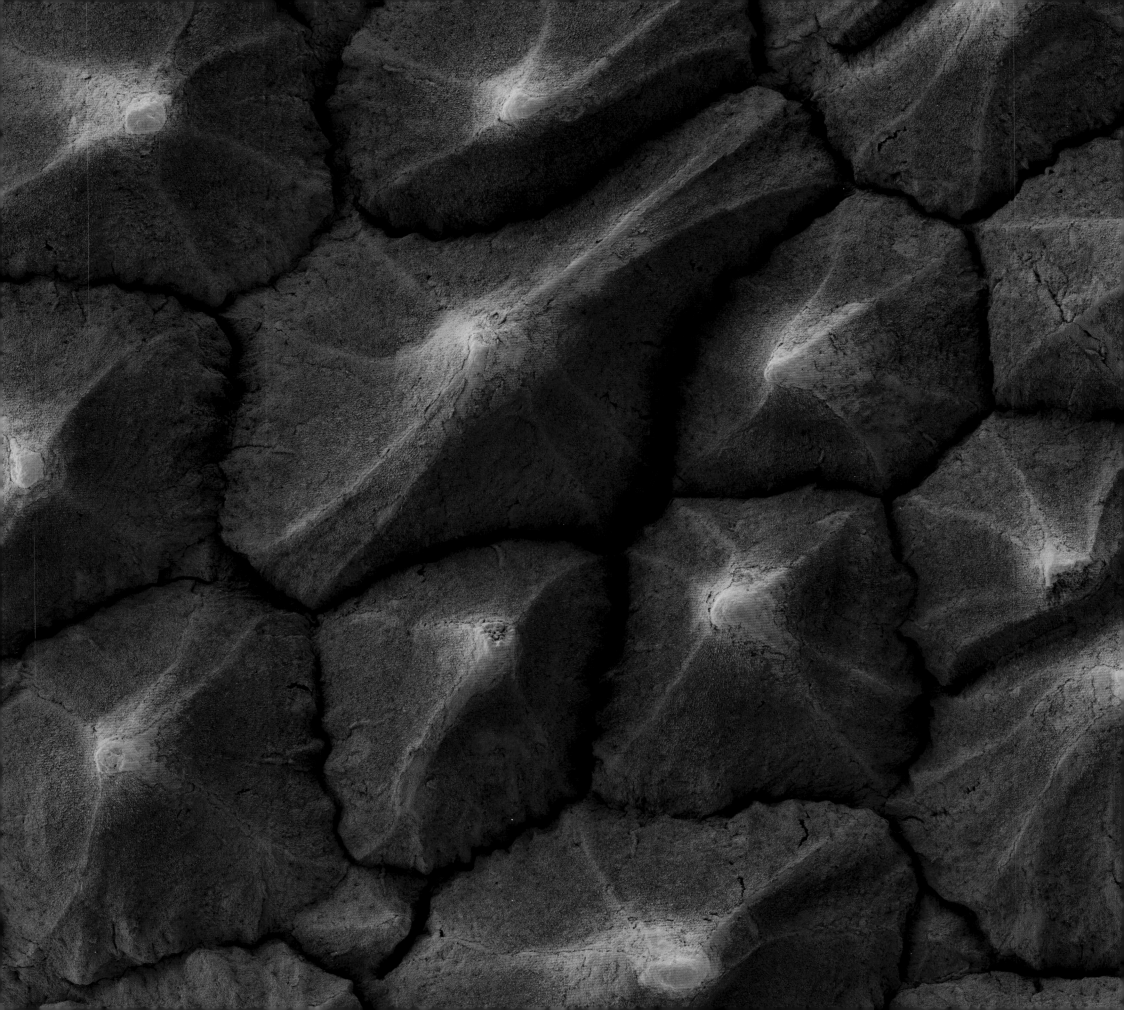

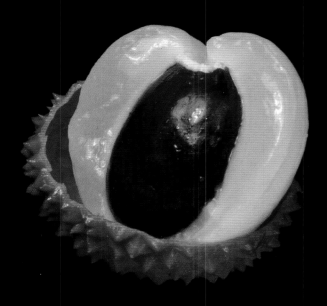

WHAT IS A FRUIT AND WHAT IS A VEGETABLE?

"What is a fruit?" may seem a trivial question but even when shopping for groceries, we unknowingly enter a world of contradictory and inconsistent concepts. We are in no doubt that apples, oranges and bananas are fruits. They offer everything we expect from a "proper" fruit, such as soft or crunchy, succulent flesh with a sweet taste that is best enjoyed raw and on its own. Professional (and even not so professional) fruit farmers are also conscious of the fact that in order for a plant to produce fruits, it has to flower first. Therefore, the golden rule of fruit farming is: no flowers, no fruits! But what about fruits without flowers? When shopping in a supermarket we may fancy a jar of jam made from carrots (the taproots of *Daucus carota* subsp. *sativus*, Apiaceae) or rhubarb (the leaf stalks of *Rheum* x *hybridum*, Polygonaceae). Provided the jar is correctly labelled as prescribed by the European Union Council Directive 2001/113/EC of 20 December 2001, somewhere on the label it will tell us about the percentage of *fruit content*.

The botanically aware reader may discern a degree of inconsistency here. After all, jam makers are no botanists and they usually neither know, nor need to know, the difference between a fruit and a root or leaf stalk. But there is an alternative interpretation. Returning to our weekly grocery shopping we will undoubtedly also choose edible plant parts that we confidently label as "vegetables."

Unlike fruits, vegetables offer a different – albeit no less delicious – culinary experience. With some exceptions, their taste is generally not sweet but savoury. Some of them, such as lettuce and radish, are best enjoyed raw but more commonly vegetables are used in cooking and require seasoning in order to enhance their otherwise somewhat uninteresting taste. Keen allotment gardeners will also know that many vegetables are not produced by flowers. Very often they consist of other parts of a plant such as leaves (lettuce, cabbage, spinach), leaf stalks (celery, rhubarb), stems (asparagus), roots (carrots, radish), underground tubers (potatoes, Jerusalem artichoke), bulbs (onions, garlic) or even young inflorescences (artichoke, broccoli, cauliflower). However, there are also plenty of examples of vegetables that develop from the ovaries of successfully pollinated flowers, such as cucumbers, squashes, pumpkins, green beans, sugar snaps and tomatoes. At the more exotic end of this kind of vegetable we find avocado, aubergine, bitter gourd and chayote. But are these really "proper" vegetables? If we recall our earlier definition of a fruit as expressed in the golden rule of fruit farming, should these vegetables not be labelled as fruits? After all, they develop from flowers and, what is more, they often contain seeds, just as fruits, and only fruits, do.

The classic question as to whether a tomato is a fruit or a vegetable has at times been highly contentious. Once, this seemingly inconsequential distinction held such great importance that it was even brought in front of the United States Supreme Court in the famous court case Nix *versus* Hedden (149 U.S. 304). The final verdict, proclaimed on 10 May 1893, ruled that a tomato is to be classified as a vegetable, at least in the sense of the Tariff Act of 3 March 1883, which at the time prescribed a tax for imported vegetables but not for fruit. Despite this authoritative decision, denying a tomato the status of a fruit reflects political rather than scientifically logical considerations.

To escape from this dilemma we can seek the objective opinion of scientists. Indeed, from a scientific point of view, "vegetable" is a culinary and not a scientific term. Its definition is subjective, arbitrary, and hence bound to be unclear. Otherwise, how would it be possible for greengrocers to classify mushrooms, which are not even plants, as vegetables? Obscured by such ambiguity, the word "vegetable" is entirely omitted from a botanist's scientific vocabulary. But even botanists are not free from conceptual worries when it comes to naming and classifying the various organs of plants. Their plight is not so much the distinction between fruit and vegetable. The botanists' problem is far more fundamental. However incredible it may seem, formulating a definition that describes *precisely* what a fruit is in a botanical-morphological sense, and deciding which plants should be entitled to have their mature reproductive organs addressed as such, has troubled the minds of botanists for centuries. In order to understand the root of the dilemma we need to know more about the different kinds of plants that surround us.

ANGIOSPERMS, GYMNOSPERMS AND THOSE THAT COPULATE IN SECRET

Living seed-bearing plants or *spermatophytes* fall within two categories: *gymnosperms,* including cycads, *Ginkgo*, conifers and the Gnetales; and *angiosperms*, better known as *flowering plants*. In the evolutionary hierarchy of the plants of our planet the spermatophytes are the most advanced. Their level of organization reaches far beyond their much more primitive[1], spore-producing relatives, the *cryptogams*. The latter comprise the algae, mosses (including liverworts and hornworts), clubmosses, horsetails and ferns. Reproducing through spores rather than seeds, cryptogams generally lack obvious sexual reproductive organs such as cones or flowers and fruits, which is why they are named as they are: *cryptogams* is derived from Greek and means "those who copulate in secret" (*kryptos* = hidden + *gamein* = to marry, to copulate). Their method of sexual reproduction is much more primitive and totally dependent on the presence of water. This is why cryptogams are usually confined to water (algae) or permanently humid environments and areas where wet periods are common in an otherwise dry climate (e.g. xeric ferns and selaginellas). Seed plants have managed to overcome this limiting impediment. By turning their male spores

Solanum lycopersicum var. *cerasiforme* (Solanaceae) – cherry tomato; a smaller variety of tomato, first domesticated in Mexico but probably from South America (Andes) originally – fruits (berries)

into *pollen* and by transforming the containers of their female spores into *ovules* that, once fertilized by the pollen, develop into seeds, the spermatophytes became independent of water for their sexual reproduction. This crucial step took place some 360 million years ago, towards the end of the Devonian time period (417-354 million years ago), a few million years before the beginning of the Carboniferous (354-290 million years ago). It was the invention of the seed – which enabled their ovules to achieve fertilization in even the driest climates – that allowed the spermatophytes to conquer nearly every habitat on earth, from the hottest deserts of Africa to the frozen plains of Antarctica. Just how successful the emergence of the seed is in evolutionary terms is demonstrated by the fact that today 97% of all land plants belong to the spermatophytes.

This most exciting chapter in the history of the evolution of the Earth's land plants is discussed in detail in our earlier book on seeds. In the context of the present book, where we focus on fruits and how they develop, it will suffice for the reader to simply remember that ovules are the organs that turn into seeds once they have been fertilized. Topographically, ovules are borne on specialized fertile leaves, called *megasporophylls*. The arrangement of the ovules on the megasporophylls differs among the spermatophytes, most notably between the *gymnosperms* and the *angiosperms*.

The naked-seeded ones

Today, the gymnosperms form a heterogeneous category of seed plants that includes four very different groups: cycads, *Ginkgo*, conifers, and the enigmatic Gnetales. There are also several extinct groups of gymnosperms that are known only from fossils, such as the cycad-like *Bennettitales* (also called cycadeoids) and the conifer-like *Cordaitales,* together with the earliest seed plants, the *pteridosperms*. The early seed plants bore their ovules and seeds "naked" on branches or along the margins of megasporophylls, which is why botanists named them gymnosperms, literally "the naked-seeded ones." The most ancient seed plants alive today, the cycads, still arrange their ovules in this primitive way. Modern-day cycads are *dioecious*, which means that the pollen and ovules are produced on separate individuals. In the primitive genus *Cycas*, the apex of the female plant alternately produces normal leaves and smaller, ovule-bearing megasporophylls. This eerily old-fashioned method of simply interspersing sporophylls with normal leaves along the same shoot is found only in female specimens of *Cycas*. All other gymnosperms bear both their ovule-bearing megasporophylls and pollen bearing leaves (*microsporophylls*) within dedicated structures, specifically side-shoots characterized by determinate growth. In the case of most cycads, this simply means that they crowd their hardened, scale-like sporophylls together – separated by gender – in male pollen cones and female seed cones. Interestingly, contrary to their obviously more conservative female counterparts, male specimens of *Cycas* also arrange their microsporophylls in cones. The prehistoric appearance of cycads is fascinating but not surprising, considering the fact that they already formed an integral part of the dinosaurs' diet. Fossils of cycads, including members of the most ancient-looking genus *Cycas*, are found in sediments that date back to the earliest Permian (290-248 million years ago). During their heyday in the following Mesozoic era (spanning the Triassic, Jurassic and Cretaceous periods), cycads were so abundant and diverse that this period is often referred to as the "age of cycads and dinosaurs." Despite a steep decline during the last 200 million years, cycads have managed to survive almost unchanged, as true "living fossils" with some 290 species.

The ginkgo (*Ginkgo biloba*) is a unique tree from China with no living relatives and therefore classified in its own division (Ginkgophyta), own class (Ginkgoopsida), own order (Ginkgoales), own family (Ginkgoaceae) and own genus (*Ginkgo*). Because its fan-shaped leaves vaguely resemble the leaflets of a frond of the maidenhair fern (*Adiantum*), the ginkgo is also called the maidenhair tree. Although they belong to ancestral species long extinct, fossils of gingko leaves have been found in sediments dating back 270 million years, to the Permian, which makes the ginkgo another example of a living gymnosperm fossil. Only the modern species, *Ginkgo biloba*, survived in a small area of south-east China, where it has long been revered as a sacred tree by Buddhists and planted in temple gardens. Like cycads, the ginkgo is dioecious.

Although they do not look quite so antiquated, conifers appeared a few million years earlier than the cycads, probably in the late Carboniferous. During the Permian and Mesozoic they dominated many forest ecosystems from tropical to boreal climates. Since then, conifers have suffered a similar decline and only 630 species survive. When it comes to their reproductive organs, conifers produce cones that look similar to those of the cycads, albeit usually much smaller. Like the male cycad cone, the male conifer cone is simply a short branch carrying many tightly packed microsporophylls. However, the microsporophylls of cycads produce numerous *pollen sacs* on their underside whereas modern conifers have just two. Superficially, the female cones of cycads and conifers also look very similar, with each scale bearing just two ovules. Despite this striking similarity, female conifer cones have evolved from much more complicated branched structures, as fossils demonstrate. Regardless of the fact that they evolved along different paths, the armoured seed cones of cycads and conifers fulfil the same very important function: they protect the ovules from physical damage and hungry predators. Nevertheless, in order to capture the vital pollen without which the egg cells cannot be fertilized, the ovules must still be

opposite: *Lepidozamia peroffskyana* (Zamiaceae) – pineapple zamia; endemic to eastern Australia – underside of a male sporophyll; contrary to the male sporophylls of modern conifers, which only bear two pollens sacs (microsporangia), cycads produce a large number of pollen sacs on the underside of their male sporophylls; width of sporophyll c. 2cm

Encephalartos ferox (Zamiaceae) – Zululand cycad; native to southern Africa – microscopic detail of the underside of a male sporophyll with dehisced sporangia; the yellow granules on the walls of some sporangia are a tiny remnant of the large amount pollen they once contained; diameter of one sporangium c. 0.8mm

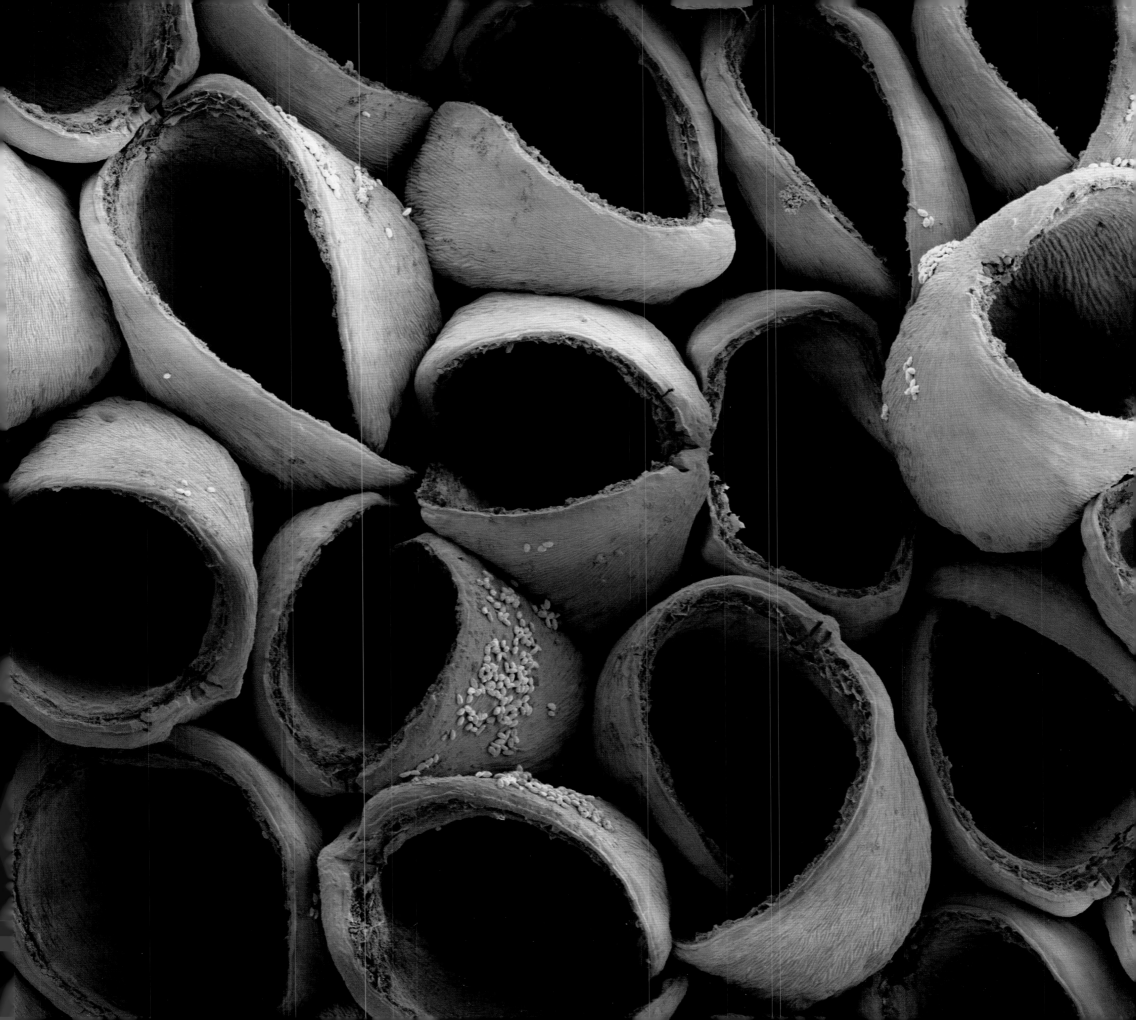

exposed to the environment. When the time for pollination comes the cone scales loosen and separate in order to provide access for airborne pollen and pollinating insects. Both cycads and conifers are therefore still deemed to be naked-seeded gymnosperms.

The fourth and most enigmatic group of gymnosperms that survived to the present day are the Gnetales. They include only three living genera, *Gnetum*, *Ephedra* and *Welwitschia*. The three are so radically different that botanists classify each of them in its own family, Gnetaceae, Ephedraceae and Welwitschiaceae. Members of the tropical genus *Gnetum* look very much like a "normal" broad-leaved tree (i.e. like an angiosperm) and nothing like a typical conifer or a cycad. One of the approximately 28 species, *Gnetum gnemon*, called "melindjo" in its native south-east Asia, is cultivated for its edible seeds, which are delicious when cooked or roasted. Flattened or ground into flour, the seeds are also used to prepare "emping" or "melindjo" crackers, a popular Indonesian snack. It was probably Sir Francis Drake who first brought the seeds of *Gnetum gnemon* to the attention of Europeans when he returned from his famous voyage around the world in 1580. Not knowing exactly what he had found on the island of Beretina in the Philippines, he named them "Fructus Beretinus."

Ephedras are distributed from the Mediterranean to China and occur also in the Americas. They are shrubby plants with green twigs that superficially resemble horsetail – cryptogams belonging to the genus *Equisetum*. Certain species of *Ephedra* contain useful alkaloids that have been used in Chinese traditional medicine for more than five thousand years. Ephedrine is the most famous of these alkaloids and is still used today in treating colds, asthma and sinusitis. It also acts as a stimulant, a side effect sometimes abusively exploited by athletes.

The third family of the Gnetales, the Welwitschiaceae, is represented by a single species, *Welwitschia mirabilis*. At home in the deserts of south-west Africa, the plant was named after the Austrian botanist Friedrich Welwitsch (1806-72), who discovered the first specimens in southern Angola in 1860. As indicated by its Latin name, which means "wondrous *Welwitschia*," the appearance of the plant is startlingly bizarre, which makes it a popular tourist attraction in the Namib Desert. Consisting of little more than a short, cup-shaped stem with two very long, opposite leaves in whose axils short, cone-bearing branches appear, *Welwitschia* is indeed one of the most unusual plants on earth. Throughout its long life – up to 1,500 years – *Welwitschia* keeps the same two strap-like leaves, which grow up to 14cm a year while withering away at their tips. Above ground, the stem can reach up to one metre in diameter. Underground, a huge taproot stretches deep down to the water table. In the extremely dry conditions of the Namib Desert *Welwitschia* has also developed the ability to absorb moisture from sea-fog dew.

The non-naked-seeded ones

If the gymnosperms have naked seeds, their spermatophyte brethren, the angiosperms, should logically have non-naked seeds, and this is indeed the case. Like many other botanical terms, the name *angiosperm* is derived from Greek (*angeion* = vessel and *sperma* = seed) and alludes to their seeds being borne inside tightly closed megasporophylls, called *carpels*. The most likely evolutionary pathway of the angiosperm carpel is via primitive leaf-like megasporophylls, similar to the kind we can still observe in *Cycas*. The transformation of such a primitive megasporophyll into a carpel can be achieved by simply folding it along the midrib and sealing the opposite edges to form a bag that encloses the ovules. But although it provides better protection, enveloping the ovules poses a significant problem: how can the pollen reach them to deliver its *sperm nuclei* for fertilization of the egg cells inside? In present-day cycads and conifers, the ovules receive the pollen grains directly via a *pollination drop*, which they exude through the *micropyle*, an opening at the apex of the ovule. The pollination drop captures pollen from the surrounding air or visiting insects. After a certain time, the pollination drop is reabsorbed, carrying the gathered pollen grains inside the ovule where they can germinate close to the egg cells. In angiosperms such a mechanism would not work because the ovules are no longer exposed to the environment. Enveloping the ovules denies pollen grains direct access. However, the angiosperms would not be the most successful and innovative group of land plants if they had failed to devise an elegant solution to this problem. Their carpels developed a *stigma* (Greek for spot or scar), a wet tissue on the surface specifically devised to receive pollen. We hypothesize that, in the beginning, a primitive stigma was formed along the seam where the opposite margins of the folded megasporophyll originally fused together[2]. During the course of evolution, the angiosperms perfected the structure of their megasporophylls by reducing the stigma to a small platform at the tip of the carpel. To ensure enhanced exposure, in some species the stigma is raised above the swollen ovule-bearing part by a slender extension of the tip of the carpels, the *style*. On the wet, glandular surface of the stigma pollen grains find the right conditions for germination. Often within minutes of their arrival, pollen tubes emerge from pre-formed openings (*apertures*) in the wall of the pollen grains and penetrate the surface of the stigma. Underneath, the pollen tubes soon enter a special canal or transmitting tissue that nourishes and guides them as they grow towards the ovules inside the *loculus*, the fertile chamber of the *ovary*.

With the evolution of the stigma, the angiosperms had not only solved the problem of providing vital access for fertilizing pollen. The stigma automatically became the carpel's single entry point for all incoming pollen and a centre for transmission of the pollen tubes

opposite: *Drimys winteri* (Winteraceae) – Winter's bark tree; native to Central and South America – flower bud with perianth removed revealing young carpels in the centre surrounded by the immature stamens; diameter 2.9mm

below: *Drimys winteri* (Winteraceae) – Winter's bark tree; flower with central cluster of separate carpels (apocarpous gynoecium); diameter 3cm

bottom: *Drimys winteri* (Winteraceae) – Winter's bark tree; fruit, a cluster of berry-like fruitlets (baccetum); diameter 2-2.5cm

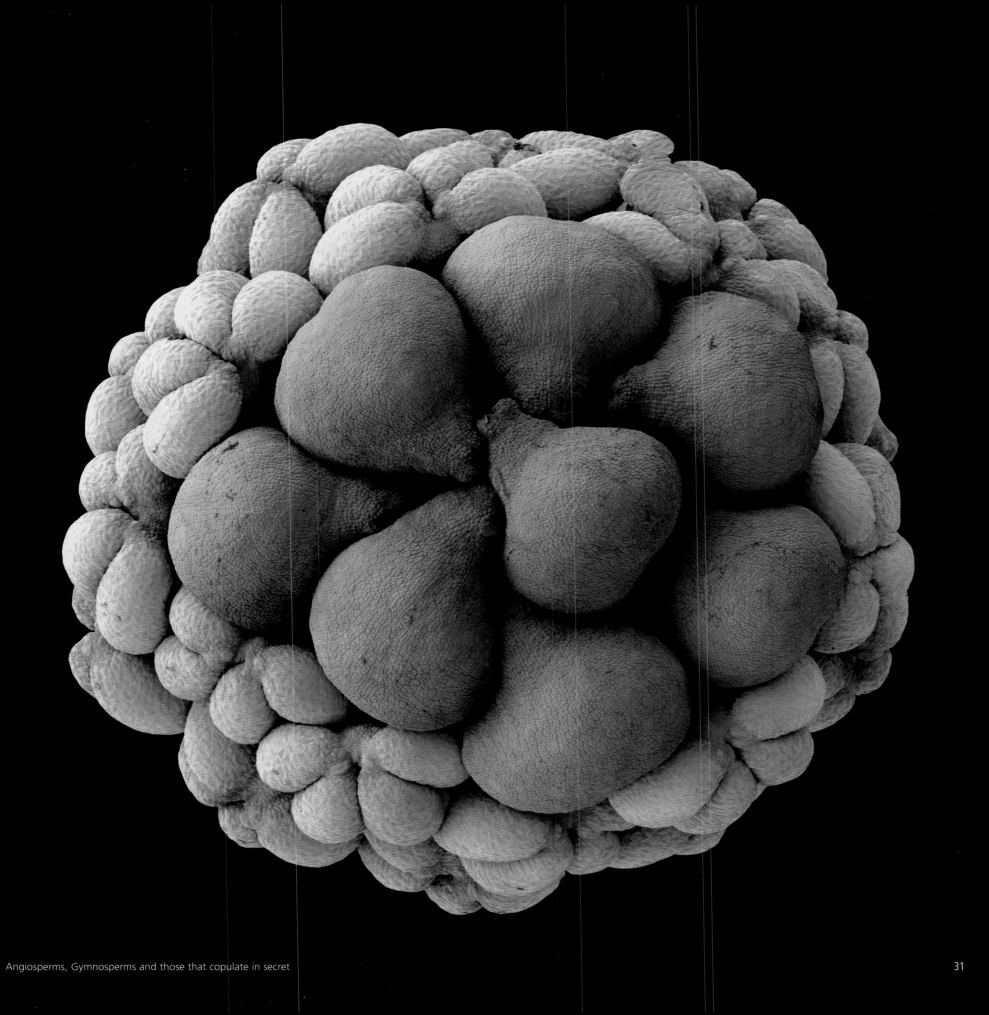

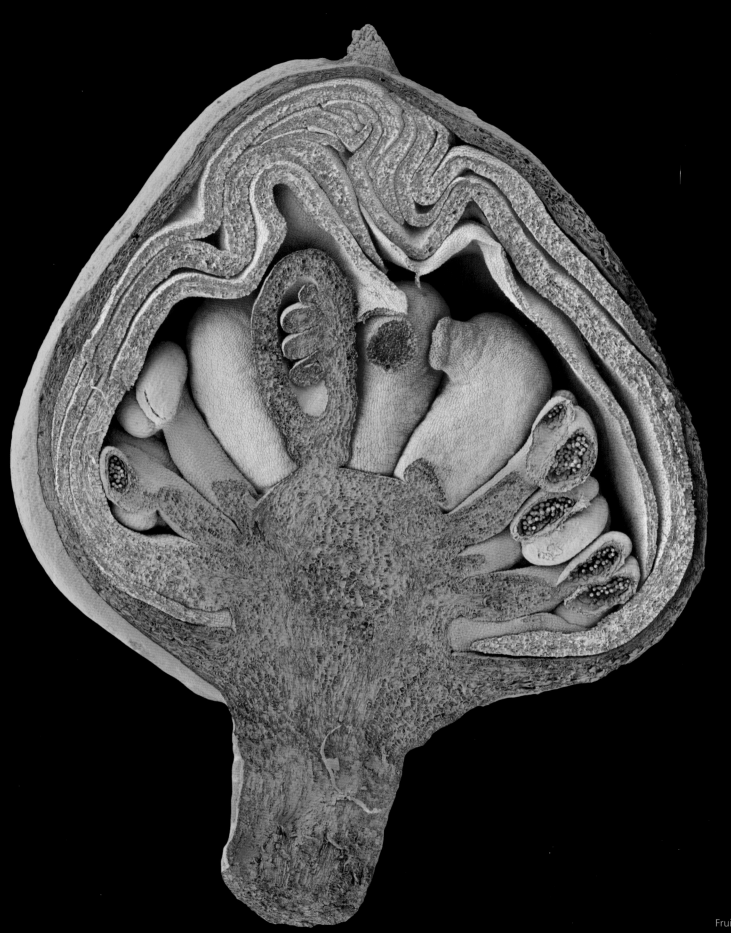

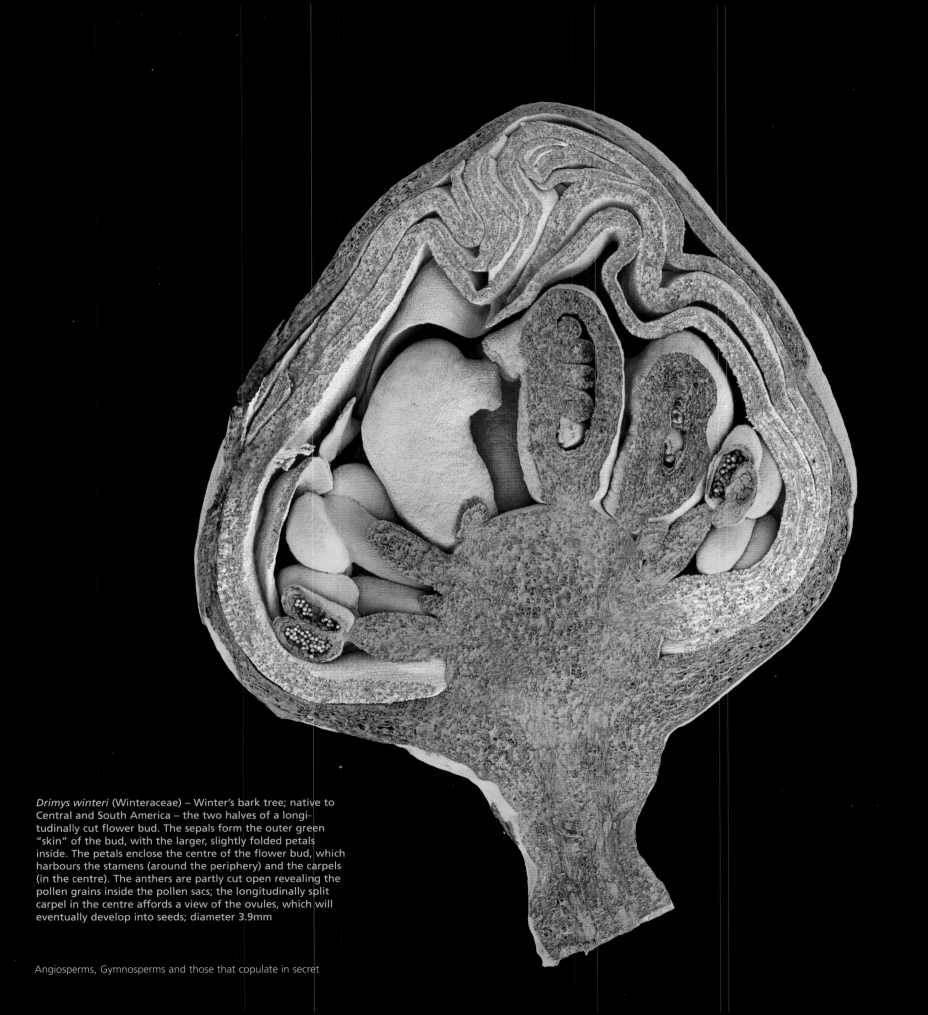

Drimys winteri (Winteraceae) – Winter's bark tree; native to Central and South America – the two halves of a longitudinally cut flower bud. The sepals form the outer green "skin" of the bud, with the larger, slightly folded petals inside. The petals enclose the centre of the flower bud, which harbours the stamens (around the periphery) and the carpels (in the centre). The anthers are partly cut open revealing the pollen grains inside the pollen sacs; the longitudinally split carpel in the centre affords a view of the ovules, which will eventually develop into seeds; diameter 3.9mm

Angiosperms, Gymnosperms and those that copulate in secret

to the ovules. With the stigma in place, a single pollination event, the visit of a pollen-laden insect for example, could potentially deliver sufficient pollen to fertilize all the ovules in the loculus. This is an incredibly efficient mechanism compared with the gymnosperms, whose naked ovules all have to be pollinated individually[3].

An abominable mystery

The first plants with carpels appeared some time between the late Jurassic (206-142 million years ago) and the early Cretaceous (142-65 million years ago), at a time when the dinosaurs were still in their prime. Although it cannot be disputed that the angiosperms must have a gymnosperm ancestry, their closest (living or extinct) relatives are still unknown and so are any intermediate forms that could help us document the evolutionary steps that led from gymnospermous to angiospermous organization. What is even more puzzling is that the angiosperms seem to have appeared abruptly out of nowhere, undergoing a remarkably rapid evolution. This observation was already baffling the greatest scientist of the nineteenth century, Charles Darwin (1809-1882), who, in a letter to the Swiss botanist Oswald Heer (1809-83) on 8 March, 1875, described the sudden appearance of the angiosperms in the fossil record as a "most perplexing phenomenon." Four years later, in a letter to the director of Kew Gardens, Joseph Dalton Hooker, dated 22 July, 1879, he famously called the rapid rise and early diversification of the angiosperms "an abominable mystery." To the present day, the question of the evolutionary origin of the angiosperms has remained unanswered. Some blame the sudden appearance of the angiosperms on a patchy fossil record. Others believe that they did indeed appear suddenly and with considerable diversity in the earth's history without any obvious antecedents. Recent research suggests an answer to this unresolved but fundamental issue in the evolutionary biology of plants. Scientists have gathered evidence indicating that the angiosperms underwent a whole-genome duplication event (i.e. a doubling of all their genes) early in their evolutionary history. Such a duplication of the entire genome yields an extra full set of genes that provides plenty of raw material for evolutionary experiments via random mutations. While most mutations have neutral or detrimental effects on the vitality of an organism and soon become eradicated by natural selection, an important few trigger the expression of new advantageous traits. Offering the maximum spectrum of opportunities for the development of new beneficial genes, whole-genome duplications are regarded as a possible mechanism that drives sudden bursts of evolution such as the origin of the angiosperms and their subsequent spread.

Whatever the answer may be, the fact remains that the carpel, together with several other significant progressions that we shall explore later, afforded the angiosperms

Ocimum basilicum (Lamiaceae) – sweet basil; originally from Asia, cultivated for more than 5,000 years – two pollen grains showing the reticulate surface pattern that is typical of pollen of members of the mint family (Lamiaceae). The smooth, longitudinal furrows are the so-called apertures, weak spots in the otherwise tough pollen wall, through which the pollen tube emerges; diameter of one pollen grain 45μm

tremendous evolutionary advantages over the gymnosperms. As a consequence, with an estimated 422,000 species, angiosperms vastly outnumber the little more than one thousand species of gymnosperms left today. It is not surprising, therefore, that the overwhelming majority of plants around us – magnolias, beeches, oaks, daffodils, roses, cacti, palms and orchids – are all angiosperms.

The great angiosperm proliferation took place around the mid-Cretaceous (some 100 million years ago). Although gymnosperms and ferns still dominated the forests, it is then that a large number of different angiosperms appeared in the fossil record. By the late Cretaceous (some 80 million years ago), the angiosperms seem to have become the predominant group of land plants in most environments (although boreal forests were still dominated by conifers), and many fossils can be identified as relatives of our modern beeches, maples, oaks and magnolias. Just as quickly, the angiosperms diversified and spread into every habitat, wherever plant life is possible, from the poles to the equator.

Today, angiosperms dominate most vegetation types. Two species have even entered the inhospitable Antarctic Circle, which is otherwise reserved for the toughest mosses, liverworts, lichens and fungi. These are the Antarctic hair grass *(Deschampsia antarctica,* Poaceae) and the Antarctic pearlwort (*Colobanthus quitensis*), a member of the pink family (Caryophyllaceae). Both occur on the South Orkney Islands, the South Shetland Islands, and along the western Antarctic Peninsula.

Some angiosperms reverted back to an aquatic lifestyle similar to that of their distant algal ancestors. Aquatic angiosperms occur abundantly in streams, rivers and fresh-water lakes (e.g. water lilies (*Nymphaea* spp., Nymphaeaceae), sacred lotus (*Nelumbo nucifera*, Nelumbonaceae), certain buttercups such as *Ranunculus aquatilis* and *R. baudotii*). A very small elite, most notably the sea grasses (*Zostera* spp., Zosteraceae), even adapted to the saline conditions of the sea where they thrive in depths of up to 50 metres. This extreme diversity of habitats and lifestyles mirrors a perplexing variety of growth forms.

Angiosperm extremists

The spectrum of angiosperm life forms begins with the tiny floating aquatic *Wolffia angusta* (Araceae) from south-east Australia. Measuring only 0.6 x 0.33mm the plant's body shows no differentiation into stem and leaves and consists of little more than a green thalloid (not differentiated into leaves, roots and stems) lump. Coincidentally, we find the opposite extreme on the same continent. One of Australia's eucalypts, the Australian Mountain Ash (*Eucalyptus regnans*, Myrtaceae), reaches a vertiginous height of nearly 100 metres. The largest living specimen, popularly named Icarus Dream, grows in the Styx Valley of

Zostera marina (Zosteraceae) – eelgrass; native to Europe – the 18 species of the Zosteraceae are the most prominent among a small elite of angiosperms that are adapted to live entirely submerged in seawater. Eelgrasses can grow in depths of up to 50m. Dried plants of *Zostera marina* are used as packaging material and stuffing for mattresses

Deschampsia antarctica (Poaceae) – Antarctic hair grass; native to southern South America and maritime Antarctica – diaspore consisting of an entire floret, comprising the mature ovary (caryposis) covered by palea (yellow), lemma (blue) and part of the feathery rachilla (spikelet axis). The Antarctic hair grass is one of just two flowering plant species that are adapted to live in Antarctica; 5mm long

Tasmania's Andromeda Reserve and currently measures 97 metres. One historic specimen of the same species felled in Gippsland, Victoria in around 1872 is said to have been 132.5 metres and some claim that it was 152.4 metres tall. In either case, it would have been the tallest tree ever found. However, the present world record for the tallest living organism is not held by an angiosperm but a gymnosperm, namely a gigantic coast redwood tree (*Sequoia sempervirens*, Taxodiaceae). Measuring 115.55 metres (379.1 feet) tall, the tree was discovered during summer 2006 in the northern Californian Redwood National Park and admiringly baptized "Hyperion" after the Titan of light. In Greek mythology, Titan was the son of the earth goddess Gaia and the sky god Uranus, and the father of Helios, the goddess personifying the sun.

Even though they currently lose the competition for the world's tallest tree, the angiosperms outnumber the gymnosperms by more than 400 times in terms of species diversity. This superiority in numbers is reflected in a much greater diversity of life forms than can be found among their extant naked-seeded cousins. Between the minuscule *Wolffia angusta* and the gigantic Mountain Ash the angiosperms enrich life on Earth with nearly half a million different herbs, trees and shrubs, creepers, climbers and scramblers, water-storing succulents, epiphytes, parasites and carnivores and many more natural wonders.

NO FLOWER, NO FRUIT?

Armed with this outline of the principal differences between the various groups of seed plants, we can now return to our original question and further explore what a fruit is in a strict scientific sense. To many botanists, defining a fruit has caused little if any headache. In 1694 Joseph Pitton de Tournefort (1656-1708) defined a fruit as "the product of a flower," as do many textbooks today (e.g. Judd et al. 2002; Leins 2000). This definition convinces through its simplicity and clarity and, after all, we arrived at the same interpretation earlier when reviewing our grocery shopping. However, in order to fully understand the consequences of this definition, we need to delve somewhat deeper into the science of botany and establish what a flower is.

Is a pine cone a fruit?

Most botany textbooks tell us that today's gymnosperms have no flowers and therefore – logically – no fruits. For example, pine cones and the often enormous seed cones of the cycads cannot be considered fruits despite the fact that they are packed with seeds. In accordance with our definition, *real fruits* have to come from *real flowers* and these are allegedly only present in the so-called *angiosperms*, which are therefore more commonly

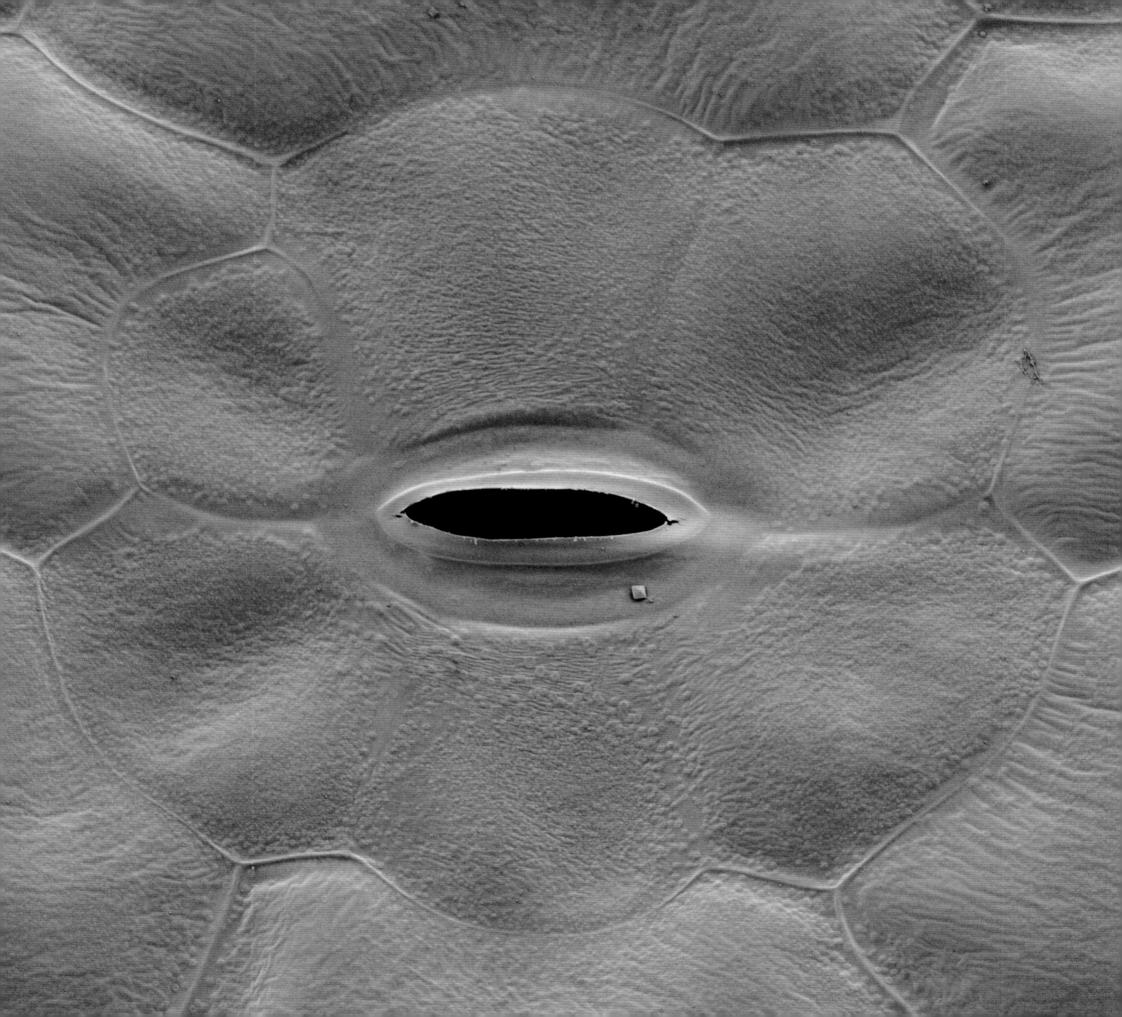

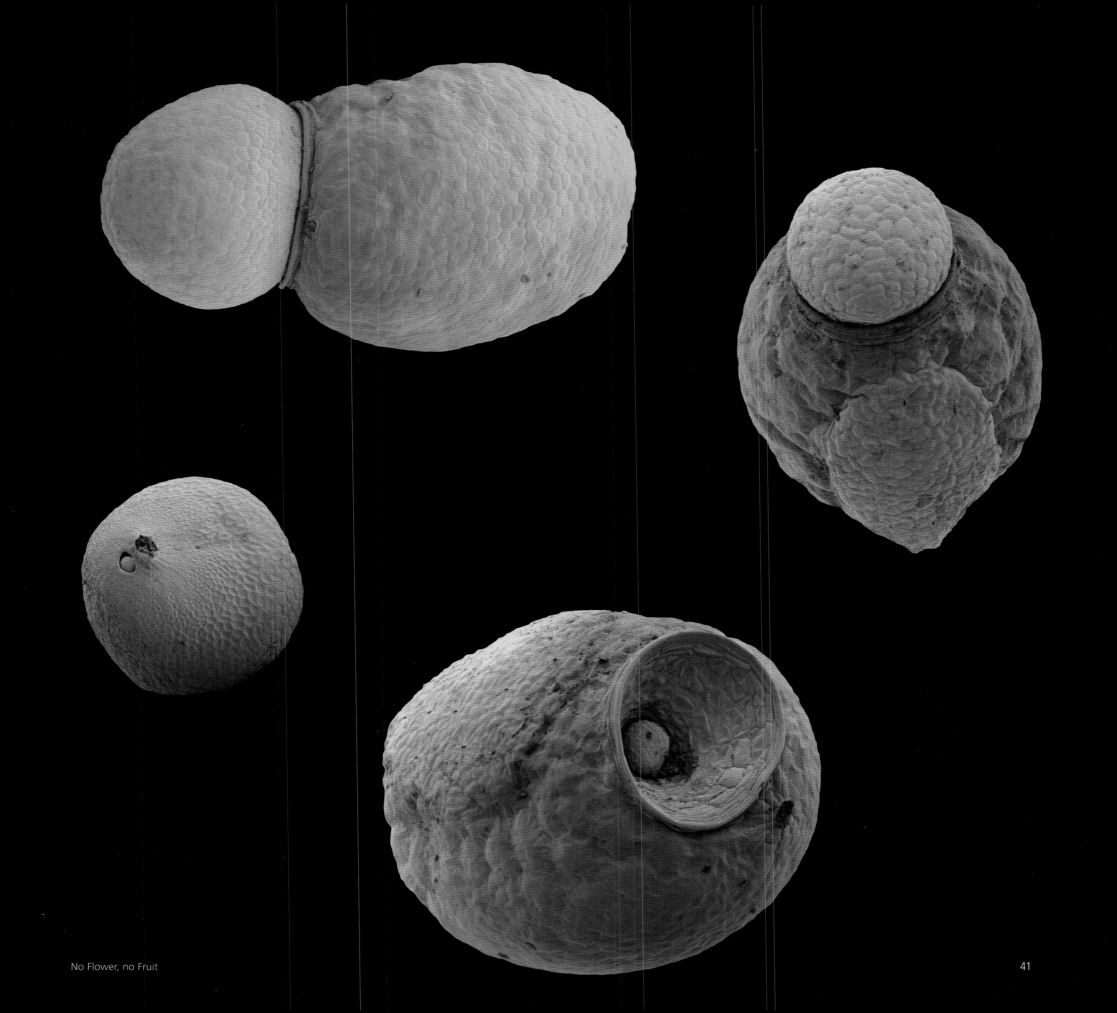

flowering plants. Scientifically, a flower is defined as a short, specialized branch whose growth terminates with the production of one or more fertile leaves, called *sporophylls*. These sporophylls bear either male or female reproductive organs, the pollen sacs and *ovules* respectively. The ovules are particularly important in our context since they are the organs from which the seeds develop; the seeds, after all, are a fruit's *raison d'être*. This far, and without any further restrictions made to the definition, the male and female cones of cycads, for example, would qualify as male and female flowers. However, cycads are gymnosperms and therefore, strictly speaking, cannot have flowers. As a logical consequence, cycads are therefore also unable to bear fruits, despite their hard-to-overlook seed cones, which, in the case of the pineapple zamia (*Lepidozamia peroffskyana*, Zamiaceae) from Australia, can be up to 90cm long and weigh more than 45kg. So just how did scientists maintain the flowering plants' exclusivity over the possession of flowers and fruits? Simply by adding a requirement that most gymnosperms find it impossible to meet. Apart from representing a specialized shoot with determinate growth bearing male and/or female sporophylls, a scientifically acceptable flower must also possess some kind of *perigone* or *perianth* in the form of additional sterile leaves surrounding the sporophylls. This may sound complicated to the non-botanist but it means nothing more than that a "proper" flower should have some more or less showy petals or similar leaves (*sepals* or *tepals*) associated with the sporophylls. And once again, most of us would agree. After all, when flowers are supposed to speak for us, who would choose a selection of humble pine cones rather than a bunch of flamboyant roses with bright red petals to catch the attention of a beautiful woman?

Nevertheless, no matter how hard botanists try to exclude the gymnosperms from the elite circle of flowering plants, there remain a few "extravagant" representatives that spoil this otherwise convincing concept. If we seek scientific truth we have to admit that the bizarre Gnetales do indeed bear real flowers. Admittedly, their flowers are very small and not at all showy but their sporophylls are clearly surrounded by a perianth, just as the scientific definition of a flower requires. In order to clearly separate what we generally call "flowering plants" from the more primitive gymnosperms, it is therefore more appropriate to call the former *angiosperms*. The name *angiosperm* is not only scientifically accurate, it also better reflects one of the crucial characters that permits a clear-cut distinction from gymnosperms, namely, the possession of closed carpels.[4]

NO CARPEL, NO FRUIT?

Successfully foiled by the Gnetales, the rule "no flower, no fruit" turns out to not entirely fulfil its purpose and limit the possession of fruits to the angiosperms. More successful in

Chamaecyparis lawsoniana (Cupressaceae) – Lawson's cypress; native to north-western North America – open seed cone with two remnant seeds (dark brown); empty cone scales show two detachment scars; like the macrosporophylls of cycads, the cone scales of conifers always bear two ovules. However, the fossil record proves that they do not represent just a single fertile leaf but rather an entire, although greatly reduced, lateral branch system; diameter 9mm

opposite: *Phytolacca acinosa* (Phytolaccaceae) – Indian pokeweed; native to east Asia – flower. Typical of dicotyledons, the Indian pokeweed possesses pentamerous flowers with five white or reddish tepals (a calyx is absent) and two whorls of five stamens (in the flower illustrated the anthers have already dropped off); only the number of carpels (eight) is deviant. Although united into a superior syncarpous gynoecium, constrictions clearly mark the boundaries between the individual carpels, each of which has its own style; diameter of flower 7.5mm

Galanthus nivalis subsp. *imperati* 'Ginns' (Amaryllidaceae) – snowdrop, garden variety; the wild form native to southern Europe – flower; typical for monocotyledons, the snowdrop possesses trimerous flowers with three large, white outer perianth segments ('sepals') and three smaller, inner perianth segments ('petals'). The green swollen part of the flower below the perianth is the inferior ovary that consists of three joined carpels

this respect is the definition adopted by many botanists that "a fruit is a mature ovary including the seeds." The inspiration for this narrow concept probably goes back to the first carpological treatment in history, Joseph Gaertner's (1732-91) well-known book *De fructibus et seminibus plantarum* (*On the fruits and seeds of plants*), published 1788-92. Gaertner distinguished between fruit ("fructus"), which he applied (*inter alia*) to the cone of *Pinus*, whereas for most angiosperm fruits he used the term *pericarpium*, which he defined as a "mature ovary." This may seem confusing today but Gaertner formulated his ideas long before 1827 when Robert Brown (1773-1858) pointed out the fundamental difference between gymnosperms and angiosperms, which until then were not treated separately by botanists. Also, at the end of the eighteenth century, the detailed structure of the gynoecium was not yet fully understood. In fact, in the thinking of Gaertner and his contemporary, the great Carl von Linné (1707-78), many angiospermous fruits (e.g. those of the sunflower family, Asteraceae) were deemed to be naked seeds.

In the nineteenth century, John Lindley (1832) summarized Gaertner's terms for pericarpia as names for fruit types and defined fruit as "*the ovarium or pistillum arrived at maturity; but, although this is the sense in which the term is strictly applied, yet in practice it is extended to whatever is combined with the ovarium when ripe.*"

Although the history is more complicated than can be explained here, little has changed over the past 170 years. Most "modern" authors still choose to define a fruit as the product of a mature ovary, although some allow other parts of the flower to be included (e.g. Raven et al. 1999; Mauseth 2003; Heywood et al. 2007), an "inconsistency" already pointed out by Lindley. Since gymnosperms have no carpels and therefore no ovaries, they are, as a logical consequence, unable to bear fruits – at least according to this popular definition.

Admittedly, carpels are a remarkable invention and have some merit. In fact, the possession of closed carpels instead of open megasporophylls is one of the most significant advantages that angiosperms have over gymnosperms. Advances in other areas, such as a more sophisticated wood structure with improved water conductivity, a refined sex life, a highly economical method of seed production[5] and greater flexibility in seedling establishment have also helped to make the angiosperms currently the ruling class among plants. However, as discussed in our previous book on seeds, sexual reproduction and seed dispersal remain the most crucial events in the life cycle of a seed plant. Therefore, probably more than anything, it was the amazing ability to adapt and perfect their flowers, fruits and seeds in so many different ways that bestowed on angiosperms their evolutionary success. Alas, the botanists' difficulty in defining a fruit does not end with the exclusion of the gymnosperms. As we are about to discover, the complexity of the problem increases within the angiosperms themselves, and the conceptual dilemma deepens. The reasons for this lie in the sheer versatility and adaptability of the angiosperms. They not only generated the most beautiful, fascinating and useful plants on Earth but at the same time present us with an incomprehensible and, as a challenge to many botanists, an almost unclassifiable diversity of flowers and fruits.

A shameless display

Applying human moral standards to the angiosperms, their flowers can only be described as a shameless and offensively flamboyant display of (plant) genitals. The first person to discover that angiosperm flowers contain a plant's male and female sexual organs was the English naturalist and physician Nehemiah Grew (1641-1712). Much of the terminology we use today to describe flowers scientifically was invented by Grew in his great work *The Anatomy of Plants*, published in 1682. In the eyes of a botanist, an angiosperm flower typically (but not universally) consists of at least four distinct whorls of specialized leaf-like structures. Initially covering and protecting the flower bud the outermost ones, called *sepals*, form the plate- or cup-shaped *calyx* (Latin: cup) of the open flower. The sepals are mostly green and smaller than the large, colourful *petals* that develop inside them, where they form the showy *corolla* (Latin: little crown) of the flower. Calyx and corolla together form what botanists refer to as the *perianth*. If all floral leaves look the same, as in a tulip, for example, they are called *tepals* and the perianth-like structure they produce is called a *perigone*. Sepals and petals often appear in threes (typical of *Monocotyledons* such as lilies, orchids and agaves) or fives (typical for *Dicotyledons* such as pinks, beans and mallows). As we move past the petals towards the centre of the flower we pass through the *androecium*, a term derived from Greek literally meaning "man's quarters." Behind what in 1826 Johannes August Christian Roeper (1801-85) humorously called the "man's quarters" are the flower's male sexual organs, represented by one or two whorls of microsporophylls or *stamens*. Each stamen consists of a slender stalk, the *filament*, which carries an *anther* at the tip. The anther is the fertile part of the stamen and contains the pollen grains distributed among four containers, the male *microsporangia* or *pollen sacs*. The carpels, representing the female reproductive organs, are located in the very centre of the flower. Once again, thanks to Roeper's creativity, the sum of all carpels of a flower is scientifically addressed as the "women's quarters" or *gynoecium* (plural *gynoecia*). More familiar terms referring to the female parts of a flower are *pistil* or *ovary*, but, as will be explained later, they are by no means synonymous.

Whilst many flowers exhibit their female sexual organs in full nudity, quite a few angiosperms have the decency to cover themselves up, limiting the "phytopornographic"

opposite: *Citrus hystrix* (Rutaceae) – kaffir lime; native to Indonesia – longitudinal section through flower bud, affording a view into the superior syncarpous ovary revealing the attachment of the ovules in the centre of the ovary (axile placentae); diameter of bud 5.8mm

Citrus hystrix (Rutaceae) – kaffir lime; native to Indonesia – flowers with four white petals, numerous stamens and a prominent superior ovary (green) consisting of several joined carpels. The aromatic leaves of the kaffir lime are a common ingredient in Thai cuisine; diameter of flowers c. 1.4cm

display to their stigmas and stamens only. They hide their female parts in a cup- or tube-like extension of the floral axis (*receptacle*), the part of the shoot that carries the floral organs. As this floral tube or *hypanthium* develops, it carries all other floral organs with it on its upper rim. The result is an *epigynous flower* (Greek: *epi* = above + *gyne* = female) in which sepals, petals and stamens sit *above* the now *inferior* and no longer visible ovary. Rather than creating a more virtuous appearance, the true evolutionary significance of epigynous flowers is that they provide better protection for their vital female parts by hiding them from potentially damaging insect visitors.

Their "exhibitionist counterpart," called *hypogynous flower* (from Greek: *hypo* = under, beneath), bears its female organs in a *superior* position where stamens, petals and sepals are attached to the receptacle *below* the exposed ovary. An intermediate condition is found in *perigynous flowers* (Greek: *peri* = around) in which a hypanthium surrounds the gynoecium but remains separate from it. Cherry blossoms, whose ovary sits in the centre of a tiny nectar-gathering, cup-shaped hypanthium, are a familiar example of a perigynous flower.

By now some readers may begin to wonder why the author is racking his brains with all these confusing terms and subtle theoretical distinctions. The truth is that a clear comprehension of the evolution and architecture of flowers, especially the structure and position of the gynoecium, is key to an understanding of how fruits are formed. For example, the wall (*pericarp*) of a fruit developing from a hypogynous flower consists of only the ovary wall, whereas an epigynous flower produces a fruit with a wall formed by the ovary wall *and* the hypanthium. Most of what we eat when indulging in an apple is tissue produced by the hypanthium – very little is contributed by the ovary wall.

Not quite the ovary of Eve

Independently of their arrangement and position relative to the other floral organs, the carpels are the womb of a flower. They bear the ovules, the organs that become seeds after fertilization. Inside the ovary, the ovules are attached to the carpel wall in special, often raised areas called *placentas*. The function of the placenta in plants is similar to the human placenta. In both cases the term refers to an organ that nourishes the developing offspring, although the analogy is not perfect. In its original meaning, "ovule" (Latin: *ovulum*, the diminutive of *ovum* = egg) suggests homology with an animal or human egg cell, but the ovule of the seed plants consists of more than just an egg cell. It is a very complicated organ – with a complex evolutionary history – that at its core hosts an egg cell. The usage of the term ovule goes back to the days of Nehemiah Grew in the seventeenth century, long before Wilhelm Hofmeister made his revolutionary discovery in 1851. By demonstrating

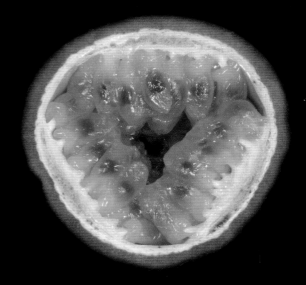

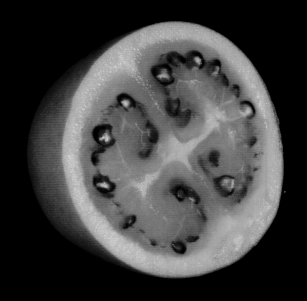

The seed contains the precious *embryo* – the miniature offspring that, upon germination, gives rise to a new plant. In terms of size, seeds can range from the enormous Seychelles nut (*Lodoicea maldivica*, Arecaceae, actually a single-seeded indehiscent fruit) which can weigh more than twenty kilos, to those of the orchid family (Orchidaceae), some of which measure less than a quarter of a millimetre in length. In such dust-like seeds the embryo is tiny and consists of just a few dozen cells. In the horse chestnut (*Aesculus hippocastanum*, Sapindaceae), avocado (*Persea americana*, Lauraceae) and mangrove (e.g. *Rhizophora* spp., Rhizophoraceae), to name but a few, the embryos are very large. The largest embryo of any plant, weighing up to one kilogram, is found in the seeds of a member of the legume family, *Mora megistosperma* (syn. *Mora oleifera*), from tropical America.

In addition to the embryo, seeds may also contain a special tissue that stores energy-rich nutrients to provide a reserve to kick-start the germinating embryo. This storage tissue, called *endosperm,* is usually large in seeds with small embryos such as those of magnolias (Magnoliaceae), buttercups (Ranunculaceae) and palms (Arecaceae). The latter include the colossal (not to say erotic) Seychelles nut palm and its distant relative, the coconut (*Cocos nucifera*). In other seeds such as the horse chestnut, the avocado and many legumes, the embryo has already absorbed the entire endosperm during its development, storing its nutrients in its own tissues. "Beefed-up" with plenty of food deposited in their swollen bodies (mostly in the *cotyledons*), seeds with such storage embryos have a great advantage. "Ready to go," they are able to germinate much faster than seeds whose embryo digests and absorbs the nutrients from the endosperm only during germination.

Assigned the vital responsibility of nourishing, protecting and finally dispersing these miraculous parcels of life, it is hardly surprising that the structure of the gynoecium has great influence on the internal and external organization of the mature fruit. In order to gain a better understanding of the tremendous diversity of fruit types, we shall explore more carefully the various types of gynoecia found among the angiosperms.

Babylonian confusion

One of the first differences we notice on a closer examination of female floral parts is that the number of carpels per flower varies among different groups of angiosperms. Strictly speaking, a single carpel is typical of members of the legume family (Leguminosae), which includes "fruit vegetables" such as beans and peas, and also many ornamentals such as sweet peas (*Lathyrus odoratus*), Chinese wisteria (*Wisteria sinensis*), black locust (*Robinia pseudoacacia*) and gorse (*Ulex europaeus*). Other familiar examples with such *monocarpellate* flowers are cherries, plums and peaches (*Prunus* spp., Rosaceae). If there are two or more carpels in a

opposite: ***Persea americana*** (Lauraceae) – avocado; native to Central America where the fruit was used as food as long as 10,000 years ago – longitudinal section of fruit (berry); clearly adapted to animal dispersal, the oil-rich avocado is one of the most nutritious and fibre-rich fruits. Its extremely large seed in combination with the green skin-colour and high oil content suggest that the avocado is adapted to dispersal by the large mammals that were part of the extinct Pleistocene megafauna that disappeared some 13,000 years ago. Today, in their native America, avocados are relished by wild cats and jaguars. Instead of relying on a hard seed coat or endocarp, the large seed protects itself with bitter-tasting toxins that discourage animals from mastication

Tahina spectabilis (Arecaceae) – a recently discovered new palm species endemic to Madagascar – longitudinal section of seed showing the small peg-like embryo embedded in a copious white endosperm. The endosperm is penetrated by ingrowths of the seed coat (so-called ruminations), possibly an adaptation against seed predators (especially insects) who are discouraged by bitter-tasting, toxic chemicals, especially tannins which are responsible for the brown colour of the seed coat. Less than a hundred individuals are thought to exist of this remarkable palm, which is over 18m tall and flowers only once in its lifetime which is estimated at 30-50 years; seed 2.1cm long

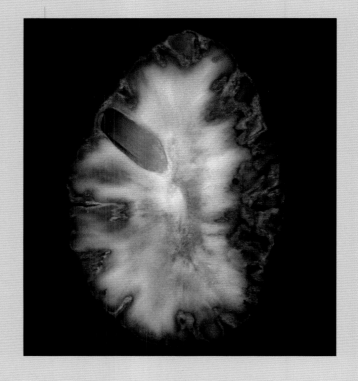

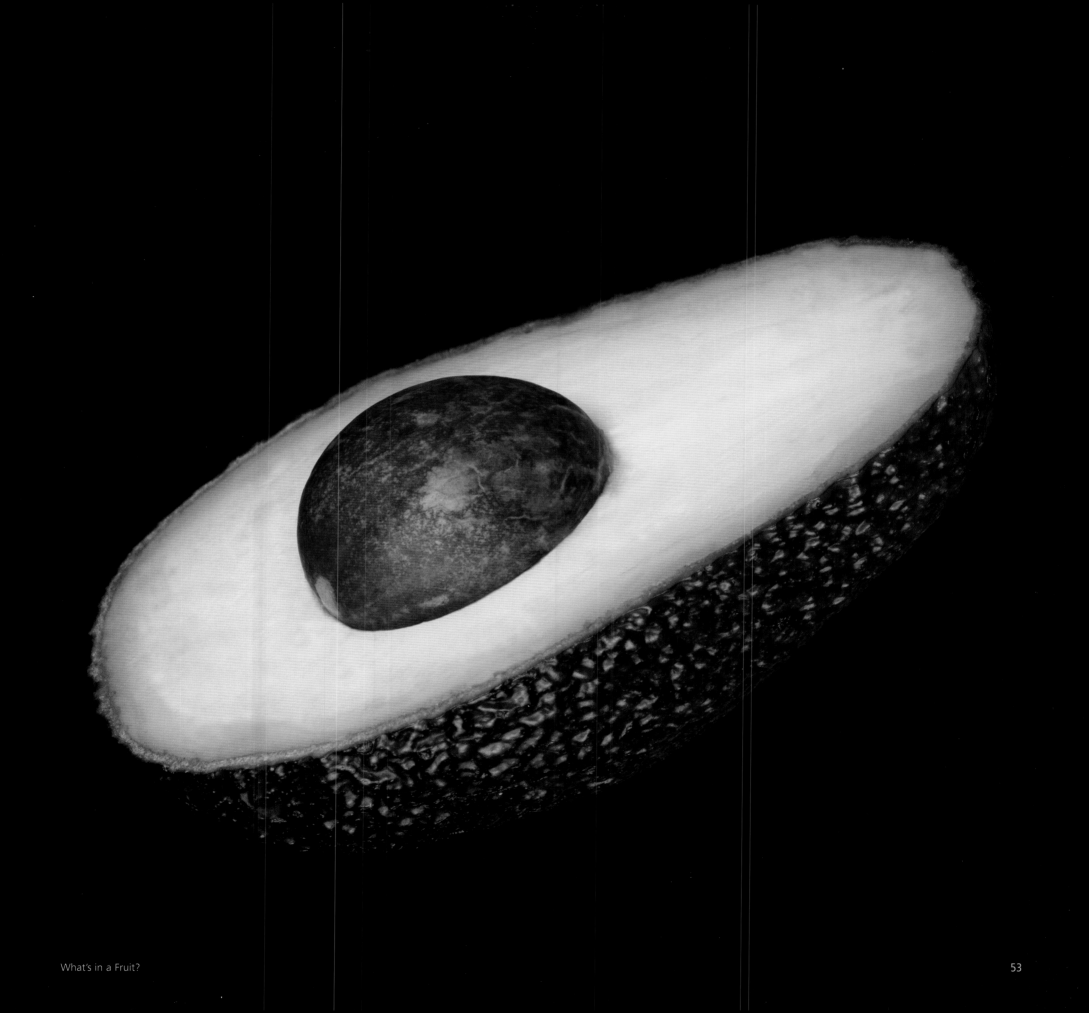

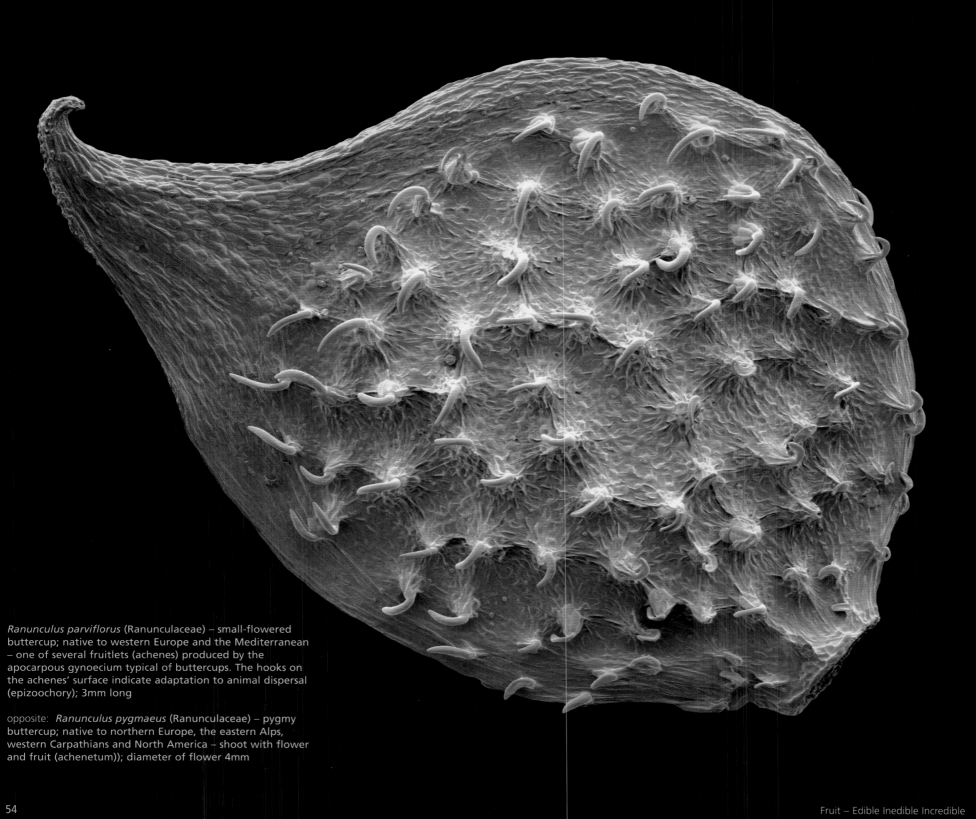

Ranunculus parviflorus (Ranunculaceae) – small-flowered buttercup; native to western Europe and the Mediterranean – one of several fruitlets (achenes) produced by the apocarpous gynoecium typical of buttercups. The hooks on the achenes' surface indicate adaptation to animal dispersal (epizoochory); 3mm long

opposite: *Ranunculus pygmaeus* (Ranunculaceae) – pygmy buttercup; native to northern Europe, the eastern Alps, western Carpathians and North America – shoot with flower and fruit (achenetum)); diameter of flower 4mm

flower, as in the majority of angiosperms, the carpels are either separate, forming an *apocarpous gynoecium,* or joined, creating a single pistil or *syncarpous gynoecium.* This may sound confusing, and indeed it is. Although there is a fine distinction between the three terms *pistil, ovary* and *gynoecium,* even many botanists use them interchangeably. The most complex-sounding term, gynoecium, is actually the simplest to explain because it refers to the sum of all the carpels of an individual flower, whether there is just one (*monocarpellate* gynoecium) or many (*pluricarpellate* gynoecium), irrespective of whether they are separate or joined. The other two are used inconsistently. In a strictly scientific sense, pistil refers either to the individual carpel of an apocarpous gynoecium or to a *syncarp* – the structure formed by several joined carpels of a syncarpous gynoecium. For example, the flower of a buttercup (*Ranunculus* spp., Ranunculaceae) with its apocarpous gynoecium has as many pistils as it has carpels. In contrast, the syncarpous ovary of tulips (*Tulipa* spp., Liliaceae) consists of a single pistil, but this is formed from three joined carpels. The term ovary has almost the same meaning as pistil except that it refers only to the swollen, ovule-containing part, and excludes both style and stigma.

Enhanced female performance

After this pernickety attempt to restore scientific precision it is time to return to the more exciting task of exploring the diversity of female floral genitalia. Whether a species has separate or joined carpels is a fundamental trait, the significance of which lies buried deep in the history of the angiosperms. The possession of separate carpels, each equipped with its own stigma, means that each carpel has to be pollinated individually. Although not entirely absent in more advanced families (e.g. Rosaceae), this condition is predominantly found among primitive living members of the angiosperms such as the Annonaceae, Lardizabalaceae, Winteraceae and Ranunculaceae. During the course of evolution, the tendency of the angiosperms to fuse their floral organs, especially the carpels, has resulted in the development of syncarpous gynoecia, in which the joined carpels share a single stigma. A stigma shared by several carpels means enhanced female performance through rationalised pollination: in progressive angiosperms with syncarpous gynoecia, all ovules in not only one but several carpels are fertilized in a single pollination event. This affords them a clear evolutionary advantage over their old-fashioned apocarpous brethren, with the result that the majority of living angiosperms are characterized by syncarpous rather than apocarpous gynoecia. Irrespective of the numerical disparity between apocarpous and syncarpous angiosperms, their fundamentally different gynoecia would lead us to anticipate major differences in their fruits.

opposite: *Citrus hystrix* (Rutaceae) – kaffir lime; native to Indonesia – flower bud with petals and stamens partly removed to allow a view of the superior ovary; after fertilization the ovary produces a small knobbly green fruit, the kaffir lime; diameter 5.5mm

Citrus sinensis (Rutaceae) – sweet orange; cultivated since antiquity; presumed origin China or India – flowers and fruits. Citrus fruits have been cultivated in south-east Asia for at least 4,000 years. With more than 60 million tons produced each year worldwide, the sweet orange is the most important of all citrus fruits

How to be a carpologist

The fact that there are flowers with single and multiple pistils is the most important character that carpologists – people who devote their lives to the study and classification of fruits – use to bring some order into the bewildering diversity of fruits. Flowers with a single pistil produce what is called a *simple fruit*. Most fruits that we eat belong in this category. Fruits developing from flowers with multiple pistils consist of a cluster of individual *fruitlets*, each representing a mature carpel. Such fruits, which include the raspberry (*Rubus idaeus,* Rosaceae) and blackberry (*Rubus fruticosus*), are called *multiple fruits*. Technically (but not evolutionarily) in limbo between simple and multiple fruits is a somewhat tricky category of fruits: they are produced by syncarpous ovaries that disintegrate into their individual carpels. This separation of the carpels can happen either soon after pollination, as in the dogbane family (Apocynaceae) and the cacao family (Sterculiaceae, now included in the mallow family, Malvaceae), or only at maturity, as in the soapberry family (Sapindaceae, for example, *Acer* spp.) and mallows (*Malva* spp., Malvaceae). In either case, the fruit is considered *schizocarpic*. Finally, mulberries (*Morus nigra*, Moraceae), although they appear similar to raspberries and blackberries, represent yet another type of fruit – a *compound fruit*. Despite the superficial resemblance, unlike raspberries and blackberries, mulberries are the product of a joint effort by several flowers. Although it might not be immediately obvious because of the small size of the flowers involved, compound fruits such as the mulberry develop from an entire inflorescence.

This was just a brief overview – many more examples of the different fruit types will be discussed later. For now, it is important to remember that simple, multiple, schizocarpic and compound fruits are the four principal categories of fruits found in the angiosperms. They form the basis for any further classification.

Once a fruit has been assigned to one of these four types, the second most important character to assess is the texture the fruit wall (pericarp) – whether it is soft or hard, juicy or dry. The third and final basic criterion of carpology distinguishes between fruits that open at maturity to release their seeds (dehiscent fruits) and those that remain closed (indehiscent fruits).

Of course, angiosperms would not be angiosperms if their universal adaptability stopped at their fruits. Among the facts of life that would-be carpologists have to face is that when studying the fruits of angiosperms, nearly every imaginable texture and structure is possible. What is most unsettling for the keen botanist is that the first three fundamental characters traditionally used to classify fruits are encountered in every conceivable combination among the angiosperms.

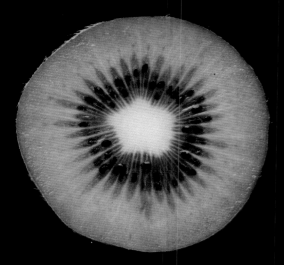

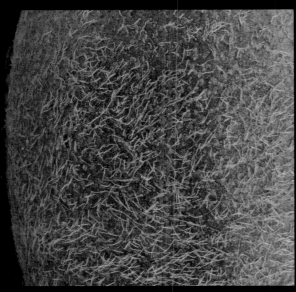

The true meaning of fruits

Knowing how fruits are formed – and which organs plants utilize to achieve their baffling variety – is a fascinating topic even if the classification of fruits according to their structure may seem like an attempt to force their overwhelming diversity into a morphological straitjacket. However, searching for general patterns and defining precise categories is an essential scientific practice. It aims to facilitate communication within the scientific community and helps to bring some order into the otherwise unmanageable chaos, no matter what aspect of the natural world is studied. As for fruits, understanding their structure and development is only the starting point for the study of all other aspects of their natural history. In fact, pure morphology is an abstraction and why fruits look the way they do can only be fully understood when taking into account their biological function, namely the dispersal of the seeds they bear.

For reasons that will be discussed later in detail, successful seed dispersal is crucial for the survival of a species. This vital function exposes fruits and seeds to highly adaptive pressures during the course of evolution. Effectively, the enormous diversity of fruit types displayed in nature is a direct consequence of these functional pressures, which often lead to astonishing similarities between unrelated species facing similar ecological challenges. Intriguing though such convergences are, they cause yet more controversy for carpologists because of the unnatural relationships among many of the taxa that are represented by a particular type of fruit.

Armed with this brief introduction to the science of carpology, and without uncovering more disconcerting truths about the real nature of fruits, we shall now begin our journey into the carpological universe, starting with the simplest fruit type of all.

SIMPLE FRUITS

Since the majority of angiosperms have flowers with a single pistil, whether it consists of just one individual carpel or several joined carpels, we are most familiar with fruits that develop from such flowers, especially if we live in the temperate northern hemisphere. Fruits that develop from a single flower with a single pistil are called *simple fruits*, irrespective of whether they are fleshy or dry, dehiscent (opening) or indehiscent (non-opening). Green beans (*Phaseolus* spp.), pea pods (*Pisum sativum*) and carob (*Ceratonia siliqua*) are all simple fruits of the legume family and, as such, develop from a single carpel. Tomatoes (*Solanum lycopersicum*, Solanaceae), oranges (*Citrus sinensis*, Rutaceae), kiwi (*Actinidia deliciosa*, Actinidiaceae), giant pumpkins (*Cucurbita maxima*, Cucurbitaceae) and papayas (*Carica papaya*, Caricaceae) are simple fruits formed from *compound pistils*. Such syncarpous gynoecia, in which two or more carpels have joined to form a single pistil or ovary, are far more common than apocarpous gynoecia and found in evolutionarily advanced families (Asteraceae, Campanulaceae, Liliaceae, Solanaceae).

If not enclosed by other flower parts, their gynoecium is visible as a single bottle- or finger-shaped pistil in the very centre of the flower (tulips, lilies, *Citrus* spp.). A syncarpous gynoecium looks broadly similar to a monocarpellate gynoecium. Often, the number of stigma lobes or branches gives a clue as to how many carpels participate in the formation of a pistil. For example, the three-lobed stigma of a tulip, a typical monocot, shows that the pistil consists of three joined carpels. However, in order to be entirely certain about the number of carpels involved in the formation of a particular ovary we need to resort to surgical means in the form of a cross-section through the middle of the ovary. The number of chambers or *locules* that can be distinguished inside the ovary generally indicates the number of carpels involved in the formation of the ovary, provided that the individual carpels retain their walls or *septae*. Take an orange or a lemon, for example. Each segment of the fruit represents one carpel and we are able to separate them so easily because the carpels retain their walls despite the fact that they are fused into a single pistil. We shall encounter more difficulties when examining a passion fruit or granadilla (*Passiflora ligularis*, Passifloraceae). Despite the fact that the fruit bears all its seeds in a single loculus, the pistil is also formed by three carpels. In a passion fruit the number of carpels joining to form the pistil is not disclosed by the number of partitions dividing the ovary but by the number of *placentae* – the areas on the inside of the pericarp where the seeds are attached.

Depending on the texture of the fruit wall and whether or not they open at maturity to release their seeds, simple fruits that consist of only the mature ovary without any other organs attached are further subdivided into berries, drupes and nuts, on the one hand, and capsular fruits on the other.

The truth about berries

In common parlance and for culinary purposes, any small edible fruit with multiple seeds passes as a *berry* (Latin: *bacca*). However, botanists apply a scientifically more rigorous definition. Only simple indehiscent fruits whose pericarp (ovary wall) becomes entirely fleshy at maturity are considered true berries, whether they contain just one or many seeds. Therefore, botanically speaking, not only blueberries (*Vaccinium corymbosum, V. myrtillus*, Ericaceae), gooseberries (*Ribes uva-crispa*, Grossulariaceae), blackcurrants (*Ribes nigrum*) and grapes (*Vitis vinifera*, Vitaceae) qualify as berries but also avocados (*Persea americana*, Lauraceae), tomatoes (*Solanum lycopersicum*, Solanaceae), aubergines (*Solanum*

opposite: *Medicago orbicularis* (Leguminosae) – blackdisk medick; native to the Mediterranean – fruit (camara), viewed from the bottom end; like the fruits of all other members of the bean family, the camara of blackdisk medick develops from a single carpel. Typical of the genus *Medicago*, the carpel is coiled into a spiral of 4-6 turns. The flat, discoid shape and narrow peripheral wing suggest that the indehiscent fruit is primarily adapted to wind dispersal; diameter 1.5cm

Dovyalis caffra (Salicaceae) – kei apple; native to the Kei River area of southern Africa – fruit (berry); despite its name, the kei apple is a berry rather than an apple (pome). The edible fruits reach a diameter of 2.5-4cm and contain 5-15 seeds. Their juicy, acidic flesh is tasty and either eaten fresh or made into jam or jelly

melongena), star fruits (*Averrhoa carambola*, Oxalidaceae) and kiwis (*Actinidia deliciosa*, Actinidiaceae). On the other hand, despite their names, strawberries (*Fragaria* x *ananassa*, Rosaceae), mulberries (*Morus nigra*, Moraceae), raspberries (*Rubus idaeus*, Rosaceae) and blackberries (*Rubus fruticosus*) are not really berries at all but a very different type of fruit that will be discussed later.

The worst misnomer of all is the "juniper berry." Although of the utmost importance when it comes to giving gin its characteristic taste, "juniper berries" are not berries at all in the eyes of a botanist, and many would claim they are not even fruits. The reason for this rejection lies in the evolutionary position of juniper. Scientifically, "juniper berries" are the fleshy cones of *Juniperus communis*, a conifer belonging to the cypress family (Cupressaceae) and hence a member of the gymnosperms, which, as we have seen, are widely deemed to lack fruits in the strictest sense. During the two to three years it takes for the aromatic seed cones of this dioecious cypress to develop, the three uppermost scales develop into a blue, fleshy cover that looks deceptively like the pericarp of a true berry.

It is worth taking a closer look at some rather unexpected examples of berries among the angiosperms.

The miraculous miracle berry

In tropical West Africa there is a shrub with small (2–3cm long), red berries that may not taste very sweet but have a quite extraordinary effect on our taste buds. A few moments after chewing the pulp of a miracle berry, as the fruit of *Synsepalum dulcificum* (Sapotaceae) is aptly called, our tongue is tricked into perceiving bitter and sour tastes as sweet, making lemons and limes taste as sweet as oranges. The miracle berry's effect is caused by the glycoprotein miraculin. Exactly how miraculin works is still unknown but it probably causes the sweet receptors on our tongue to be activated by acids. Within an hour or less, the illusion fades and our taste buds return to normal. Miracle berries are highly perishable and therefore difficult to export. However, local tribes have used the fruits for centuries to "sweeten" their food and drink.

West Africa is home to yet another remarkable fruit with an equally suggestive name. The flesh of the serendipity berry (*Dioscoreophyllum cumminsii*, Menispermaceae), which is actually a drupe, also contains an interesting protein called monellin. Unlike miraculin, monellin really does taste sweet – about two thousand times sweeter than sucrose. However, although monellin is potentially a great natural sweetener, it is expensive to extract and becomes denatured at high temperatures and so it would be unsuitable for use in processed food.

below: *Juniperus flaccida* (Cuppressaceae) – weeping juniper; native to Mexico and southern Texas – seed cone; the fleshy fruits of junipers are commonly referred to as "berries" but they do, in fact, represent conifer cones; dia. of fruits c. 1cm

bottom: *Averrhoa carambola* (Oxalidaceae) – star fruit, carambola; cultivated for centuries in south-east Asia, presumed origin between India, Sri Lanka and Indonesia – immature fruits (berries) and flowers; star fruits are eaten fresh and made into jams, pickles, juice and liqueur. The crunchy flesh has a sweet but slightly tart, acidic taste reminiscent of a mixture of melon, apple and kiwi

Hylocereus undatus (Cactaceae) – dragon fruit, pitahaya; native to tropical America – fruit (acrosarcum); although they appear to be berries, dragon fruits are not simply a mature ovary. In the Cactaceae family, the actual flower is literally sunken into a piece of shoot, which is why the ovary-bearing part and later the fruit are often covered with leaves (scale-like bracts). Inside, the dragon fruit contains hundreds of tiny black seeds embedded in an edible, sweet, white or red pulp, which consists of the seeds' fleshy funicles; fruit c. 16cm long

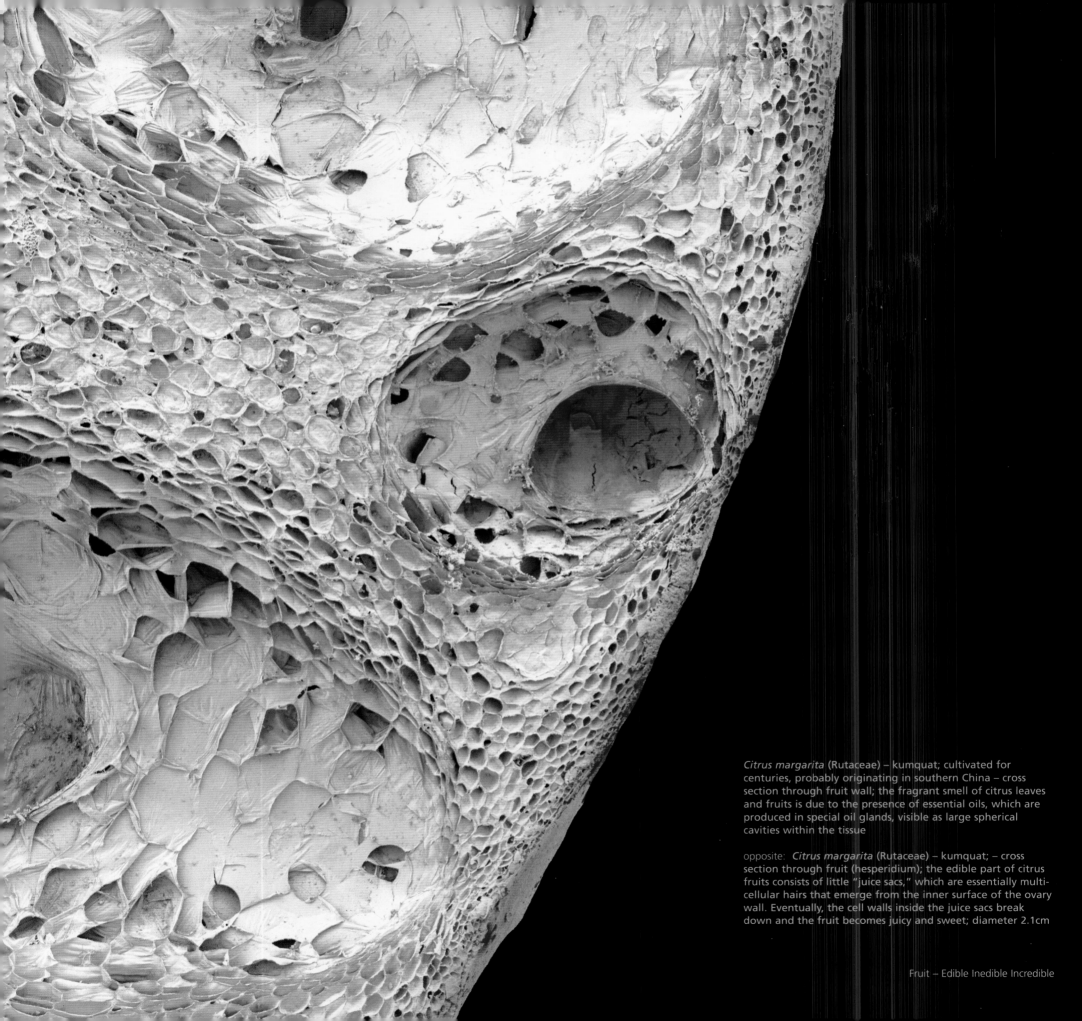

Citrus margarita (Rutaceae) – kumquat; cultivated for centuries, probably originating in southern China – cross section through fruit wall; the fragrant smell of citrus leaves and fruits is due to the presence of essential oils, which are produced in special oil glands, visible as large spherical cavities within the tissue

opposite: *Citrus margarita* (Rutaceae) – kumquat; – cross section through fruit (hesperidium); the edible part of citrus fruits consists of little "juice sacs," which are essentially multi-cellular hairs that emerge from the inner surface of the ovary wall. Eventually, the cell walls inside the juice sacs break down and the fruit becomes juicy and sweet; diameter 2.1cm

Golden apples

Although their fruit wall is predominantly soft, some berries have a particularly tough outer rind. The best-known examples of such armoured berries are citrus fruits. The twenty or so species of the genus *Citrus* occur naturally from northern India to China through south-east Asia, reaching their southernmost point in north-eastern Australia (Queensland). For sweet oranges (*Citrus sinensis*, Rutaceae), mandarins (*C. reticulata*), grapefruits (*Citrus* x *paradisi*) and other citrus fruits such as lemons (*C. limon*), pomelos (*C. maxima*), limes (*C. aurantifolia*), Seville oranges (*C. aurantium*, used for marmalade) and kumquats (*C. margarita*), botanists long ago coined a special name, *hesperidium,* an apparently unlikely name for such a harmless fruit, but not to those who enjoyed a classical education as did the botanists of yore. The orange was the golden apple of Greek mythology, and *Hesperides* was a garden in the west in which golden apples grew. The name hesperidium is simply the result of the Latinization of its Greek root. Despite this glorious name bestowed on them by the ancient Greeks, oranges are not always orange in colour. When travelling in the tropics one can look in vain for orange oranges, only to be surprised by the sweet taste of the very unripe-looking, dark-green fruits on offer everywhere. In tropical countries where it never gets cool, oranges remain green, even when mature. This is because the orange pigments (carotenes) are produced only at lower temperatures. If ambient conditions fluctuate between warm and cold, the fruits may well change colour accordingly. The edible part of an orange has a peculiar origin. Close examination of the segments of an orange, grapefruit, lemon or other citrus fruit reveals that the flesh is formed by multicellular hairs, the distal part of which becomes enlarged. These hairs emerge from the surface of the inner ovary wall and fill the entire space of the loculi around the seeds. When the cell walls inside the hair finally break down, the cavity is filled with juice, creating the "juice sacs" that are so enjoyable to eat.

Fragrant citrons

Citrons are prized for their thick aromatic peel rather than their fleshy pulp. Although most of us have never seen the fresh lemon-like fruits of *Citrus medica*, we have almost certainly consumed their candied peel, which is widely used in the food industry, especially as an ingredient in confectionery and fruit cakes. Like its relatives the lemon and lime, the citron probably originated in India, but because it has been cultivated by humans for thousands of years, its exact origin may never be known. Seeds of the citron dated to 4000 BC have been found in Mesopotamia (today's Iraq). In ancient times, citrons were used mainly for religious and medicinal purposes, serving as a remedy for sea

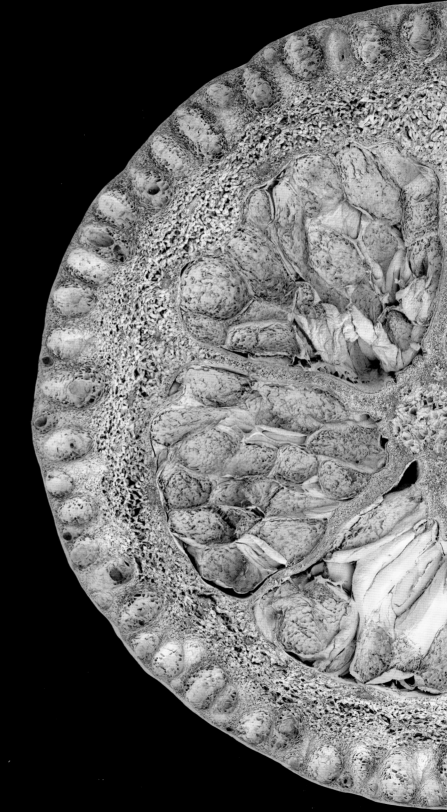

sickness, pulmonary and intestinal ailments, dysentery, and other health problems. With their pleasant fragrance, fresh citrons were also used as a source of perfume and served as natural air fresheners, a purpose for which they are still employed today in central and northern China. Long known to the people of the Orient, the citron finally reached Europe in about 300 BC, when the armies of Alexander the Great introduced the fruit to the Mediterranean, where it is still grown today.

Buddha's hand

The first time that citrus fruits are mentioned in Chinese writings is during the Chou dynasty (1027–256 BC). However, the citron reached China only in around 300 AD. Either here or, as some scholars believe, in northern India, a "freak" variety emerged that has been named *Citrus medica* var. *sarcodactylis*. The epithet "sarcodactylis" (literally, fleshy fingers) aptly describes the bizarre-looking fruit in which the segments appear to have separated into finger-like lobes. Buddhist monks in China and Japan thought it looked like the praying hands of Buddha and adopted this graceful oddity of nature. They have revered it as a symbol of happiness, wealth and longevity for more than a thousand years. The fruit itself is usually green and consists of little more than the spongy rind with hardly any flesh or seeds. Used in the same way as other citrons, Buddha's hands are esteemed for their curious shape and exquisite, non-bitter aroma. Today, the Buddha's hand or "fingered citron" is grown commercially for its strongly aromatic zest, which is almost exclusively used for the manufacture of candied peel. Occasionally the fruits are sold in Western supermarkets where their eccentric appearance catches the attention of unsuspecting grocery shoppers. Fashionable chefs like to add fresh slices of this exotic delicacy to salads and fish dishes to give them a subtle yet familiar lemon taste.

Sizeable pepos

The "pepo" is another special case among berries. Like the hesperidium, it is a berry with a thick, leathery rind. A hint as to where to find pepos in nature gives us the name itself, which, once again, is a Latin word that has its roots in ancient Greek (*pepon* originally meant "ripe"). It was used by Galen, Theophrastus and Hippocrates to describe ripe fruits, for example *sikuopepona* (ripe cucumber). The word *pepon* became *pepo* (= large fruit) in Latin. In about 79 AD, Pliny noted that *cucumeres* (cucumbers), when of excessive size, are called *pepones*. During the course of history, the Latin *pepo*, used for any melon, gourd or pumpkin, became the French *pompon*, which became *pompion* or *pumpion* in English. Finally, early colonists of the New World somewhat inappropriately added the Old Dutch

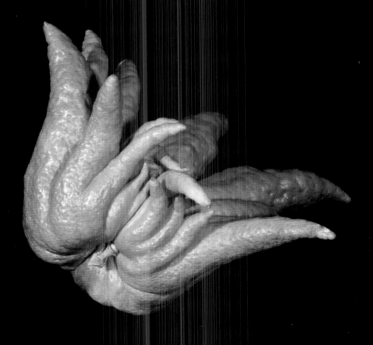

Citrus medica var. *sarcodactylis* (Rutaceae) – Buddha's hand, fingered citron; ancient cultigen, originally from northern India – fruit (hesperidium); the uniquely bizarre shape of this fruit results from the carpels partly separating into finger-like segments. In China and Japan the fruits are esteemed for their strong fragrance and ornamental shape. The fingered citron contains little or no flesh but the peel is candied and used to aromatize food; 16cm long

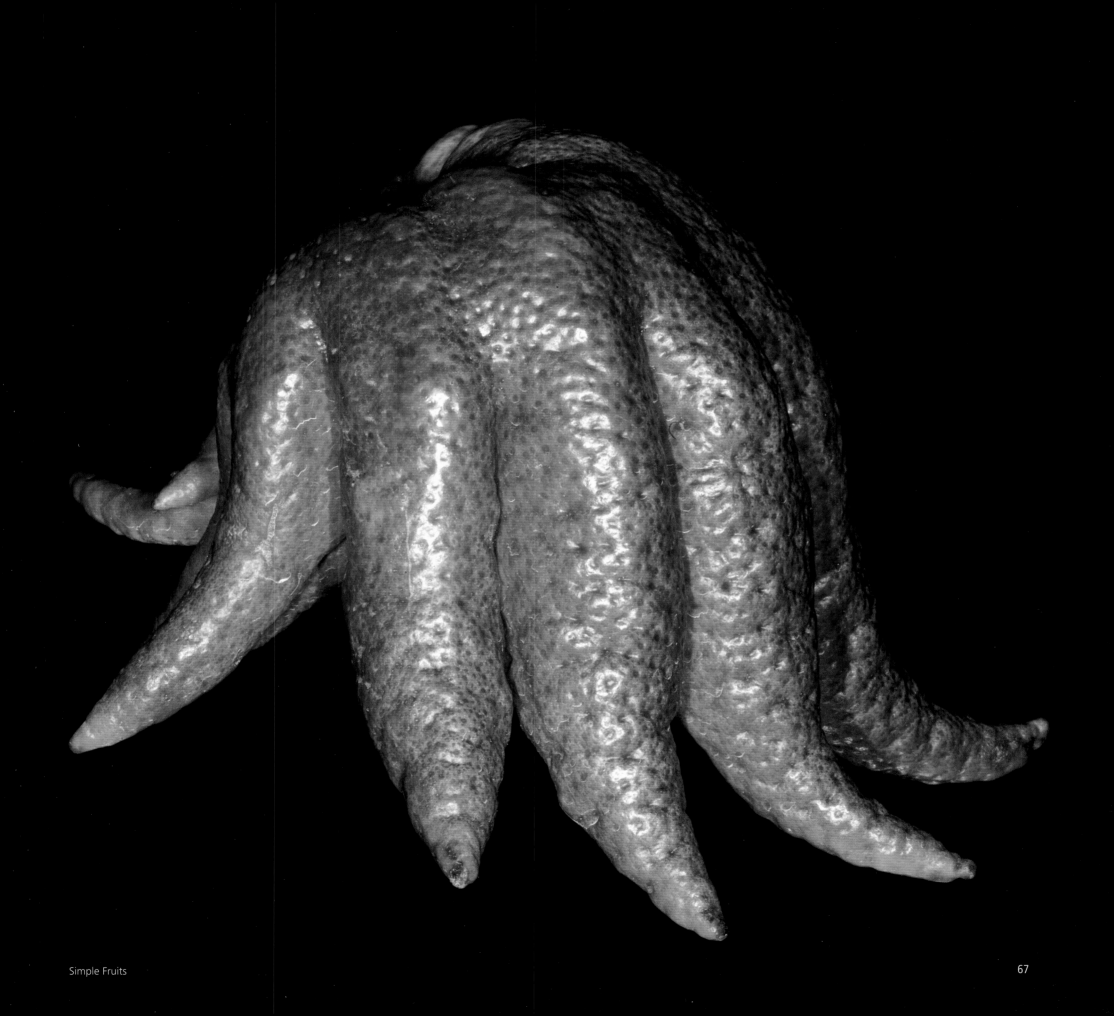

diminutive ending "-ken" to create the word "pumpkin." Pumpkins – members of the
genus *Cucurbita* (Cucurbitaceae) – originated in the tropical and warm climates of the
Americas where they were already an important part of pre–Columbian culture. The fruits
of *Cucurbita maxima*, better known as giant pumpkins, are the subject of fierce competition
in both the United Kingdom and the United States. Every autumn ambitious gardeners
jealously compare the size of their pumpkins. The largest pumpkin ever grown weighed
681.3kg, a world record established on 7 October, 2006 by Ron Wallace of Greene, Rhode
Island, USA. This extraordinary berry (or, strictly speaking, pepo) was not only the biggest
pumpkin but also the largest fruit of any angiosperm ever recorded. Besides the truly
gargantuan giant pumpkins, many other fruits of Cucurbitaceae (the gourd, melon and
pumpkin family) are pepos, including squashes, cucumbers, and the Mexican vegetable
pear or chayote (*Sechium edule*), not to mention the much smaller fruits of many wild
species. There are also several perhaps unexpected examples of pepos that do not belong
to the pumpkin family, such as passion fruit (*Passiflora* spp., Passifloraceae), papayas (*Carica
papaya*, Caricaceae) and bananas (*Musa acuminata*, Musaceae).

Although they have a thick, leathery rind in common, the hesperidium and the pepo
differ in one feature that can only be observed in a cross–section of the fruit. Most
commercial citrus fruits are bred to be almost seedless, but we sometimes find the odd seed.
If seeds are present, they are always found in the middle of the fruit, attached to the centre.
A pepo lacks the septae between the carpels, which in an orange create its segments and, as
a result, the seeds are attached to the inside of the surrounding fruit wall.

Soft shell, hard core *or* how to be a drupe

In contrast to berries, where the entire pericarp (the ovary wall of a ripe fruit) is more or
less soft, stone fruits or *drupes* are characterized by an indehiscent pericarp that is
differentiated into three distinct layers: the thin skin (*epicarp*), the fleshy pulp (*mesocarp*) and
the stone formed by the hard woody inner layer of the pericarp (*endocarp*). A typical
familiar example of a drupe is the fruit of *Olea europaea* (Oleaceae), better known as the
olive. Apart from being immensely useful and delicious, the olive is particularly significant
in the context of drupes. A closer look at the etymology of the word "drupe" reveals that
it is derived from the Latin *drupa*, which itself stems from the Greek *dryppa*, the name of
the olive in ancient Greece.

The stone of a drupe is often mistaken for the seed itself but usually houses a single seed
at its core. In a kind of "evolutionary takeover," the hard endocarp provides for the physical
protection of the seed, a role normally performed by the seed coat. Having lost its function

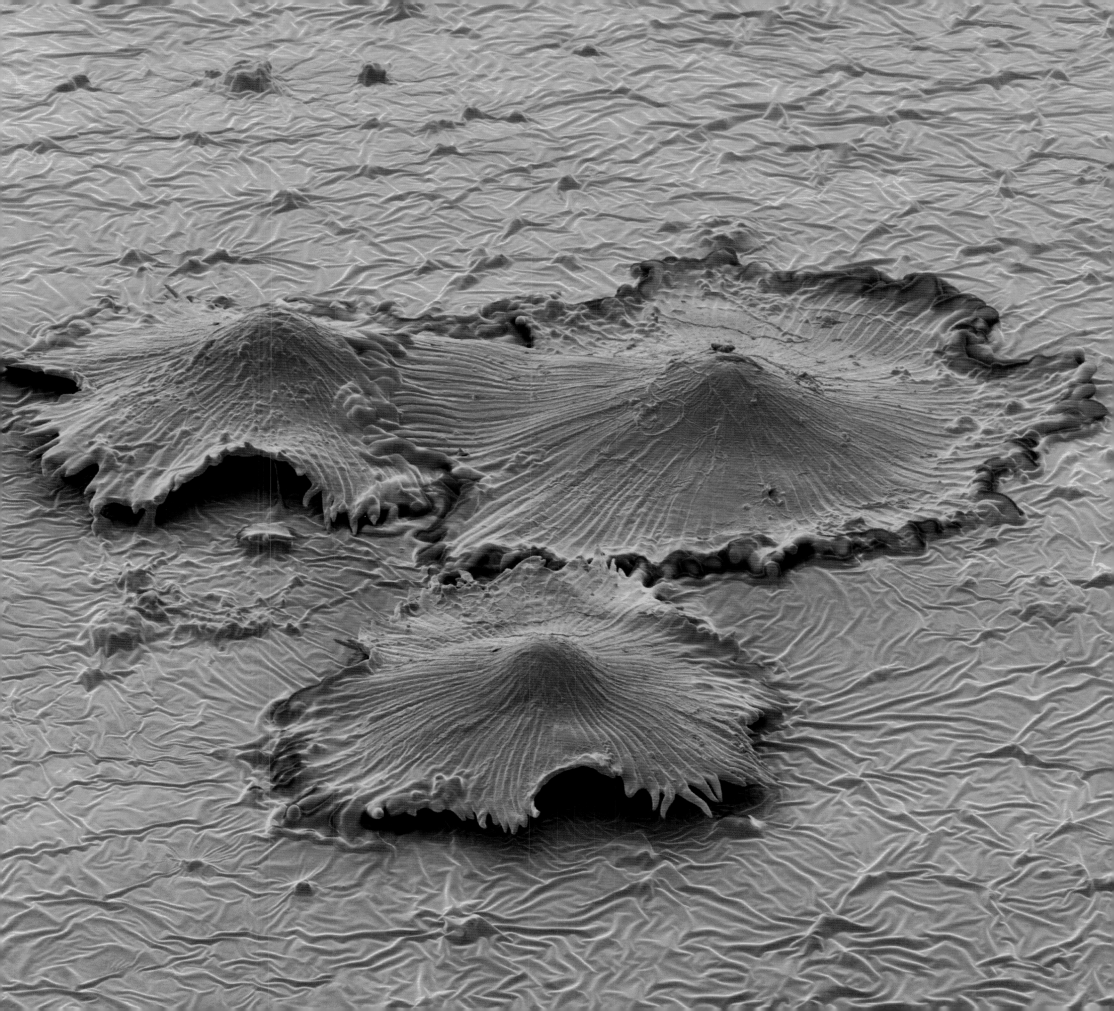

the coat of a drupe-borne seed is usually thin and rudimentary, especially in phylogenetically old drupes. For example, the brittle, brown skin that covers shelled pistachios (*Pistacia vera*, Anacardiaceae) represents the remnants of the seed coat.

The most common edible stone fruits in temperate climates belong to members of the rose family (Rosaceae). Most popular among these are cherries (*Prunus avium*), plums (*Prunus* x *domestica*), apricots (*Prunus armeniaca*), peaches (*Prunus persica* var. *persica*) and nectarines (*Prunus persica* var. *nucipersica*). Their stones contain only a single seed, which makes good sense from a dispersal point of view. Just like berries, drupes are adapted to be eaten by animals, who chew the fruit and swallow the pulp together with the seeds or stones. However, unlike drupes, berries often contain many small seeds, in line with the principle of safety in numbers. Even if some seeds get squashed by the grinding molars of the dispersing animal, at least some will pass through the mouth unharmed and end up in the gut, from where they will eventually be released unscathed and equipped with a pile of instant fertilizer. Drupes have evolved a slightly different strategy. Because their seeds are generally much larger, the chances of getting past the beak of a bird or the teeth of a mammal without being damaged are low. To ensure a safe passage, the fruit encapsulates its seed in an armoured stone, which is swallowed or, if too big, simply discarded by the feeding animal.

Although most drupes contain only a single stone with a single seed, there are some exceptions. The West African sugar plum (*Uapaca guineensis*, Phyllanthaceae) contains three separate single-seeded stones (*pyrenes*) in one fruit, whereas the Australasian Burdekin plum (*Pleiogynium timoriense*, Anacardiaceae), bush tucker of the Aborigines in Queensland, has a single large stone with several seeds, each locked in its own individual compartment.

Nuts about nuts

By now we should be getting used to the fact that botanists use familiar fruit terms in a very different and much more rigorous sense than we do in our everyday language. This incongruence between culinary and botanical usage could hardly be greater when it comes to nuts. For the food industry, chefs and "regular" consumers who enjoy a tasty nibble, any large edible kernel that requires forceful liberation from a hard shell before consumption is unscrupulously addressed as a nut. In a botanical sense, a *nut* is only a nut if it consists of nothing but the mature ovary of an indehiscent simple fruit with a hard, dry pericarp, usually harbouring a single seed. This is true of hazelnuts (*Corylus avellana*, Betulaceae), sweet chestnuts (*Castanea sativa*, Fagaceae), walnuts (*Juglans regia*, Juglandaceae), pecan nuts (*Carya illinoiensis*, Juglandaceae), beechnuts (*Fagus* spp., Fagaceae), acorns (*Quercus* spp., Fagaceae),

below: *Juglans regia* (Juglandaceae) – walnut; native to Eurasia – fruit (pseudodrupe); the fruit appears to be a drupe, but its fleshy husk is not derived from the outer layers of the pericarp but is formed by a series of fused bracts. The actual ovary develops into a nut with a hard, dry pericarp

bottom: *Prunus dulcis* (Rosaceae) – almond; native to Western Asia – fruit (nuculanium); the fruit is similar to a drupe but the epi- and mesocarp are dry and split open to disperse the stone. A genuine "carpological troublemaker" that is often oxymoronically called a "dry" or "dehiscent" drupe

below: *Castanea sativa* (Fagaceae) – sweet chestnut; native to south-eastern Europe and the Mediterranean – fruit (trymosum); what appears to be a simple capsule is actually a compound fruit. Up to three female flowers are surrounded by a cupule, which later forms the spiny husk around the chestnuts, each of which represents a mature ovary

bottom: *Quercus robur* (Fagaceae) – English oak; native to Europe and the Mediterranean – fruit (glans); the open cupule of the oaks holds a single mature ovary – the acorn – unlike other Fagaceae such as beech (*Fagus sylvatica*) that hold several

and unshelled peanuts (*Arachis hypogaea*, Leguminosae[6]), although the latter often contain more than one seed. Other culinary "nuts" such as unshelled almonds (*Prunus dulcis* var. *dulcis*, Rosaceae), pistachios (*Pistacia vera*, Anacardiaceae), and, at least technically, cashew nuts (*Anacardium occidentale*, Anacardiaceae) are, in fact, the stones of drupes, whereas Brazil nuts (*Bertholletia excelsa*, Lecythidaceae), macadamias (*Macadamia integrifolia* and *M. tetraphylla*, Proteaceae), ginkgo nuts (*Ginkgo biloba*, Ginkgoaceae) and pine nuts (*Pinus pinea*, Pinaceae) simply represent seeds. We should not forget that ginkgos and pines are gymnosperms and as such are not officially entitled to bear fruits, not to mention real nuts.

Because of their ambiguous meaning, even in the scientific literature, the most recent comprehensive carpological classification by Richard Spjut (1994) rejects "nut" and "nutlet" as scientific terms and replaces them with several more precisely defined terms (*achene, camara, carcerulus, caryopsis* and *cypsela*). Whilst we agree with Spjut, for the purposes of this book we have nevertheless chosen to use "nut" and "nutlet" in the strict sense defined above in order to keep the complexity of the subject at a level suitable for a broader, non-specialist audience.

Walnuts or waldrupes?

Those who hoped that this was the end of the confusion over nuts may be amused to learn otherwise. Many of the fruits that we just classed as proper nuts qualify only if nothing but the qualities of the mature ovary are taken into account. For example, fresh walnuts look more like drupes. They are covered by a fleshy green husk that peels off easily when the fruits are ripe. However, the bony shell of the walnut represents the pericarp (ovary wall) in its entirety, whilst the fleshy green husk is formed by a series of fused *bracts* (modified leaves) that initially subtend the ovary. As the fruit develops, the fused bracts grow until they cover the ovary completely and create the illusion of an epi- and mesocarp, rendering the walnut not a drupe but a *pseudodrupe*. This may seem to be a rather exceptional case but pseudo-drupes are also typical of members of the oleaster family (Elaeocarpaceae) such as sea buckthorn (*Hippophae rhamnoides*), where the persistent hypanthium forms a loose, fleshy cover around the achene formed by the ovary. However, the walnut and the Elaeagnaceae are only two examples of a large group of carpological troublemakers assembled in a category labelled "anthocarpous fruits." We shall meet more of these interesting eccentrics later.

Glans quercus

Although they are real nuts, beechnuts and sweet chestnuts pose a different problem. Borne in groups of usually two (beechnuts) or three (sweet chestnuts), they appear to be shed from

pary capsules, which would mean that they are seeds and not nuts. However, the spiny husk
s not derived from the ovary wall but rather represents a very peculiar structural character
of the beech family (Fagaceae), in which it is called a *cupule* (Latin *cupula* = little cask). The
simplest cupules are found in oaks (*Quercus* spp.) in which the cup holds a single nut, the
acorn. So unusual is the cupule of Fagaceae that its developmental origin has long been a
mystery. Some have argued that it is derived from lobed extensions of the stalk beneath each
female flower; others have demonstrated quite convincingly that the cupule represents a
modified perianth. Whatever the origin of the cupule, it forms an integral part of the fruit
and renders the acorn a rather special nut, aptly called *glans* (Latin = acorn) by hardcore
carpologists. In beeches (*Fagus* spp.), chestnuts (*Castanea* spp.) and other members of
Fagaceae, the cupule contains two or more nuts and effectively creates a kind of "super-
fruit" or, in scientific terms, a compound fruit, correctly addressed as a *trymosum*.

Two fruits in one – cashew nut and cashew apple

Arguably the tastiest and botanically most interesting nut is the cashew nut. Nowadays
cultivated and naturalized almost all over the tropics, the cashew tree (*Anacardium occidentale*,
Anacardiaceae) is originally native to the coastal plains of north-eastern Brazil, where it
forms part of the so-called *restinga* vegetation. Its fruits were utilized by Brazilian Indians
long before European colonization in the sixteenth century. Called "acajú" by the members
of the Tupi tribe, the name was converted by the Portuguese into "cajú" and eventually
became "cashew" in English.

Not quite creating a compound fruit, but nevertheless complementing the "nut" that it
carries with a very special accessory, the stalk (*pedicel*) of the cashew nut (*Anacardium
occidentale*) produces a large, fleshy, pear-shaped swelling known as the "cashew apple."
Dangling from the base of the cashew apple is the kidney-shaped seed-bearing part of the
fruit, the actual cashew nut. Interestingly, upon closer examination of a fresh fruit, the
cashew nut turns out not to be a real nut after all but a drupe, the type of fruit that is typical
of so many of its close relatives in the sumac family (Anacardiaceae), such as the mango
(*Mangifera indica*). Although hardly recognizable as a drupe, the leathery pericarp of the
cashew nut does indeed display the three defining layers: an outer skin (epicarp), a very thin,
quick-drying but nevertheless soft middle layer (mesocarp), followed by the dominating
thick, woody endocarp. Whilst the pericarp of the cashew nut is poisonous owing to an
acrid phenolic oil that causes dermatitis, the harmless, extremely juicy cashew apple can be
enjoyed by drinking the juice and discarding the fibrous residue. Because of the problems
caused by the toxic shell of the "nut," Latin Americans, West Indians and West Africans have

opposite: *Exocarpos sparteus* (Santalaceae) – broom ballart,
native to Australia – fruit (glans); although only very distantly
related to the sumac family (Anacardiaceae), the broom
ballart's small, bird-dispersed fruits are structurally very
similar to the cashew. In the dried fruit shown here the
wrinkly, shrunken part represents the once round, smooth,
fleshy flower stalk; 8mm long

Anacardium occidentale (Anacardiaceae) – cashew nut; native
to north-eastern Brazil, widely cultivated throughout the
tropics – fruit (glans); the cashew nuts we eat represent only
the storage embryos borne inside the single-seeded mature
ovary which, although superficially nut-like, turns out to be a
drupe with a very thin layer of flesh. The entire fruit also
includes a large, fleshy, pear-shaped swelling of the flower
stalk, the "cashew apple"

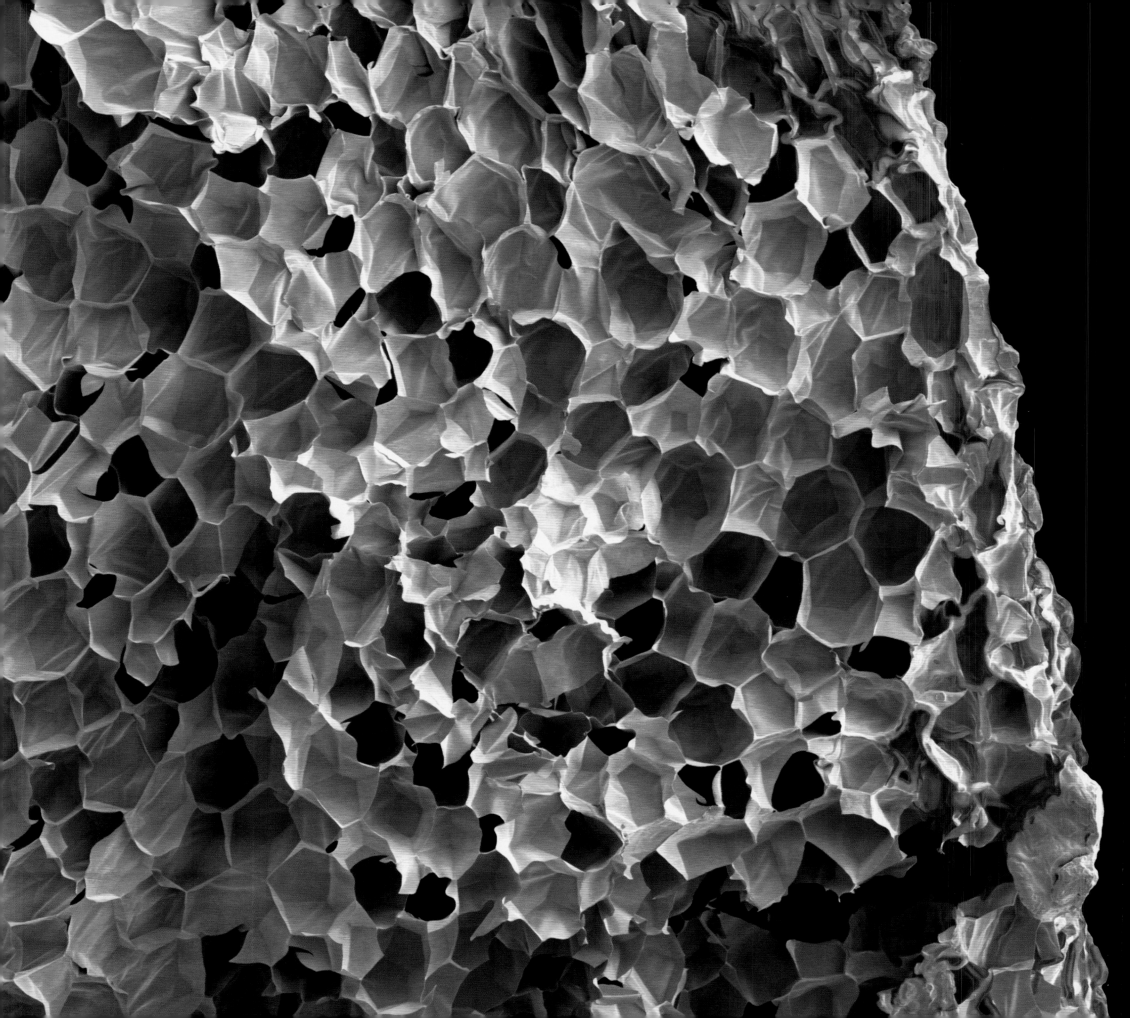

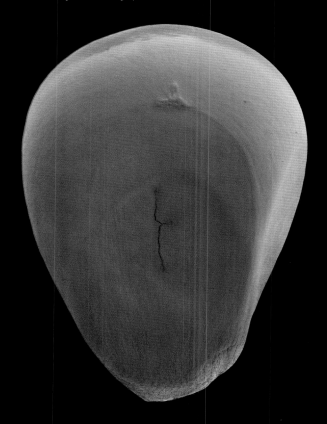

long used only the succulent "apple," making it into wine and refreshing beverages, similar to lemonade, such as the Brazilian "cajuado." However, on a worldwide scale the seed of the cashew tree is still the main commercial product, despite the laborious cleaning process that makes cashew nuts the most expensive of all nuts (in a culinary sense). In the wild, the brightly coloured, 5-10cm long cashew apple acts as a tasty reward for the animals it needs for dispersal. Fruit bats and monkeys pick the fruits to feed on the yellow- to scarlet-coloured "apple" but discard the poisonous "nut," leaving the seed inside unharmed.

Although not closely related to the cashew tree, the Australian broom ballart (*Exocarpos sparteus*), a hemiparasitic member of the sandalwood family (Santalaceae), has evolved the very same type of fruit. Admittedly on a much smaller scale, the broom ballart offers a round, fleshy, bright red, miniature "apple" to entice birds into dispersing its fruits.

Wheat "grain" and sunflower "seed" – caryopsis and achene

We have already revealed that some of our most cherished nuts are seeds. However, the "nutty" confusion works both ways. With the pericarp having taken over the physical protection of the seed from the seed coat, many nuts look and behave deceptively like seeds. Mere mortals may therefore be forgiven for calling some nuts "seeds" in common parlance, especially if the respective specimens are very small and not at all nut-like in that they do not appear either chunky or delicious. The most common examples are the fruits of the grass family (Poaceae), to which most of our cereal grains belong. Each whole grain of wheat (*Triticum aestivum*), oat (*Avena sativa*), rye (*Secale cereale*), barley (*Hordeum vulgare*), rice (*Oryza sativa*), and maize (*Zea mays*) represents a single nut-type fruit, traditionally called a *caryopsis*.

Although sunflowers and other members of the same family (Asteraceae) have little in common with grasses, their "seeds" are likewise single-seeded nuts. For nuts like these, the correct scientific label is *achene*. With various definitions scattered throughout the botanical literature, its precise meaning is obscure to most botanists who generally use it to categorize small nuts with a pericarp that is contiguous with the seed coat and soft enough to be squashed between one's finger nails. Admittedly, this does not sound particularly scientific, and the distinction between *achene* and *caryopsis* can only be made with the help of additional requirements such as the degree of cohesion between the pericarp and the seed coat or the position of the ovary (superior or inferior). The truth is that, in botany, the fruit of the grasses has long been called a *caryopsis*. Old habits die hard, so it is simply easier to tailor definitions around established terms rather than breaking with long-standing tradition. Attentive readers may already suspect that there are yet more compromises of this kind to be revealed.

Samaras – nuts gone airborne

There remains one elite group of nuts, namely those that have developed sophisticated aerodynamic structures that allow them take advantage of air currents for their dispersal. The structures that enable these fruits to go airborne are usually wings or feathery appendages of various origins. Depending on the organ that contributes the flight aid, carpologists distinguish between various kinds of flying nut.

The ovary wall can produce flat outgrowths acting as wings that are delicately balanced around the centre of gravity of the (usually) single-seeded fruit. In recognition of this ingenious "invention," which it took humans centuries to design, botanists bestowed on them the name *samara*. Samaras can have a single, one-sided wing, as in the fruits of ash trees (*Fraxinus* spp., Oleaceae) and the fruitlets of maples (*Acer* spp., Sapindaceae), two of which jointly form a whole fruit (a *samarium*) until they break up at maturity. Surprisingly similar fruits have evolved independently in the legume family, which is otherwise characterized by bean-like pods. The place to look for the most spectacular leguminous samaras is Brazil. Here we find the tipu tree (*Tipuana tipu*), which has become a popular street tree throughout the tropics, the Pau de Moco (*Luetzelburgia auriculata*) and the Brazilian zebra wood tree (*Centrolobium robustum*), whose samaras consist of a viciously spiny, golf-ball sized structure bearing a gigantic wing up to 30cm in length.

The samaras of elm (*Ulmus* spp., Ulmaceae), hop tree (*Ptelea trifoliata*, Rutaceae), and Christ's thorn (*Paliurus spina-christi*, Rhamnaceae), to name but a few, are equipped with a continuous wing surrounding the central, seed-bearing part. Once again, the same model is found in some members of the legume family, most notably in wild teak (*Pterocarpus angolensis*). The large samaras of this African tree are similar to the fruits of the Brazilian *Centrolobium robustum* and cover their seed-bearing part with long soft spines. Since even a strong gust of wind would carry the heavy fruits only a few metres away from the parent tree, the spines developed as part of a two-pronged dispersal strategy, aimed at becoming entangled in the fur of passing mammals. Another example where a close relative has fruits with a peripheral rather than a unilateral wing is the Chinese *Dipteronia*, which together with *Acer*, formerly constituted the small family Aceraceae until both genera were subsumed in the Sapindaceae.

A crosswise set of four wings occur along the sides of the nuts of the southern African raasblaar (*Combretum zeyheri*) and other *Combretum* species. Although the raasblaar boasts sizeable samaras of up to 8cm in diameter, it is beaten by one extraordinary member of the mallow family (Malvaceae). The fruits of the cuipo tree (*Cavanillesia platanifolia*), a giant tree that grows in the rainforests of Central America, are surprisingly similar but larger and bear

below: *Dipteronia sinensis* (Sapindaceae) – dipteronia; native to China – fruit (samarium); the schizocarpic fruit breaks into two wind-dispersed fruitlets (samaras) that bear a peripheral wing around the central seeded portion; fruit c. 5-6cm wide

bottom: *Centrolobium ochroxylum* (Leguminosae) – amarillo de Guayaquil; native to Ecuador – fruit (samara); slightly smaller, but otherwise very similar to those of the Brazilian zebra wood tree (*Centrolóbium robustum*), the samara of this species consists of a viciously spiny, golf-ball-sized structure with a 20-25cm long wing. The wing and the spines indicate a double dispersal strategy (wind and animal dispersal), although the fierce spines on the seed-bearing part could represent an adaptation to deter seed predators

Dipterocarpus grandiflorus (Dipterocarpaceae) – keruing belimbing (Malay); native to the lowland rainforests of south-east Asia – fruit (pseudosamara); in members of the genus *Dipterocarpus*, only two of the five sepals enlarge and develop into wings as the fruit matures. Because the wings are not formed by the ovary wall as in true samaras, the fruits of the Dipterocarpaceae are called *pseudosamaras*; fruit 25cm long

opposite: *Scabiosa columbaria* (Dipsacaceae) – small scabious; native to southern Europe, western Asia and north-east Africa – "flower;" like the sunflower family (Asteraceae) the members of the teasel family arrange their flowers in head-like inflorescences (capitula); diameter 4cm

Scabiosa columbaria (Dipsacaceae) – fruit (cypsela); the cypselas of the teasel family possess a collared "air bag" derived from four united bracts as well as a set of stiff awns derived from the calyx of the flower. The collar assists wind dispersal whereas the awns are suitable for hooking on to the fur of passing animals; 6mm long

more wings than the raasblaar. Even more impressive are the flying nuts of the Dipterocarpaceae. With either two, three or five large apical wings, they gracefully helicopter from the canopy to the ground of the lowland rainforests of south-east Asia. However, since their "rotor blades" are formed by the persistent sepals of the flower and not by the ovary wall, the fruits of the Dipterocarpaceae qualify only as *pseudosamaras* rather than genuine samaras.

Cypselas – achenes gone airborne

In order to get their plump, heavy nuts airborne, samaras need large aerodynamic wings. Tiny achenes can get by with just a plume of hairs or a feathery appendage that affords them high wind resistance. Many of the sunflower's relatives in the family Asteraceae add delicate parachute-like structures to their achenes that enable them to hitch a ride on even the lightest breeze.

The "seed head" of a dandelion (*Taraxacum officinale*) or a meadow salsify (*Tragopogon pratensis*), for example, is a densely packed crowd of "parachute" fruits waiting to go airborne. In the sunflower family a large number of tiny flowers are typically arranged together in inflorescences that mimic an individual flower, hence their densely clustered fruits. When it is present, the feathery appendage, called a *pappus*, develops from the highly modified calyx of the tiny flower. Its origin from sepals is more obvious in the less delicate sessile pappuses of the fruits of *Galinsoga brachystephana* and the Texas sleepy-daisy (*Xanthisma texanum*) than in the stalked feathery parachutes of dandelion and meadow salsify. A pappus significantly changes a fruit's appearance and function, which is why in the eyes of the botanists it has become a *cypsela*.

Similar but more elaborate cypselas than those of the Asteraceae are found in the related teasel family (Dipsacaceae). Here, the pursuit of a double dispersal strategy has resulted in a more sophisticated fruit structure. Contrary to the asteraceous pappus, the wind-catching parachute of a dipsacaceous cypsela is not produced by the flower's calyx but by the collar of an external "air bag." The air bag, which is formed by four laterally fused bracts surrounding the flower, helps to lower the specific weight of the fruit, thereby enhancing its buoyancy in the air. In addition, the calyx of the flower of the Dipsacaceae develops into a set of stiff awns, suitable for hooking onto the fur of a passing animal. The airborne qualities of the cypselas of the teasel family are far weaker than those of the sunflower family owing to their larger size and weight. Perhaps the pursuit of a double dispersal strategy – anemochory and epizoochory – compensates for the Dipsacaceae's limited air-travel skills.

opposite: *Scabiosa crenata* (Dipsacaceae) – native to the central and eastern Mediterranean – fruit (cypsela); like the small scabious (previous page), the cypsela of *Scabiosa crenata* pursues a double dispersal strategy: the papery collar facilitates wind dispersal whereas the rough calyx awns are poised to hook on to the fur of a passing animal; diameter 7.2mm

bottom: *Scabiosa crenata* (Dipsacaceae) – a cluster of fruits (infructescence) as it develops from the flower-like inflorescence (capitulum)

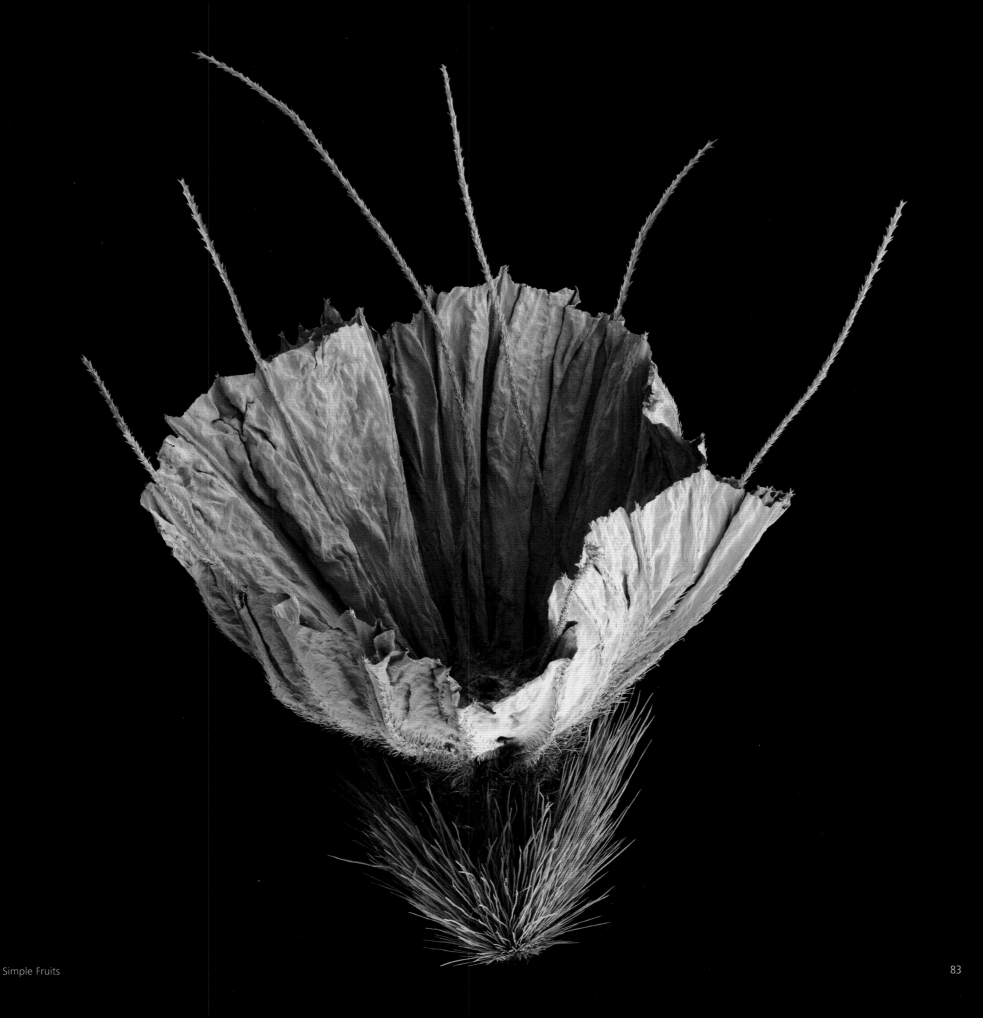

*Galinsoga brachyste*phana (Asteraceae) – native to Central and South America – fruit (cypsela); in the tiny shuttlecock-like cypselas of this species the rays of the modified calyx perform as tiny feathery wings; 2.5mm long

Xanthisma texanum (Asteraceae) – Texas sleepy-daisy (name alluding to the capitula, which close at night), the only species of the monotypic genus; native to the south-eastern United States – fruit (cypsela); although its spreading pappus rays are able to assist wind dispersal, they are very narrow with teeth along the margin that make them even better adapted to animal dispersal (epizoochory); 7mm long

Simple Fruits

Pods and such like

We intuitively call any dry fruit that encloses one or many loose seeds within an air space a *capsule* or a *pod*, especially if it rattles when shaken. Botanists are less impressed by acoustic qualities and take a slightly different view. Many believe that every capsule can pass as a pod but not every pod qualifies as a capsule. Some even limit the usage of the term "pod" to fruits of the legume family (Leguminosae). We consider "pod" a colloquial rather than a scientific term, since it is commonly used for all dry fruits with a firm fruit wall surrounding a cavity that contains one or more seeds, irrespective of whether the underlying ovary is composed of one or several carpels and whether or not the fruit opens (dehisces) at maturity. Carefully distinguished from monocarpellate and indehiscent pods, which will be discussed later, botanically correct *capsules* are defined as *simple dehiscent fruits* formed by *syncarpous gynoecia*, that is, at least two or more joined carpels. Despite this rather narrow definition, the capsule is one of the most frequently encountered fruit types among the angiosperms.

Capsules *or* seven ways to open a fruit

To qualify as dehiscent fruits, capsules must somehow open at maturity to release their seeds. With respect to the opening (dehiscence) of the pericarp, there are several possible strategies. Most commonly, capsules open along pre-formed lines of dehiscence. These lines can run along the middle of each loculus, creating a *loculicidal capsule*, or coincide with the septae, resulting in a *septicidal capsule*. Loculicidal capsules are far more common than septicidal capsules. They are found in many species of both monocotyledons and dicotyledons. Typical *monocots* with loculicidal capsules are agaves (*Agave* spp., Agavaceae), aloes (*Aloe* spp., Aloeaceae), irises (*Iris* spp., Iridaceae), lilies (*Lilium* spp., Liliaceae) and the truly amazing bird-of-paradise flower (*Strelitzia reginae*, Strelitziaceae) from South Africa. The loculicidal capsule of *Strelitzia* is typical of monocots in that it is formed by three joined carpels and hence splits, when ripe, into three large valves, each consisting of two halves of adjoining carpels. Upon opening, the fruits reveal the most curious seeds. Lined up along the placenta, marked by the ridge (*septum*) that runs along the middle of each valve, are two rows of pea-sized black seeds, which attract the attention of birds by means of an appendage (*aril*) that looks like a shaggy, bright orange wig. The fact that it is almost impossible to find an open capsule with any seeds left is proof of the great success of this advertising strategy. Interestingly, the related traveller's palm from Madagascar (*Ravenala madagascariensis*, Strelitziaceae) has very similar seeds but with a blue wig.

Loculicidal capsules, which are obviously a very efficient method of dehiscence, are also common in *dicots*. One example that everyone remembers from childhood is the horse

chestnut (*Aesculus hippocastanum*, Sapindaceae). Unlike its distant relative the sweet chestnut (*Castanea sativa*, Fagaceae), which bears true nuts, horse chestnuts are seeds, borne individually or as a pair, within large, spiny, loculicidal capsules.

In septicidal capsules, where the carpels separate along their septae, each valve consists of one entire carpel. This is beautifully illustrated by the boat-shaped valves of the fruits of the eastern Australian crow ash (*Flindersia australis*), a member of the large citrus family (Rutaceae).

A loculicidal capsule is easily distinguished from a septicidal capsule when septae are present. However, if the capsule has only a single loculus and the seeds are either attached to the inside wall of the pericarp (*parietal placentation*) or to a central column (*central placentation*), then the position of the dehiscence lines relative to the placenta(s) has to be determined.

Teeth, fissures, cracks and lids

Apart from these two mainstream types of capsules, there are some interesting variations on the theme. Some capsules open regularly along longitudinal *sutures* but only near the apex, not along their entire length. Their short valves look like little teeth, which has earned them the name *denticidal capsules*. To satisfy the requirements of a strict carpologist, a denticidal capsule is not supposed to split along more than one-fifth of its length. The capsules of the purple loosestrife (*Lythrum salicaria*, Lythraceae), campions (*Silene* spp., Caryophyllaceae) and primroses (*Primula* spp., Primulaceae), to name but a few, all comply with this rule. Their seeds are gradually dispersed through the narrow opening at the top of the capsule like salt from a saltshaker as the fruits sway back and forth in the wind on their flexible, slender stalks.

Fissuricidal capsules open regularly along sutures between a closed apex and base. This seemingly impractical type of dehiscence has evidently evolved from regular types of capsules as a consequence of their development from an inferior ovary. Fissuricidal capsules are the typical fruit of orchids (Orchidaceae) and Cannaceae (e.g. Indian shot, *Canna indica*).

When loculicidal or septicidal capsules open in a way that causes their septae to break near the central axis of the fruit leaving a persistent column in the centre, they become what botanists refer to as *septifragal capsules*. Some of the most beautiful fruits of this type are found in the mahogany family (Meliaceae). For example, the impressive capsules of the tree *Swietenia mahagoni*, the economically important Cuba mahogany of the trade, shed 40–50 large, winged seeds after the heavy pericarp has split and dropped off the stalk high up in the canopy. Birds rather than the wind are invited to carry away the seeds of the brightly

Papaver rhoeas (Papaveraceae) – corn poppy; native to Eurasia and North Africa – fruit (poricidal capsule)

left: side view of capsule; the seeds are flung out as the capsule sways in the wind on its long, flexible stalk. The upper protruding rim prevents rain water from entering through the pores

right: top view of the capsule; the downy rays represent the persistent remains of the stigma; diameter 6.5mm

coloured septifragal capsules of *Hedychium horsfieldii*, a member of the ginger family (Zingiberaceae). When ripe, the vivid orange and – for a capsule – unusually fleshy pericarp peels back to reveal the central column, which carries three ridges of densely packed, deep red seeds, signalling an edible reward in the form of a fleshy seed coat.

The two remaining ways in which a fruit can open to release its seeds do not follow the pre-formed longitudinal lines of dehiscence along the septae or down the middle of the loculi. One possibility is that the pericarp opens with a number of pores, each marking a loculus. For example, poppy capsules (*Papaver* spp., Papaveraceae) open with a ring of pores around the top, whereas bellflowers (*Campanula* spp., Campanulaceae) open with three pores at the base. Seed dispersal in such *poricidal capsules* follows the same strategy as in the denticidal capsules of the pink family (Caryophyllaceae), where the seeds are expelled as the fruits sway in the wind. The second unconventional method of capsular dehiscence is "self-circumcision:" a transverse split parallel to the equator cuts across all the carpels resulting in the formation of a lid. Fruits of this kind are called *circumscissile capsules* or *pyxidia* (singular *pyxidium*, New Latin, from Greek: *pyxidion* = small box). They are found in a variety of angiosperms, including the temperate scarlet pimpernel (*Anagallis arvensis*, Myrsinaceae), the twinleaf (*Jeffersonia diphylla*, Berberidaceae), plantains (*Plantago* spp., Plantaginaceae), and the tropical monkey pot (*Lecythis pisonis*, Lecythidaceae) from South America.

Follicle and coccum

In addition to this cornucopia of capsules, there are various pods that are considered "non-capsular" because they consist of only a single mature carpel and/or remain closed even when ripe. Familiar pods formed by hypogynous flowers whose gynoecium consists of a single carpel are from the legume family (Leguminosae) and the protea family (Proteaceae). Bearing in mind that a carpel is simply a folded, fertile leaf, we can distinguish between its ventral meridian, which represents the line along which the opposite edges of the leaf originally fused, and the dorsal meridian, which coincides with the midrib. These meridians are the two preferred longitudinal lines of dehiscence (sutures) along which pods open. A monocarpellate pod that splits only along the ventral suture is called a *follicle,* whereas one that splits in half along both the ventral and dorsal suture is called a *coccum.*

Carpels that open in a follicle-like manner are often found in primitive angiosperms with multiple fruits such as the Ranunculaceae (e.g. *Caltha palustris*), Paeoniaceae (*Paeonia* spp.), and Illiciaceae (e.g. *Illicium verum,* star anise), but these will be discussed later. A rare example where a follicle truly represents a simple fruit is the Australian macadamia nut (*Macadamia integrifolia* and *M. tetraphylla,* Proteaceae). Each macadamia follicle contains a

below: *Strelitzia reginae* (Strelitziaceae) – bird-of-paradise flower; native to South Africa – fruit (loculicidal capsule) with two remnant seeds. The black, pea-sized seeds have a nutritious, bright orange aril to attract birds for dispersal; diameter of fruit c. 6-8cm

bottom: *Ravenala madagascariensis* (Strelitziaceae) – traveller's palm; native to Madagascar – fruit (loculicidal capsule); a close relative of the bird-of-paradise-flower, the traveller's palm has very similar fruits but the seeds carry blue instead of orange arils

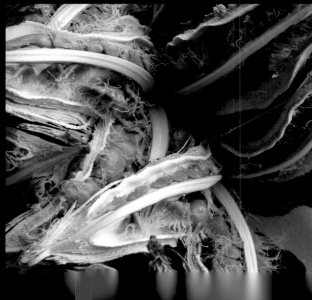

below: *Hedychium horsfieldii* (Zingiberaceae) – Java ginger; native to Java (Indonesia) – fruit (septifragal capsule); the three leathery, bright orange valves reveal the seeds embedded in a fleshy red aril to attract birds for dispersal; c. 3cm long

bottom: *Hakea orthorhyncha* (Proteaceae) – bird-beak hakea; native to Australia – fruit (coccum); the fruits of hakeas consist of just one carpel that splits along the dorsal and ventral side. Their thick, woody pods provide protection against both fire and predators, especially heavy-billed, seed-eating members of the parrot family (Psittacidae); fruit 4-5cm long

large single seed (the macadamia "nut") and opens with a narrow split along the ventral suture. The gap in the pericarp is far too narrow for the seed to escape and appears to be a rudimentary reminder of its evolutionary past when the seeds were probably set free from the follicle as in some of its close relatives, such as grevilleas (*Grevillea* spp.).

In south-western Australia we find representatives of the protea family, which possess some extraordinary examples of coccum-type fruits. Adapted to the dry, fire-prone environment of the Australian outback, many hakeas (*Hakea* spp.) retain their fruits for years after they have reached maturity, keeping them tightly closed to protect the precious seeds. Their often heavily armoured pods (cocci) open only when the water supply to the fruit is interrupted, which occurs naturally when the plants die through fire, disease or insect damage. The strategy of accumulating the entire crop on the plant year after year is called *serotiny*. Serotiny is a common adaptation of many woody plants in habitats where wildfires are a frequent, or even seasonal event, such as the Australian bush, African savannahs, the fynbos of South Africa's Cape region and some forests in North America. In such environments, the safest time for seed dispersal is immediately after a fire. With hardly any organic matter left to burn, new fires are unlikely in the near future, providing a safe window of opportunity for seed dispersal and germination. Serotiny creates a natural canopy-stored seed bank, ensuring the survival of a species in an area even after the mature individuals have died in a fire.

Considering the high temperatures they have to withstand, it is not surprising that the fruits of serotinous species are often equipped with a thick, hard pericarp. Research has shown that in strongly serotinous *Hakea* species, which retain their fruits for at least five years, the woody fruit wall is more likely to be thicker and harder than in non- or weakly-serotinous species. Although it is obvious that a thicker pericarp provides better insulation from heat, the amount of material some strongly serotinous hakeas invest in the physical protection of their fruits seems grossly exaggerated (e.g. *Hakea orthorhyncha, H. sericea, H. platysperma*). After all, the need to survive fire is just as critical for weakly-serotinous hakeas with thin-walled fruits. The true adaptive significance of the heavy body armour seems to be greater protection from the heavy-billed, seed-eating members of the parrot family (Psittacidae), which abound in Australia.

Pods as in "pea pods"

The legumes (Leguminosae) are the third largest family of angiosperms and second only to the grasses (Poaceae) in their economic importance. There are some 19,000 species of legumes, all of which have just one carpel per flower available to turn into a fruit. That this is by no means a handicap is shown by the astonishing array of shapes, sizes and dispersal strategies of legumes. The typical fruit of the legume family is a dry, dehiscent pod that opens along the dorsal *and* ventral suture, splitting the fruit in half. The alert reader will recall that this describes precisely what we have just identified as a coccum, and this is indeed so. Nevertheless, a coccum produced by a member of the Leguminosae is – aptly, one may argue – called a *legume* (Latin: *legumen* = bean). The reason for this dual terminology that permits two names for the same thing is yet another concession made by carpologists to accommodate older usage. Since the Leguminosae are such a large and economically important plant family, their familiar fruits have traditionally been called "legumes" since Carl von Linné introduced the term in 1751.

We are particularly familiar with the edible members of the Leguminosae, such as peas (*Pisum sativum*), beans (*Phaseolus* spp.), and popular ornamentals such as sweet peas (*Lathyrus odoratus*), lupins (*Lupinus* spp.), gorse (*Ulex europaea*), Chinese wisteria (*Wisteria chinensis*), black locust (*Robinia pseudoacacia*) and flamboyant (*Delonix regia*). Pods of this type often explode to eject their seeds. Their catapult mechanism is based on cross-textured fibres in the pericarp that cause the carpel halves to twist in opposite directions as they dry out. Eventually, the tension is released when the dorsal and ventral sutures burst suddenly. Lupins, gorse and sweet peas all use this self-dispersal mechanism effectively, but they can hardly compete with some of their tropical relatives. *Tetraberlinia morelina*, a legume that grows in the rainforests of west Gabon and south-west Cameroon, manages to shoot its seeds over a distance of up to 60 metres, certainly helped by its great height. It holds the world record for the longest ballistic dispersal distance of any plant.

Sweet bean pods
Other members of the Leguminosae produce similar but indehiscent pods. The best-known example of such a *single nut* or *camara* (Greek: *kamara* = vault, chamber) is the peanut (*Arachis hypogaea*). Peanuts are grown primarily for human consumption and represent the world's third major oilseed crop, after soybean and cotton. China alone produces more than 10 million tons per year, and the fact that Americans eat about 4 million pounds (unshelled weight) each day is further proof of their popularity. Other interesting camaras include the pods of the carob tree (*Ceratonia siliqua*), the tamarind (*Tamarindus indica*) and the tropical American ice-cream bean (*Inga* spp., especially *I. edulis, I. feuilleei*). Unlike peanuts, it is the fleshy pulp and not the seeds that constitute the edible part of these pods.

Nut-like on the outside, but fleshy inside, the tamarind or "Indian date" contains a brown, sticky, sweet-sour, acidic pulp, rich in vitamin C and citric acid. It has long been

opposite: *Acacia vittata* (Leguminosae) – Lake Logue wattle; native to south-western Australia – fruit (legume); the fruits of the members of the legume family develop from a single carpel that, when ripe, typically splits along the dorsal and ventral suture; 2.1cm long

Acacia vittata (Leguminosae) – Lake Logue wattle – seed; like many wattles (Australian acacias), the seeds of the Lake Logue wattle are equipped with a "bait" (elaiosome) to attract ants for dispersal; 3.8mm long

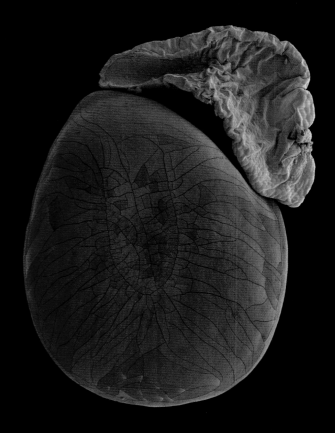

cultivated by humans (Marco Polo referred to it in 1298), particularly in semi-arid regions, where it grows luxuriantly into one of the most beautiful ornamental trees of the tropics. The fruit was first thought to be produced by an Indian palm, hence its old Arabic name "tamar-u'l-Hind" which means "date of India." The natural origin of the tamarind tree is probably tropical Africa. From there Arabian seafarers introduced it to the Mediterranean and south-east Asia, where its fruits have become an integral part of the culture. Tamarind pulp is widely used to flavour drinks, chutneys and a variety of dishes, especially in Asian cuisine, and is an essential ingredient for an old English favourite, Worcester sauce. Adding to the tamarind's domestic value is its high content of citric acid, which makes the juice of overripe fruits one of the best agents for polishing copper and brass. The natural dispersers of tamarinds are more interested in the fruit's nutritious rather than its practical values. They are mainly ruminants such as deer and gazelles but in south-east Asia monkeys are among the chief dispersal agents.

Edible pods of biblical significance develop from the allegedly semen-scented flowers of the "kharoub" (Arabic) or carob tree, *Ceratonia siliqua*, from the eastern Mediterranean. Referred to as "locust beans" in the Bible, the brown, leathery fruits of this small to medium-size tree and wild honey were supposedly the only food of John the Baptist while he lived in the desert (Mark 1, 6), hence their other name "St. John's bread." However, some believe this interpretation is wrong and that John might indeed have lived on migratory locusts (probably dipped in honey). In any case, he would have done well eating the fruits of the carob tree. High in sucrose (almost 40 per cent) and other sugars, as well as gum, the soft, dry pulp, called "carob" like the tree itself, is sweet, though a hint of butyric acid adds a slightly unpleasant smell. Carob tastes very similar to chocolate, but is hypoallergenic and contains only one third of the calories, just half the amount of fat, no theobromine or other psychoactive substances, and lots of protein and pectin, the latter being an excellent colon cleanser. All this seems almost too good to be true. Carob flour is widely used in health foods as a substitute for cocoa. If not turned into chocolate substitute, locust beans serve as a traditional feed for livestock in the Middle East, where carob trees have been in cultivation for at least 4000 years. The thickish pods, 15–30cm long, are also chewed raw, especially on the Jewish holiday of Tu Bishvat. They are edible only after the pericarp has changed colour from green to dark brown. When fully ripe, the pods emit a heavy scent to attract natural dispersers, including the very assiduous Egyptian fruit bat (*Rousettus aegyptiacus*). The round flat seeds are rock hard and can pass through the jaws and gut of most animals largely unharmed. Because of their remarkably uniform weight, carob seeds were used in ancient times as units for measuring small quantities of

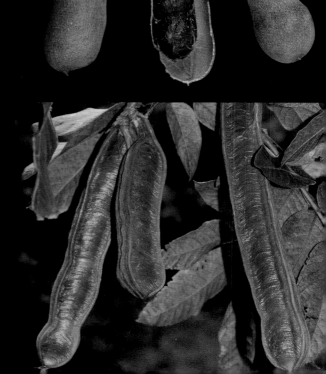

below: *Tamarindus indica* (Leguminosae) – tamarind, known only in cultivation, probably originating in tropical Africa – fruits (camaras); the hard seeds are embedded in a sticky brown pulp, which is eaten fresh and used in curries; dia. 2.5cm

bottom: *Inga feuilleei* (Leguminosae) – ice cream bean; known only in cultivation but probably native to Bolivia and Peru – fruits (camaras); the fruit contains a delicious sweet pulp that resembles vanilla ice cream; fruit c. 20-30cm long

Entada sp. (Leguminosae) – monkey ladder vine; photograph taken by H.J. Schlieben in 1935 in Tanzania (then Tanganyika); the species shown could either be *Entada gigas* or *Entada rheedii*, both producing gigantic pods (craspediums) reaching a width of 8-15 cm and a length of up to 1.8m in *E. gigas* and allegedly even 2m in *E. rheedii*. The fruits of *Entada* are the largest pods of any legume

precious gemstones. The system was eventually standardized and today the internationa unit for the weight of diamonds, the "carat" (derived from "carob"), is precisely 20 milligrams, the equivalent of a typical carob seed.

The third and last delicious legume pod we want to discuss here is the ice-cream bear The cylindrical and often spirally twisted pods of *Inga edulis* grow up to a metre in lengt and contain numerous large green seeds embedded in an edible, translucent, white pulp Contrary to the tamarind and the carob, where the pulp is derived from the inner layers c the pericarp, the white juicy envelope encapsulating the seeds of the ice-cream bean consist of the fleshy outer epidermis of the seed coat (*sarcotesta*). Because of their delicious taste, ice cream beans are very popular in many parts of Central and South America, where they are almost always eaten raw. The sweet flavour of the pulp resembles vanilla ice cream, hence the name. During the tropical wet season when fruits are abundant, monkeys and birds feas on the sweet pulp and scatter the soft seeds.

The world's largest bean pod

Among the most interesting curiosities of natural history are the superlatives. When it comes to the largest "bean" of all, the dimensions of the pods of the monkey ladder vine (*Entada* spp.) are truly staggering. Growing as lianas in the tropical forests of Central and South America, Africa, Asia and Australia, the flattened, twisted shoots writhe high up into the forest canopy. Like spiral staircases, the astonishingly strong vines of the monkey ladder provide natural canopy walkways for many animals, including snakes, sloths and, of course monkeys. When fruiting, *Entada gigas* and *E. rheedii* produce gigantic pods. With a width o 8-15cm and a length of up to 1.8m in *E. gigas* and allegedly even 2m in *E. rheedii*, they are the largest pods of any legume species. Unlike normal beans, their spiral (*E. gigas*) or straigh (*E. rheedii*) pods (botanically *craspediums*) are subdivided into 10-20 segments, each containing a single very hard, chestnut-brown seed, 5-6cm in diameter. At maturity, the seed-bearing segments separate transversely from each other and vertically from a wooden frame that runs along the periphery of the fruit. Dropped out of their frame and littered across the forest floor, many of these "seed packets" are washed into streams and rivers by the tropical rain, if they did not land there in the first place. The seeds are kept afloat by ai trapped within the fruit segments and between the cotyledons of the embryo. They finally reach the ocean where one of nature's most amazing seed dispersal stories begins. Once in the open sea, the fragile wall of the fruit compartment soon withers away but the hard waterproof seeds remain buoyant for at least two years. During this time they can travel for thousands of miles, drifting on the ocean's surface currents. One of the most striking examples is the Gulf Stream, which carries tons of exotic drift seeds from South America and the Caribbean to the shores of north-western Europe every year. Among the most frequent arrivals on European beaches are the seeds of the predominantly neotropical *Entada gigas*, whose suggestive shape explains their popular name "sea hearts." The more rectangular seeds of the paleotropical *Entada rheedii*, known to drift seed collectors as "matchbox beans" or "snuffbox beans," are mainly found on beaches in south-east Asia and the Pacific region. Both sea hearts and matchbox beans have been used in games, as baby teethers and, hollowed out and hinged, as snuff and matchboxes.

Seeds in prison

At the close of our carpological exploration of "pods," one kind remains to be discussed, namely those multicarpellate pods that would be perfectly acceptable capsules if only they would open. These rattling pods have an air space around the seeds inside their hard shell (which can be fleshy or dry) and have long been a carpological nuisance.

Among the various troublemakers are the ferociously spiny fruits of *Uncarina grandidieri* and the devil's claw (*Harpagophytum procumbens*), both belonging to the sesame family (Pedaliaceae), chilli and bell peppers (*Capsicum annuum*, Solanaceae), and the rose apple (*Syzygium jambos*, Myrtaceae).

The most remarkable of all indehiscent pods must be those of the Brazil nut tree (*Bertholletia excelsa*, Lecythidaceae), a huge tree up to 60m tall from the rainforests of South America. Whereas its relatives such as the monkey pot (*Lecythis pisonis*) have large capsules that open with a wide lid, the fruit of the Brazil nut tree appears to have lost the ability to produce an escape hatch for its seeds. Harder even than the rock-like seeds (Brazil "nuts"), the fruit of *Bertholletia excelsa* is a large spherical woody pod, 15cm on average in diameter and weighing up to 2.5kg. Inside, the fruit bears 15-25 seeds embedded in a yellow pulp. Whilst providing the precious seeds with maximum protection against predators, the almost impenetrable pericarp, which needs an axe to crack it, constitutes a major obstacle to germination. This problem is solved by the Brazil nut tree's natural ally, the agouti (*Dasyprocta agouti*). Only the teeth of this medium-sized rodent are sharp enough to gnaw a hole in the pod through which to extract the nutritious seeds. The agoutis, like squirrels, are scatter-hoarders and tend to eat only part of their bounty while burying the leftover seeds somewhere in the forest floor, up to 400m away from the parent tree. Either death or forgetfulness on the part of the agouti ensures that new Brazil nut trees grow in the wild.

However intriguing the story of the Brazil nut tree, multicarpellate pods that remain stubbornly closed even at maturity pose a conceptual challenge. Botanists were unable to fit

opposite: *Hippocrepis unisiliquosa* (Leguminosae) – horseshoe vetch; native to Eurasia and Africa – fruit (camara); adapting to the shape of the fruit, the seeds are bent around the elliptic invaginations of the fruit wall. Although the adaptive strategy behind the curiously shaped pods is difficult to interpret, its flat and very lightweight construction may assist wind dispersal. Moreover, the overlapping margins of the invaginations and the peripheral bristles may help hook the fruits to the fur of animals (epizoochory); diameter 1.8cm

Harpagophytum procumbens (Pedaliaceae) – devil's claw, grapple plant; native to southern Africa and Madagascar – fruit (carcerulus); the large woody grapples of this devil's claw are adapted to cling to the feet and fur of animals, who may suffer terrible wounds. The Khoisan peoples of the Kalahari Desert have used the tuberous root of the devil's claw for thousands of years to treat pain during pregnancy and to prepare ointments to heal sores, boils and other skin problems. Extracts from dried roots are today sold as a natural remedy against pain and inflammation caused by arthritis and other painful ailments; fruit 9cm long

into any of their carefully crafted definitions and had to resort to "semantic violence," referring to them as "indehiscent capsules," despite the fact that their own definition of a capsule requires the respective fruit to be dehiscent. In an attempt to put some scientific logic into the chaos of fruit classification, Richard Spjut (1994) reviewed and revised the terminology created during the eighteenth and nineteenth centuries, the heyday of carpology. As it turned out, without having exactly the same meaning in mind, in 1813 Charles François Brisseau de Mirbel (1776-1854) had already created a scientific term that offers a more plausible alternative to "indehiscent capsule," namely the Latin *carcerulus*, meaning "little prison."

Inside-out drupes

To complicate matters further, there are some "little prisons" where the seeds are not surrounded by air but are embedded in a juicy pulp. The flesh beneath the hard outer shell can belong to the seeds themselves if they possess a sarcotesta, as in cacao pods, the fruits of the chocolate tree (*Theobroma cacao*, Malvaceae). Mostly, the pulp originates from the inner layers of the pericarp. This is the case in Old World baobabs (*Adansonia* spp., Malvaceae), the New World calabash tree (*Crescentia cujete*, Bignoniaceae), and the Asian bael fruit or Bengal quince (*Aegle marmelos*, Rutaceae).

A particularly impressive example of this type of fruit grows on a tree named *Couroupita guianensis*, a member of the monkey pot family (Lecythidaceae). The large size and spherical shape of its woody pods gave this tropical curiosity from the Guianas the name "cannonball tree." Unlike most species of the monkey pot family whose fruits open with a lid, those of the cannonball tree remain closed. Within their hard, woody shell the fruits contain numerous hairy seeds embedded in a white pulp. When ripe, the heavy fruits drop to the ground where the impact can cause them to crack. Intact fruits remain untouched underneath the tree until they are broken up and eaten by peccaries, ground-dwelling *frugivores* (fruit-eating animals) that play an important role as seed dispersers in neotropical rainforests. Local people also feed the pulp to domestic pigs. Although not poisonous, the pulp has a rather foul, offputting smell so that few people are tempted to taste it. Those who prefer to miss out on this doubtful culinary experience can still use the hard exterior of the fruit as a container.

In a way, fruits of the kind of the cannonball act like "inside-out drupes" with their hard outer shell and soft core. As so often, western textbooks are lost for words in the face of such tropical extravagance and resort to unspecific ("indehiscent pod") or Procrustean ("indehiscent capsule") terminology. Thankfully, scientific logic has been

Syzygium jambos (Myrtaceae) – rose apple; native to southeast Asia; fruit (carcerulus), crowned by the incurved calyx; inside the large cavity of the fleshy pericarp lie one or two large seeds. The edible fruits, which are naturally dispersed by fruit bats and possibly also monkeys, have a subtle aroma of rose water, hence the name; diameter c. 4cm

below: *Couroupita guianensis* (Lecythidaceae) – cannonball tree, native to tropical America – flower and fruit (amphisarcum); inside the hard shell the woody pods contain numerous hairy seeds in a white pulp; diameter c. 20cm

bottom: *Theobroma cacao* (Malvaceae) – cacao; originally from the Amazonian rainforest – fruits (amphisarca); the heavy fruits grow directly on the trunk and major branches. Under the tough rind lies a sweet, juicy pulp, typical of primate-dispersed fruits; fruit c. 20-30cm long

restored by the brave carpologists who coined the term *amphisarcum* to describe the structural essence of these armoured fruits.

To be or not to be a drupe

Numerous oxymoronic expressions similar to "indehiscent capsules" have been created by botanists in a Procrustean attempt to provide conformity by violent means. They include "dry drupe" or "drupaceous nut" (for the coconut), "dehiscent drupe" (for the almond), and "dehiscent berry" (for the nutmeg). Each played its part in discrediting the scientific merits of carpology.

As for the mistreated coconut (*Cocos nucifera*, Arecaceae) and almond (*Prunus dulcis*, Rosaceae), their fruits are indeed very drupe-like in that they possess a thin epicarp, a soft mesocarp, and a hard endocarp. Hence, they would be perfect drupes if only they had the required fleshiness of the mesocarp, but instead it is dry and fibrous. In the case of the coconut, the dryness of the mesocarp can easily be explained by the preferred dispersal strategy. Whereas "normal" drupes produce a fleshy outer layer to attract animal dispersers, the coconut – which is, in any case, too large to be swallowed by an animal – has evolved to become a true seafarer. Rather than providing an edible reward, the thick fibrous husk is extremely resistant to seawater and keeps the coconut buoyant in the ocean for months. The average maximum distance that a live coconut can travel is 5,000 kilometres. It is alleged that viable fruits have been found on Norwegian beaches.

In the case of the almond, matters are slightly more complicated because the soft outer layers (epi- and mesocarp) of the fruit split along the ventral suture to expose the single-seeded stone, a carpological misdemeanour responsible for their paradoxical branding as "dehiscent drupe." The natural dispersers of almond stones are birds such as jays and woodpeckers but also small mammals, especially rodents.

MULTIPLE FRUITS – SEVERAL FRUITLETS FROM A SINGLE FLOWER?

So far, we have considered only *simple fruits* that develop from flowers with either just one individual or several joined carpels. But what do fruits look like when they develop from flowers with two or more separate carpels? Flowers with such apocarpous gynoecia are predominantly found in a relatively small number of primitive families such as the Annonaceae, Illiciaceae, Schisandraceae, Magnoliaceae, Paeoniaceae, Ranunculaceae and Winteraceae. Their numerous separate carpels can easily be recognized in the centre of their flowers, where they can be arranged spirally around a central column as in magnolias (*Magnolia* spp.), the related tulip tree (*Liriodendron tulipifera*) and certain buttercups

opposite: *Rubus phoenicolasius* (Rosaceae) – Japanese wineberry; native to northern China, Korea and Japan – fruit (drupetum); like its relatives the raspberry (*Rubus idaeus*) and the blackberry (*Rubus fruticosus*), the Japanese wineberry has an apocarpous gynoecium that produces a multiple fruit in the form of a cluster of tiny drupelets. Like the rest of the plant, the calyx is covered in sticky, bristly hairs. The sweet, juicy fruits are edible; diameter c. 1cm

Stones from the drupelets of *Rubus fruticosus* (bottom left, blackberry; 3mm long), *Rubus phoenicolasius* (bottom right, Japanese wineberry; 1.6mm long), *Rubus laciniatus* (top, cutleaf blackberry, parent of some garden blackberries; 1.6mm long)

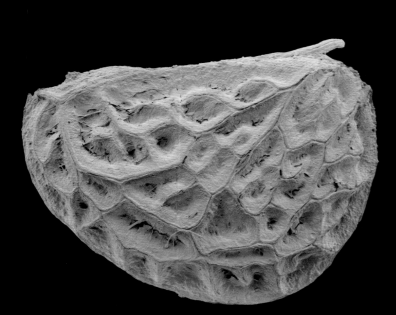

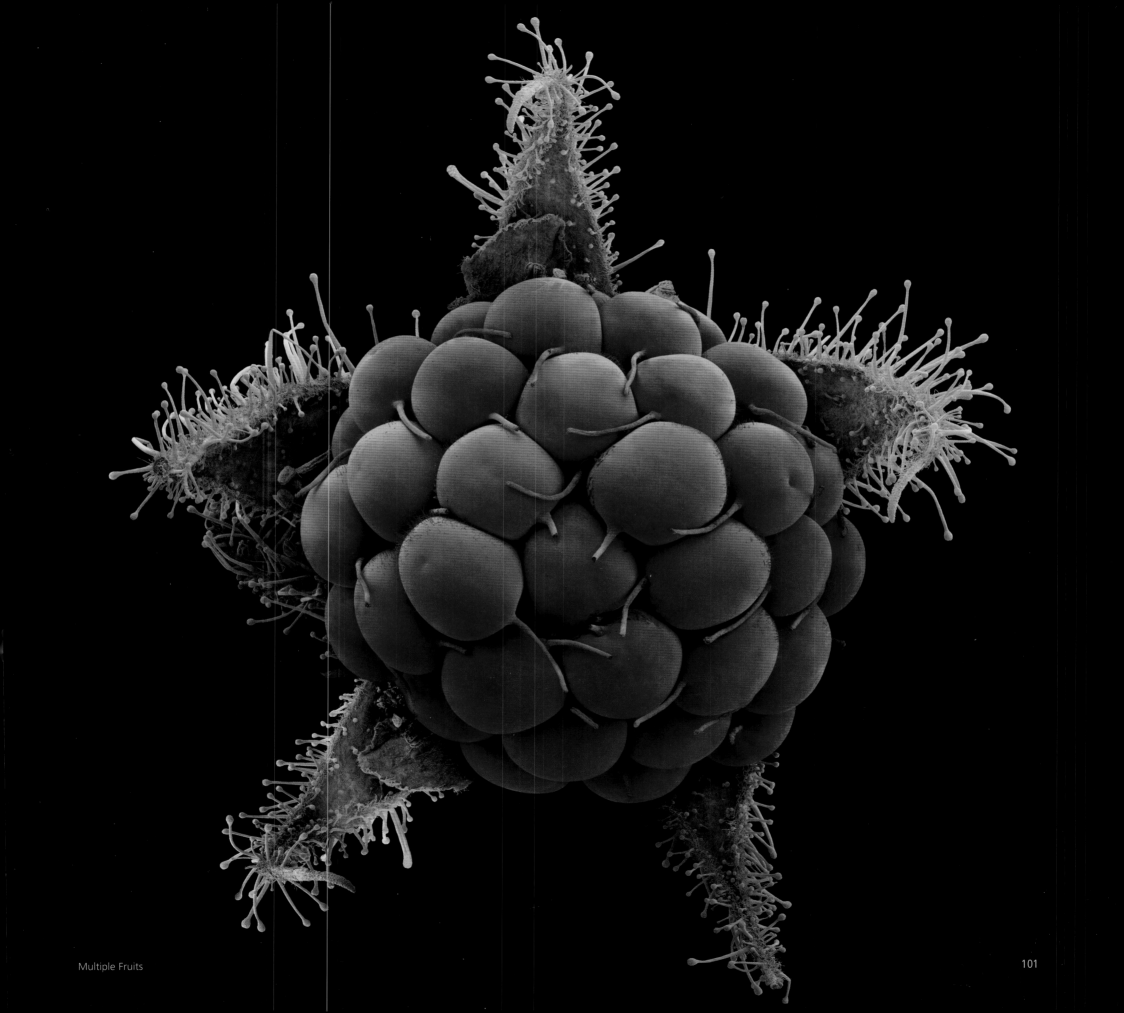

(*Ranunculus* spp., Ranunculaceae); grouped into a loose cluster (e.g. *Drimys winteri*, Winteraceae); or, alternatively, arranged like the spokes of a wheel as in star anise (*Illicium verum*, Illiciaceae).

When these primitive flowers develop into a fruit, each carpel produces a fruitlet of its own. The result is a *multiple fruit* consisting of a cluster of individual fruitlets. Depending on the differentiation of the pericarp, these fruitlets can be follicles, cocci, nutlets (or achenes), drupelets or berrylets. Clusters of follicles (so-called *follicetums*) are produced by a variety of families, not all of which are closely related. The best known among them are star anise, magnolias (*Magnolia* spp., Magnoliaceae), stonecrops (*Sedum* spp., Crassulaceae), peonies (*Paeonia* spp., Paeoniaceae), and many members of the buttercup family (Ranunculaceae) such as marsh marigolds (*Caltha palustris*), larkspurs (*Delphinium* spp.), monkshoods (*Aconitum* spp.), columbines (*Aquilegia* spp.) and bugbanes (*Cimicifuga* spp.). In the buttercups themselves (*Ranunculus* spp.) the carpels are single-seeded and remain closed at maturity. As a result, each flower produces a cluster of nutlets or achenes, forming what in carpology is defined as an *achenetum*. And so the naming of multiple fruits continues by simply adding the ending *-etum* to the name of the corresponding simple fruit, a quite sensible approach first suggested in 1835 by the French politician and botanist Barthélemy Charles Joseph Dumortier (1797-1878). Following Dumortier's approach, the politically correct name for multiples of drupes and berries would be *drupetums* and *baccetums* (Latin: *bacca* = berry), respectively.

In our temperate climate the most popular drupetums are produced by two members of the rose family (Rosaceae), *Rubus idaeus* and *Rubus fruticosus*, better known as raspberries and blackberries. Plants with flowers that give rise to a cluster of berries are rather rare in temperate climates. Nevertheless, such baccetums do exist. One illustrative example is the Winter's bark tree, *Drimys winteri* (Winteraceae), which occurs from Mexico to Terra del Fuego. The Winter's bark is named in honour of Captain John Winter who was the first to introduce the plant to Europe in 1578. Commander of the *Elizabeth* and Vice Admiral to Sir Francis Drake on his famous voyage around the world, Winter used a tonic made from the pungent bark of *Drimys winteri* to cure scurvy among his crew. The fruits of the Winter's bark tree also have a peppery taste and are not very palatable.

A far more pleasant culinary experience is offered by the fruits of some of the Winter's bark tree's cousins in the annona family (Annonaceae). Names such as cherimoya (*Annona cherimola*), bullock's heart (*A. reticulata*), soursop (*A. muricata*), sweetsop (*A. squamosa*) and biriba (*Rollinia mucosa*) depict a selection of deliciously tasting tropical fruits that all come from flowers with apocarpous gynoecia. Given the latter, one would expect a cluster of

below: *Cocos nucifera* (Arecaceae) – coconut; found in tropical regions worldwide, precise origin unknown – fruits (nuculania); the fruit is similar to a drupe but the spongy mesocarp is dry and fibrous so it floats. The correct carpological term for such a "dry drupe" is "nuculanium"

bottom: *Rollinia mucosa* (Annonaceae) – biriba; native to South America – fruit (syncarpium); the fruit is formed by many separate carpels which amalgamate in the fruit. Each tubercle represents a single carpel. A most delicious fruit that tastes like lemon meringue pie; diameter c. 10-15cm

fruitlets as in the related Winter's bark tree. Yet, surprisingly, the apple- to melon-sized fruits come in a single homogenous structure. The solution to this conundrum provides a glimpse of the events that take place after pollination. As they grow the numerous carpels fuse together and with the underlying swollen receptacle form a fleshy, deceptively berry-like fruit that is scientifically called a *syncarpium* (Greek: *syn* = together + *karpos* = fruit). Its surface is smooth in the cherimoya and bullock's heart or covered with conical or spiky protuberances (one per carpel) in the sweetsop, soursop and biriba, respectively. These fruits have in common a thin, green to brownish skin covering a juicy white pulp into which numerous extremely hard, bean-size seeds are embedded. Only the curious markings on the surface reveal the true multiple nature of the fruits. They create a scale-like pattern, reminiscent of the skin of a reptile, the outlines of the "scales" marking the boundaries between the individual carpels. Curiously, one of the fruits of the annona family, the unsavoury alligator apple (*Annona glabra*) from Florida and the West Indies, was so named because of this striking similarity.

SCHIZOCARPIC FRUITS *OR* HOW TO EMULATE THE MULTIPLE EXPERIENCE
In the custard apple and its relatives, primitive angiosperms have found a way to overcome the separation of their carpels allowing them to produce a uniform structure that looks and acts deceptively like a multicarpellary simple fruit. From a human perspective the ambition of trying to be more modern by emulating advanced syncarpous angiosperms is perfectly understandable. Besides, in the case of fleshy, animal-dispersed fruits such as the custard apple and its relatives, it is advantageous to offer all seeds in a single package that can be harvested on one visit. However astonishing it may seem, the copying works both ways. Despite being blessed with advanced syncarpous gynoecia, some families have found ways to emulate the archaic "multiple experience." Their fruits separate into their carpellary constituents, a process that can take place during development but occurs most often only at maturity. Because of their disintegrating behaviour, these fruits are described as *schizocarpic* (Greek: *skhizein* = to split + *carpos* = fruit). Depending on the number of carpels involved, schizocarpic fruits split into two or more fruitlets, each fruitlet consisting of either a whole or half a carpel. Mostly, though, the fruitlets themselves are dry, indehiscent and single seeded, which technically renders them almost true nuts. However, since they are *fruitlets* rather than *fruits*, they are aptly referred to as *nutlets* instead of *nuts*.

Dry schizocarpic fruits are typical of the carrot family (Apiaceae), which includes common kitchen spices such as caraway, cumin, coriander, aniseed and fennel. Formed by two joined carpels, the fruits of the Apiaceae split longitudinally into two nutlets. The nutlets, each of which represents an entire carpel (*monocarp*), initially remain attached to the

central column (*columella* or *carpophore*) of the fruit until they are blown away by the wind (e.g. *Artedia squamata*) or become entangled in the coat of a passing animal (e.g. *Daucus carota*). The very distantly related stickywilly (*Galium aparine*, Rubiaceae) has similar bicarpellate schizocarpic fruits but the two nutlets separate in a clean break without leaving behind a columella. The winged nuts of maples (*Acer* spp.) and the related *Dipteronia* from China also develop from two joined carpels and split in a very similar way when ripe. More than two carpels are involved in the schizocarpic fruits of certain Malvaceae, especially from the tribe Malveae (for example, *Abutilon* spp., *Alcea* spp., *Malva* spp.). Their ripe fruits, which look like wagon wheels, disintegrate into numerous (three or many more), often beautifully ornamented nutlets with sophisticated surface patterns. The most spectacular among them are also adorned with hairs, bristles and spines.

The members of the mint family (Lamiaceae) and the borage family (Boraginaceae) take the division of their ovary a step further. Once more, two carpels form a syncarpous gynoecium but the ovary is deeply lobed so that each carpel is longitudinally divided into half. The result is a bicarpellate ovary with four single-seeded compartments. At maturity, the four compartments break apart into single-seeded nutlets, each representing half a carpel or a *mericarp*. The mericarps of the Lamiaceae, which include well-known kitchen herbs such as sage (*Salvia officinalis*), oregano (*Origanum vulgare*), thyme (*Thymus vulgaris*) and basil (*Ocimum basilicum*), look deceptively like true seeds, and this is what most people consider them to be.

Schizocarpic fruits of the exotic kind are those of the genus *Ochna*, namesake of its family, the Ochnaceae, also known as the "Mickey Mouse plant family." In the flower the carpels, of which there are between three and twelve, are joined only at the base whilst sharing a common style. After pollination, the receptacle becomes fleshy and each individual carpel develops into a black, oily drupelet that attracts birds. The schizocarpic ovary of the Indian pokeweed (*Phytolacca acinosa*, Phytolaccaceae) from East Asia (India to China) develops into a black, segmented fruit that resembles a miniature pumpkin. Unlike *Ochna*, the individual carpels separate into single-seeded berrylets. Since in carpological nomen-clature, the ending *-arium* indicates schizocarpy, the schizocarpic berry of *Phytolacca acinosa* is correctly addressed as *baccarium* (Latin: *bacca* = berry).

As the previous examples indicate, most schizocarpic fruits separate into single-seeded nutlets, drupelets or berrylets. In some cases, though, the fruitlets are many-seeded, dehiscent follicles, as in the dogbane family (Apocynaceae) and certain Malvaceae such as *Brachychiton* and *Sterculia*. It is easy for their carpels to split since they are almost separate from the beginning, joined only at the tip by a shared style.

page 104: *Cimicifuga americana* (Ranunculaceae) – American bugbane; native to eastern North America – fruit (follicetum) with a few (aborted) seeds; as in other members of the buttercup family, the fruit of the American bugbane de-velops from an apocarpous gynoecium. Of the 3-8 carpels of a flower, 2-4 ripen into follicles. The seeds are dispersed by the wind or when raindrops hit the spoon-shaped follicles causing the seeds to be catapulted away; 1.25cm long

page 105: *Cimicifuga americana* (Ranunculaceae) – American bugbane – seed; the bizarre lobes of the seed coat are most likely an adaptation to wind dispersal; they increase the air resistance of the seed; 4.3mm long

opposite: *Ocimum basilicum* (Lamiaceae) – sweet basil – fruit (microbasarium); the syncarpous gynoecium of the members of the mint family consists of two carpels that are deeply lobed into four single-seeded compartments (half-carpels). When the fruit is ripe, the four compartments break into single-seeded nutlets, each representing half a carpel or a mericarp. In basil flowers the upper lobe of the calyx is expanded into a spoon-shaped cap which, in the fruit, acts as a springboard for raindrops, which help to expel the mericarps from the deep calyx; 8.5mm wide

Ocimum basilicum (Lamiaceae) – sweet basil, cultivated for more than 5,000 years, originally from tropical Asia – glandular hair on the persistent calyx of the fruit (pointed non-glandular hairs in background); the mint family is rich in aromatic herbs and medicinal plants. Many species, including basil, possess glandular hairs on the surface of their aerial parts that secrete essential oils, the volatile chemicals that are responsible for the typical taste and smell of the plants. The essential oil accumulates between the outer wall and the cuticle of the 1-4 cells forming the head of the glandular hair; diameter of head 110μm

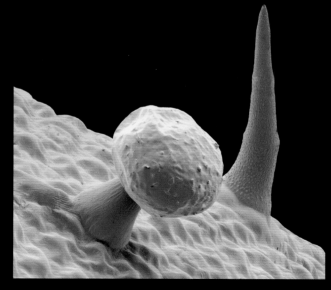

Daucus carota (Apiaceae) – wild carrot; native to Europe and south-western Asia

right: whole plant; the wild carrot is a biennial plant; during the first year it grows a rosette of leaves while building up a taproot that stores energy (mainly sugars) to produce the flowers in the second year. Edible carrots (*Daucus carota* subsp. *sativus*) are a domesticated form with greatly enlarged and better tasting taproots

below: close-up of infructescence; diameter 5cm

opposite: whole fruit (polachen arium); the fruits of the carrot family (Apiaceae) develop from an inferior ovary composed of two joined carpels. At maturity the two carpels separate into two individual single-seeded indehiscent fruitlets. The fruitlets of the wild carrot are covered in long spines with recurved hooks at their tips, an adaptation to facilitate dispersal by animals (epizoochory); 5.5mm long

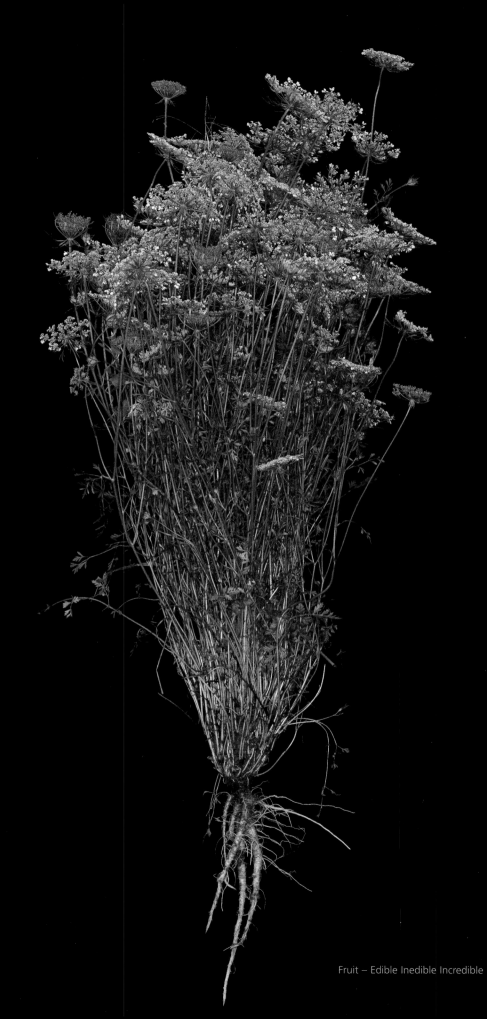

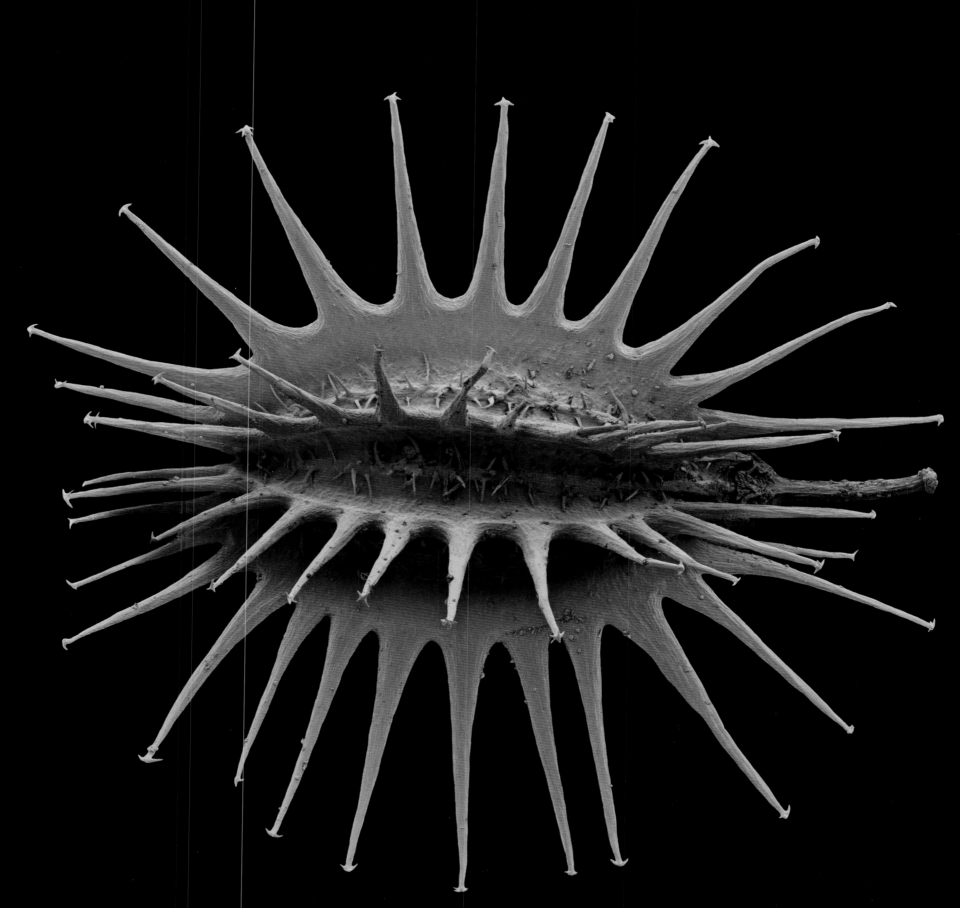

opposite: *Phytolacca acinosa* (Phytolaccaceae) – Indian pokeweed; native to east Asia – inflorescence; the superior syncarpous ovary of the flower has 7-8 lobes, each representing a single carpel; diameter of flower 7.5mm

Phytolacca acinosa (Phytolaccaceae) – Indian pokeweed; native to east Asia – fruit (baccarium); when fully mature, the individual carpels separate into single-seeded berrylets, hence the name *baccarium* (Latin: *bacca* = berry) for this type of schizocarpic fruit; diameter 7.8mm

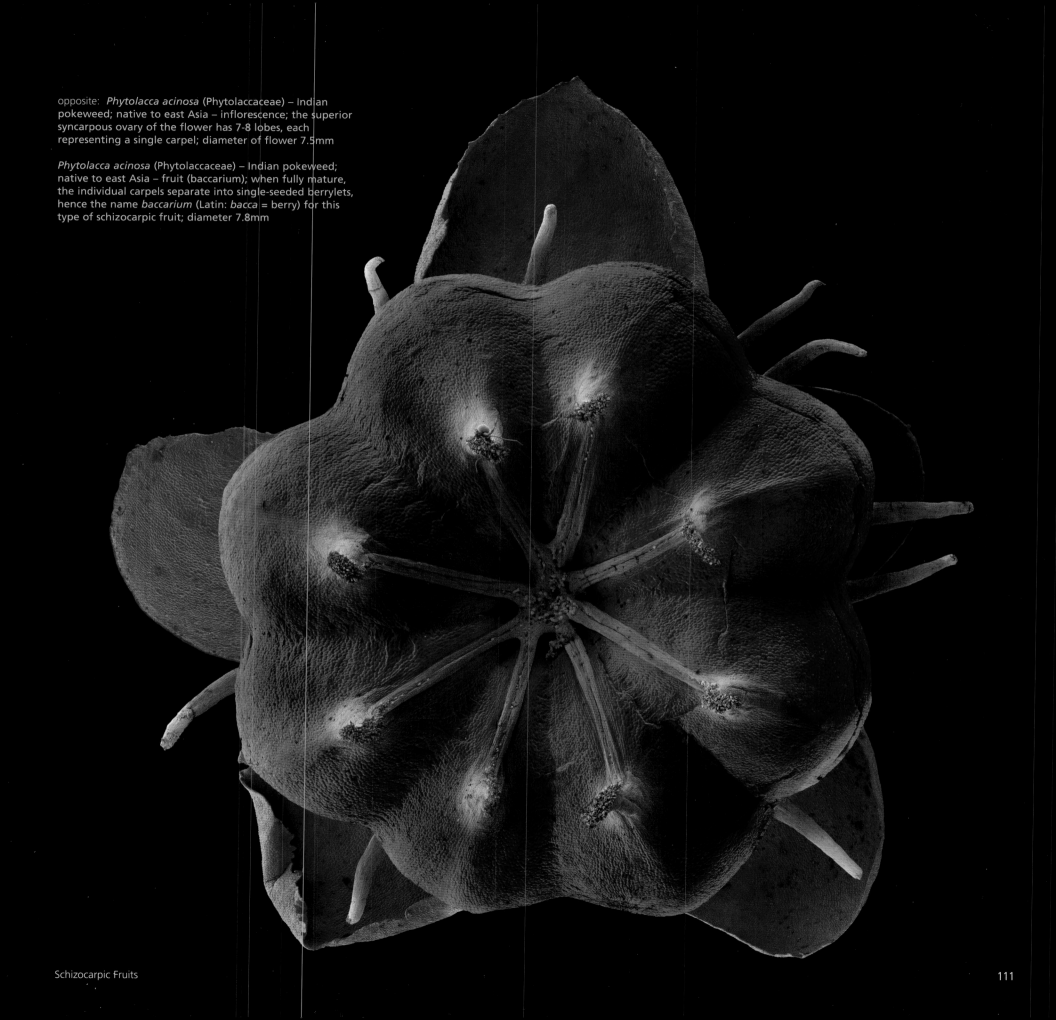

Schizocarpic Fruits

Pyrus pyrifolia (Rosaceae) – Chinese pear; native to east Asia
– fruit (pome); in pears (*Pyrus* spp.), apples (*Malus pumila*)
and quinces (*Cydonia oblonga*), the bulk of the fleshy part is
formed by the hypanthium which is why they are classified as
anthocarpous fruits; diameter 8cm

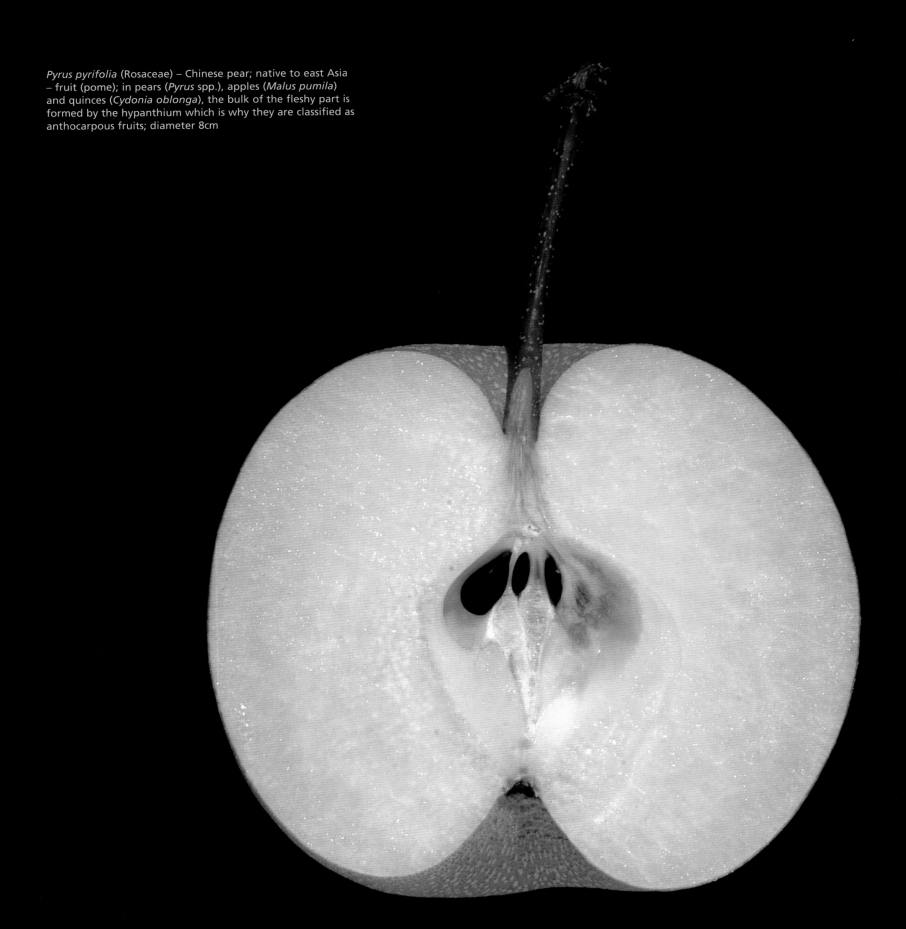

ANTHOCARPOUS FRUITS – THE CARPOLOGISTS' TOUCHSTONE

Classifying fruits according to their structure is a difficult task when only the diverse
products of the gynoecium are considered. The carpologists' real challenge begins when
parts of the flower other than the ovary alone participate in the formation of yet more
complex fruits. Literally, any floral organ (petals, sepals, receptacle, pedicel) and even
accessory structures such as bracts can persist and develop after fertilization to join the
mature (apocarpous or syncarpous) ovary in order to form what is respectfully called an
anthocarpous fruit (Greek: *anthos* = flower + *karpos* = fruit). Strictly speaking, this is already
the case in fruits developing from flowers with inferior ovaries, a character typical of the
pumpkin family (Cucurbitaceae), banana family (Musaceae), and coffee family (Rubiaceae),
to name but a few. In these groups, the fruit wall is formed by both the hypanthium and
the ovary wall. Most of the time, the two layers are so perfectly joined that the boundary
between them cannot be distinguished, even under a microscope. Although early
carpologists such as the Frenchman Nicaise Auguste Desvaux (1784-1856) and the
Englishman John Lindley (1799-1865) considered ovary position an important criterion for
fruit classification (Desvaux 1813; Lindley 1832; Takhtajan 1959), most modern carpologists
have in the meantime conveniently dropped this idea. For example, fruits with an entirely
fleshy fruit wall are deemed to be berries, irrespective of whether they are derived from a
superior ovary (kiwi, *Actinidia deliciosa*, Actinidiaceae) or an inferior ovary (mistletoe berries,
Viscum album, Santalaceae). In apples (*Malus pumila*) and other Rosaceae-Maloideae such as
medlar (*Mespilus germanicus*), loquat (*Eriobotrya japonica*), quince (*Cydonia oblonga*) and pear
(*Pyrus communis*), it is all too obvious that the bulk of the fleshy part of the fruit is formed
by the hypanthium and not the ovary wall. The demarcation line between the different
tissues is marked by a ring of vascular bundles that is clearly visible in a cross-section as a
dotted circle surrounding the apple core (the green dots representing the sectioned vascular
bundles). The seed-bearing core of apples and their Maloideae brethren, is formed by the
hardened, parchment-like endocarp of the inferior ovary, the crucial character that, together
with the prominent hypanthium, defines this anthocarpous fruit type as a *pome*. Contrary to
what is often stated in textbooks, the carpels of the Maloideae are not apocarpous but
syncarpous, even if their ventral sides are joined only for a very short distance, leading to
the star-shape of the apple core. Other members of the rose family (Rosaceae), including
the rose (*Rosa* spp.) itself, have apocarpous gynoecia and consequently produce multiple
fruits. Despite the fact that their carpels turn into nutlets, rosehips are very similar to apples
in that the nutlets are embedded in a fleshy, urn-shaped hypanthium. In accordance with
Dumortier's suggested principle of naming multiple fruits, rosehips represent an antho-

carpous fruit type called *pometum*. In the genus *Fragaria*, yet another member of the rose family, things are very different. Instead of being enclosed in a cup-shaped hypanthium, the carpels constituting the apocarpous gynoecium are arranged on a convex receptacle. As the flower turns into a fruit, the receptacle greatly expands and becomes one of the most popular of all fruits, the strawberry. The carpels themselves do not contribute much substance to the fruit. They turn into minute nutlets, visible as the tiny brown granules on the surface of the strawberry.

Far more obvious than the participation of the hypanthium or receptacle in a fruit is the involvement of other floral parts, as we have already seen in the cashew apple (swollen pedicel), walnut (fused bracts), the cypselas of the Asteraceae (calyx) and Dipsacaceae (calyx and bracts), and the pseudosamaras of the Dipterocarpaceae (enlarged sepals). However, the final and ultimate challenge is posed by those fruits that are formed by entire inflorescences. Not only do they include non-gynoecial organs such as bracts, perianths and inflorescence axes, they also break the idealistic "one flower – one fruit" dogma, shaking to its foundations the botanists' morphological concept of a fruit.

COMPOUND FRUITS – A SINGLE FRUIT FROM SEVERAL FLOWERS?

Having had to come to terms with the fact that a single flower can give rise to several fruit(let)s by generating a multiple or schizocarpic fruit, bold readers – turned apprentice carpologists by now – will remember that the reverse is also possible. The dehiscent trymosa of beeches and sweet chestnuts, for example, were exposed as genuine "nut bags." They look deceptively like a "normal" capsule but what appear to be the seeds are actually single-seeded nuts borne inside a cupule.

The angiosperms owe much of their evolutionary success to the tremendous diversity of their fruits and seeds, and the joining of several flowers to form a single *compound fruit* is yet another carpological concept that they pursued with some fascinating results. The ovaries of compound fruits can either fuse together or stay separate. What defines a compound fruit is that an entire inflorescence turns into an *infructescence* that disarticulates as a single unit and functionally acts like a single fruit. In the simplest case just two ovaries join partly to form a "double berry," as in the Eurasian fly honeysuckle (*Lonicera xylosteum*, Caprifoliaceae) or a "double drupe," as in the North American partridge berry (*Mitchella repens*, Rubiaceae). Whilst the honeysuckle berries fuse only at their base, the union of the two ovaries of the partridge berry is so complete that the fruit would appear to be a "normal" (simple) drupe or berry if it was not for the two dimples at the apex that betray its origin as two ovaries. The small, blueberry-size partridge berries are edible and both

fruits and leaves were commonly used by Native Americans as a parturient to promote easy childbirth. Alas, despite their medicinal qualities, partridge berries are not one of the most luscious compound fruits.

More complex and impressive are the compound fruits of certain dogwoods (*Cornus* spp., Cornaceae). In the subgenus *Syncarpea*, for example *Cornus kousa* subsp. *chinensis* from China, the inconspicuous, greenish flowers are crowded together in dense, spherical clusters, which are surrounded by striking large, white petal-like bracts. After pollination the bracts drop and the flower cluster turns into a bright red head of coalescent drupes with a diameter of approximately 2cm. The sweet, juicy, custard-like pulp of the fruits is edible but the bulk of the fruit consists of the stones of the individual drupes and the skin tends to be rather tough and unpleasant.

As is so often the case, a visit to the tropics or, alternatively, the local supermarket provides us with a far more impressive insight into the culinary potential of compound fruits. One of the most delicious examples of a fruit that is the result of the concerted effort of an entire bunch of flowers must be the pineapple. After several years of vegetative growth the pineapple plant (*Ananas comosus*, Bromeliaceae) produces a spike-like inflorescence, the thickened axis of which carries numerous inconspicuous, sessile flowers, each subtended by a bract. The uppermost bracts are sterile. They look very similar to the normal leaves of the pineapple plant and form the familiar tuft of green leaves at the apex of the fruit. The hard core of a pineapple represents the fibrous inflorescence axis, whereas the edible part is formed by the basal parts of the fertile bracts, the flowers, in which the sepals persist, and the inflorescence axis. All these different organs merge with hardly any recognizable line of demarcation into a soft, sweet, juicy "superberry" or *sorosus*, to use the scientifically correct term. The hard rind is formed by the persistent sepals, the apices of the ovaries and the tissue of the bracts whose free, distal parts form the triangular, parchment-like scales on the surface. After fruiting, the entire plant dies. Since cultivated pineapples are bred to be seedless, propagation is exclusively vegetative and the tuft of green leaves crowning the fruit yields a perfect cutting for propagation.

With few exceptions, the Moraceae (mulberry family) are an entire plant family in which compound fruits are typical. Despite the fact that the black mulberry (*Morus nigra*) looks almost like a blackberry, it is formed by an entire female inflorescence in which the tiny perigones (four decussate *tepals*) and the underlying inflorescence axis become succulent. The ovaries themselves turn into small, single-seeded drupes, the tiny stones of which form the hard bits of the fruit. Black mulberries are delicious when eaten fresh and were used in the Middle Ages to add colour and flavour to wine. It is thought that the black

mulberry is derived from the white mulberry (*Morus alba*) through millennia of human domestication. Today, both species are cultivated for their edible fruits and as ornamental trees in temperate Asia, Europe and North America. In its native China, *Morus alba* gained enormous economic importance for its leaves rather than its fruits. The leaves of the white mulberry tree are the sole diet of the silkworm (*Bombyx mori*), and have for centuries been the lynchpin of China's famous silk industry. But the mulberry is not the only member of the Moraceae with a long history of human exploitation. In the same family we find economically important species with much larger edible compound fruits such as the breadfruit (*Artocarpus altilis*), the jakfruit (*Artocarpus heterophyllus*) and the fig (*Ficus carica*).

The breadfruit and the Mutiny on the Bounty

Better known in temperate regions for its connection with the much-romanticized mutiny on the Bounty, the breadfruit has long been a staple crop for the people of the Polynesian islands. It is believed to have originated in the Indo-Malayan Archipelago where the breadfruit tree, *Artocarpus altilis* (Greek: *artos* = loaf of bread + *carpos* = fruit), has been cultivated for its starchy, nutritious fruits since antiquity. Similar to a mulberry, the spherical to oblong breadfruit develops from a female inflorescence bearing a large number of tiny flowers whose tubular perigones form the fleshy part. Unlike a mulberry, the breadfruit's ovaries turn into large achenes and the entire compound fruit grows up to 30cm in diameter and weighs up to 4kg. Its high starch content, especially when unripe, earned it its name, which is the same in nearly all languages. There are two different kinds of breadfruit. One is seedless but the other one, sometimes called "breadnut," has large edible seeds enclosed in achenes. The seedless variety can only be propagated vegetatively and is almost certainly the result of long-standing cultivation by humans. Green, unripe breadfruits are eaten baked, boiled or fried while the pulp is still white and mealy. The immature flesh contains 30–40 per cent starch and resembles potatoes in taste and texture. As breadfruits ripen, they turn yellow and part of the starch is converted into sugar. Hence, fully mature fruits have a sweet flavour and are used as an ingredient in puddings, cakes and sauces. In most regions of the Pacific, especially Polynesia and Micronesia, the seedless forms of the breadfruit are preferred. People in New Guinea and the rest of Melanesia prefer the "breadnut." The flesh is of little value but the seeds, which are similar in size and taste to chestnuts, are highly relished, being eaten roasted or boiled.

It is assumed that migrating Polynesians spread this valuable source of food throughout the Pacific region. The first time Europeans encountered the impressive trees, up to 20m tall, with their large, lobed, dark glossy green, leathery leaves and extremely sticky, white

opposite: *Morus nigra* (Moraceae) – black mulberry – microscopic detail of fruit showing the inidividual fruitlets with the remains of their withering stigma; fruitlet 5.3mm wide

Morus nigra (Moraceae) – black mulberry; cultivated since antiquity, probably from China originally – fruit (sorosus); although it looks similar to a blackberry, the black mulberry is formed by an entire female inflorescence in which the tiny perigones (four decussate tepals) and the underlying inflorescence axis become fleshy. The ovaries themselves turn into small, single-seeded drupes, whose tiny stones form the hard bits of the fruit; fruit c. 2.5cm long

latex was probably in the Marquesas in 1595, and a decade later in Tahiti. However, it was only after Captain James Cook and his crew experienced its edible qualities during their visit to Tahiti in 1769 that English explorers became enthusiastic about the breadfruit. Also travelling on board HMB *Endeavour* were Joseph Banks, one of the ablest young botanists of the time, and his friend Dr. Daniel Solander, himself a highly gifted scientist and former student of Carl von Linné. Back in England the two botanists greatly praised the extraordinary qualities of the breadfruit tree. At that time the owners of the sugar cane plantations in the British West Indies had long been looking for a cheap source of food for their slaves. After the American War of Independence (1775-83) goods could no longer be imported from the United States into Jamaica, where repeated crop failures, caused by hurricanes and droughts, had led to several periods of famine between 1780 and 1786. A new cheap, reliable source of food was needed, and the easily grown breadfruit tree with its large carbohydrate-rich fruits seemed ideal. Realizing the economic potential as well as the humanitarian need, the Royal Society petitioned King George III to send an expedition to collect breadfruit plants in the South Seas and take them to the West Indies. The King's scientific advisor, Joseph Banks, who had eaten and enjoyed the breadfruit, backed this grand plan and made the necessary arrangements. On 23 December, 1787 the King dispatched Lieutenant William Bligh, Captain James Cook's navigator on his third and fatal voyage, to command HMS *Bounty* with a crew of 45 men on a mission to take breadfruit plants from Tahiti to the Caribbean. After sailing 27,000 miles, the *Bounty* reached Tahiti on 26 October 1788. The crew spent nearly six months collecting and propagating breadfruit plants. Many of them fell in love not only with the idyllic life-style in this tropical paradise but also with some of the Tahitian girls. When the *Bounty* left the island on 5 April 1789 bound for the Caribbean, it had on board 1015 breadfruit plants and a very disgruntled crew. The *Bounty* was a remodelled merchant ship formerly named *Bethia*, which was both too small and too understaffed for its mission. The famous mutiny was led by the 24-year-old Fletcher Christian who longed to escape the overcrowded conditions on board and return to his hedonistic life-style. It resulted in William Bligh (he was made "Captain" later) and eighteen others being set adrift in a 7m longboat on 28 April, 1789. Bligh's party first landed on the nearby island of Tofua but after a violent confrontation with the islanders, decided to sail to the distant Dutch colony of Timor. Despite having food and water for only five days, all the men miraculously survived the heroic voyage. After forty-one days and 3,618 nautical miles (some 6,700 kilometres) they reached the beaches of the island of Timor on 13 June, 1789.

Upon his return to England, Bligh was promoted to Captain in the Royal Navy. In 1791 he was given command of HMS *Providence* by the British Admiralty and dispatched on a

opposite top: *Artocarpus heterophyllus* (Moraceae) – jakfruit; cultivated for centuries, originally probably from India (western Ghats) – longitudinal section of fruit (sorosus); like its relative the breadfruit, the jakfruit is the product of an entire inflorescence. The tastiest parts are the soft, slightly rubbery "bulbs" that are formed by the perigones of the fertilized flowers. Each bulb encloses an egg-shaped, light brown achene, 3cm long, which is also edible. The jakfruit still enjoys its greatest popularity in India, where it is the third most popular fruit after mangoes and bananas

opposite bottom: *Artocarpus heterophyllus* (Moraceae) – jakfruit – up to 90cm in length, 50cm in diameter, and weighing up to 40kg, it is the largest fruit borne by any tree. As is typical of mammal-dispersed fruits, a fully ripe jakfruit emits a musty, sweet odour with a hint of rotten onions

Artocarpus altilis (Moraceae) – breadfruit; native to the Malay Peninsula and the western Pacific – fruit (sorosus); like a mulberry, the breadfruit develops from an entire female inflorescence bearing a large number of tiny flowers, whose tubular perigones form the fleshy part of the compound fruit. The breadfruit, so called because of its high starch content, has been a staple crop for the people of the Polynesian islands since antiquity; diameter up to 30cm

second mission to Tahiti in order to take the promising breadfruit to the West Indies. This time the mission succeeded and in February 1793 Bligh delivered 1,200 breadfruit plants and other valuable trees to Jamaica. But, contrary to everyone's expectations, the slaves did not share the Polynesians' enthusiasm for breadfruit. They refused to eat it.

The largest fruit a tree can bear

Among the other plants that travelled on board HMS *Providence* were specimens of a very close relative of the breadfruit, the jakfruit (*Artocarpus heterophyllus*, Moraceae), which Captain Bligh had collected on the island of Timor. Although usually spelled "jackfruit," the name is an English adaptation of the Portuguese "jaca" which in turn is derived from the Malayalam name for the fruit "chakka." The correct spelling of the common English name should therefore be "jakfruit."

Like the breadfruit, the jakfruit has long been cultivated in the tropical regions of Asia. Its presumed origin lies in the rainforests of the Western Ghats in India, where it still enjoys its greatest popularity, ranking as the third most popular fruit after mangoes and bananas. It is assumed that migrating peoples carried it eastwards from India to the Malay Archipelago. Unlike the breadfruit tree, the jakfruit tree has smaller, entire leaves and bears its fruit-producing female inflorescences only on the main trunk and the thickest branches. The reason for this peculiar arrangement of the female inflorescences is very obvious at harvest time: at its best, a jakfruit can be as long as 90cm, with a diameter of 50cm, and can weigh up to 40kg. With such gargantuan dimensions, the jakfruit is the largest fruit borne by any tree on Earth. Although its size never fails to impress, opinions on its edible qualities diverge considerably. To begin with, the very sticky latex in its rind makes it difficult to peel. In addition, a fully ripe jakfruit emits a musty, sweet odour with a hint of rotten onions, and so is likely to make a negative first impression on the uninitiated. However, careful removal of the rough, green to yellowish-brown rind, which is formed by the hardened distal parts of the perigones of the flowers, reveals the delicious golden-yellow flesh. Its pleasant fruity aroma and rich flavour, which has been variously described as a mixture of melon, pineapple, mango, papaya and banana, compensate for the first disagreeable olfactorial impression. The soft but slightly rubbery "bulbs" formed by the perigones of the fertilized flowers constitute the tastiest part of the pulp. Each bulb encloses an egg-shaped, light brown achene, 3cm long. Rich in carbohydrate and protein, the chestnut-flavoured seeds, called jaknuts, can be eaten raw, boiled or roasted. In between the bulbs lie strands of tougher, fibrous tissue called "rags." The chewy rags are derived from the perigones of unfertilized flowers and yield an excellent jam-setting agent.

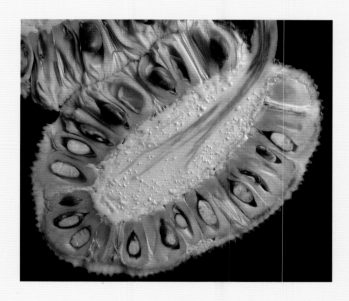

Figs, gnats and sycophants

It may seem rather an anticlimax to those still in awe of the humongous dimensions of the jakfruit, but a closer look at the fig (*Ficus carica*, Moraceae), a less well-endowed cousin of *Artocarpus heterophyllus,* proves once more that size is not everything in life. Whereas its cousins, the breadfruit and jakfruit, had a more adventurous past, the historic record of human indulgence in the common fig (*Ficus carica*) is much more impressive. Although it owed its early fame to its leaves, which, according to the Bible, were used by Adam and Eve to cover up their nakedness, it is for its fruit that the fig has been cultivated and held in the greatest esteem by humans around the Mediterranean for thousands of years. The Egyptians and Greeks admired them as one of the most delicious of foods. They were the favourite fruit of queen Cleopatra, and the Athenians were so proud of their figs that it was forbidden to take them out of Attica. Those who reported the unlawful export of figs were called "*sykophantai*," meaning "fig informers." As some *sykophantai* abused their influence in a manipulative manner, the name was later generally applied to all informers, liars, flatterers, impostors and parasites, hence the English word sycophant. The fig was also much admired by the Romans, who spread it throughout their empire. Pliny the Younger (c. 61–112 AD), who knew of 29 different varieties of fig, found that "figs increase the strength of young people, preserve the elderly in better health and make them look younger, with fewer wrinkles." Because of its long domestication the exact natural origin of the fig is not known but it is generally assumed to be somewhere between the Near East and the Black Sea. It has long been known that the Egyptians were cultivating the fig more than six thousand years ago. However, the remains of edible figs have been discovered recently at the site of an ancient village in the Jordan Valley dating back to 11,200 to 11,400 years ago. The fact that these figs were seedless means that they were formed parthenogenetically (without prior fertilization), a clear indication that they came from cultivated trees. If the interpretation of this sensational discovery is correct, the common edible fig would be the first plant species domesticated by humans in the Neolithic Revolution, more than a thousand years before barley and wheat. The fact that *Ficus carica* can most easily be propagated by cuttings supports the idea of its early domestication.

Although it belongs to the same family as the jakfruit, *Ficus carica* and the other c.750 species of the predominantly tropical to subtropical genus *Ficus* are distinctly different from all other Moraceae. They are therefore classified in their own separate tribe, Ficeae. This taxonomic separation is based mainly on their peculiar inflorescence, called a *syconium,* which, after pollination, matures into the fruit that we commonly call a fig (Greek: *sykon* = fig). In simple terms a fig could be described as a miniature inside-out jakfruit. In scientific

opposite: *Ficus carica* (Moraceae) – common fig; ancient cultigen, probably from south-west Asia originally – microscopic detail of the entrance (ostiole) to the syconium; the entrance of the fig cavity is closed by numerous tightly packed bracts which, at the time of pollination, give way to a narrow passage through which the fig's pollinators, tiny wasps of the family Agaonidae, can enter the flower-lined cavity; diameter 1.1cm

Ficus carica (Moraceae) – common fig – fruits (syconia); the common fig is only one of the worldwide c. 750 members of the genus *Ficus* (figs). All figs bear their tiny flowers inside a peculiar inflorescence called a *syconium*, which after pollination matures into the fruit we commonly call a "fig"; diameter of fruit c. 4cm

terms, a syconium represents an invaginated spherical, oblong or pear-shaped pulpy inflorescence axis, the inside of which is lined with two or three (in small figs) and several thousand (in the largest figs) tiny male and/or female flowers. To illustrate how a syconium is theoretically formed the idea of a sunflower head curving up its margin to form first a bowl and then an urn with a small opening at the top can be used. For the botanically better versed, the hypothetical sunflower analogy is superfluous. Certain Moraceae, namely the members of the tribe Dorstenieae (e.g. *Dorstenia contrajerva*), present their minuscule flowers and fruits openly on a flat, plate-like inflorescence axis, just like the sunflower. Morphologically, their inflorescence or infructescence (when mature) provides a real-life model of an open fig. Real-life figs are closed, apart from a small apical opening (ostiole) that is lined with tightly overlapping bracts. Certain tropical *Ficus* species even bear their syconia up to 10cm underground, which means that we need to explain how the flowers are pollinated. The answer is one of the most astonishing examples of co-evolution and describes the mutual dependence between a plant and an animal in a relationship that is vital for both partners. During the course of evolution tiny insects belonging to the fig wasp family Agaonidae (Hymenoptera) have adapted their reproductive cycles to take place inside figs. In return for the fig's provision of shelter and food, the 1–2mm long wasps provide an essential service to the fig by pollinating its flowers. The cycle of fig wasp reproduction and fig pollination begins with a gravid female wasp forcing its way through the narrow ostiole into the fig, often breaking its wings on the way. Inside, the wasp pollinates the long-styled (fertile) female flowers with pollen brought from another fig and lays its eggs in the ovaries of short-styled female flowers that the fig specifically provides for this purpose. In the literature the short-styled flowers are often called "gall flowers" and are considered sterile as opposed to the fertile, long-styled flowers. However, unlike typical insect galls, ovaries hosting a developing fig wasp show no aberrant tissue formation and, if not infected, short-styled flowers are just as capable of producing a normal seed-bearing drupe as long-styled flowers. The latter, though, have the advantage of being safe from infestation for the simple reason that the wasp's ovipositor is only long enough to reach down the stylar canal and into the ovary of the "sacrificial" short-styled gall flowers, each of which is inoculated with one egg. After laying its eggs, the female wasp weakens and dies inside the fig. Upon inoculation with a wasp egg the ovules in the short-styled flowers are stimulated to produce endosperm tissue but no embryo. For the next two or three weeks, the endosperm tissue, which must be initiated parthenogenetically (i.e. without prior fertilization of the ovule), will provide the food for the developing wasp larvae. Unlike the fertile female flowers that develop into small single-seeded drupes (the reason for the tiny stones inside a fig), the

opposite: *Blastophaga psenes* (Agaonidae) – female fig wasp; pollinator of the common fig (*Ficus carica*) – only a few pollen grains stick to the almost glabrous body of a fig wasp. Pollen is transported in special pollen pockets on the thorax, in pollen-holding cavities (so-called "corbicula") on the front legs, on the antennae, and in the folds between the segments of the abdomen. The female fig wasps lay their eggs in the ovaries of the female flowers where the larvae develop and also pupate. The wingless male fig wasps hatch first and mate with the females whilst they are still inside the fig's ovaries; 1.7mm long

Ficus carica (Moraceae) – common fig; ancient cultigen, probably from south-west Asia originally – pollen grains on a fig wasp (*Blastophaga psenes*, Agaonidae); 10.4μm long

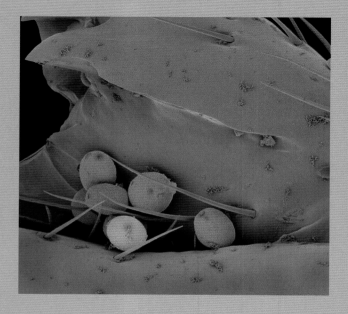

short-styled flowers die once the wasps have hatched. The hatching of these tiny insects begins after two to three months and follows an astonishingly sophisticated protocol, probably controlled by the high carbon dioxide content inside the fig. The male wasps emerge first. Unlike their female counterparts, they will never leave the confinement of the fig and hence are borne wingless with reduced legs and eyes. Facing a very short lifespan, and suffering from what in the outside world would be considered severe gender inequality, the male wasps' existence is nevertheless a sweet one. Male wasps are equipped with much stronger jaws than female wasps. Their mission is to use this asset to free the still unhatched young females from their ovarial confinement and, in doing so, also impregnate them. During their short lives, in many fig species the male wasps also have to gnaw holes in the wall of the syconium through which the females — roused by the influx of fresh air — can leave. The change of atmosphere inside the syconium also causes the fig to swell and ripen. In other species, including the common edible fig (*Ficus carica*), the female wasps leave through the ostiole, which by then has widened to allow them passage. Whatever the exit strategy, the winged females pass the ripening male flowers on their way out and pack their pollen baskets (corbiculae) with pollen before they take off for the next fig, where their reproductive cycle starts anew.

This pollination strategy applies to about half the species of the genus *Ficus*, namely those that are monoecious. The other half consists of dioecious species with a slightly more complex distribution of their male and female flowers between different individuals. Monoecious figs bear male flowers and both kinds of female flowers (short- and long-styled) within the same syconium. Dioecious figs produce two different kinds of syconia on separate male and female trees. The syconia of so-called male trees are only functionally male since they also contain short-styled female flowers (to host their fig wasps) and pollen-producing male flowers. Female trees are purely female with syconia lined exclusively with long-styled female flowers. This particular type of dioecy where the individuals of a species are either *hermaphrodite* or female is called *gynodioecy*. Pollination and seed production follow a slightly different pattern in gynodioecious figs. After leaving her native "male" syconium, a female, pollen-laden fig wasp will try to enter the ostiole of another syconium of its host fig species. She may find herself in another "male" syconium where she can successfully deposit her eggs in the short-styled flowers. However, if she is unlucky, she will end up in a female syconium and find herself unable to lay her eggs because her ovipositor is too short for the long-styled flowers. Looking in vain for short-styled flowers, the wasp at least fulfils her duty to the fig tree and pollinates its flowers before her short life expires inside the syconium. Although her life may seem wasted, other wasps will be laying their eggs

below: *Ficus carica* (Moraceae) – common fig; ancient cultigen, probably from south-west Asia originally – longitudinal section of fruit (syconium); an edible fig is lined with female flowers whose ovaries develop into tiny drupes; diameter 3.5cm

bottom: *Dorstenia contrajerva* (Moraceae) – contrayerva; native from southern Mexico to northern South America – fruit (trymosum); the compound fruit technically represents an "open fig." Lots of tiny drupes are sunken into the flattened inflorescence axis. When ripe, their stones are expelled with force; diameter of fruit 3.5cm

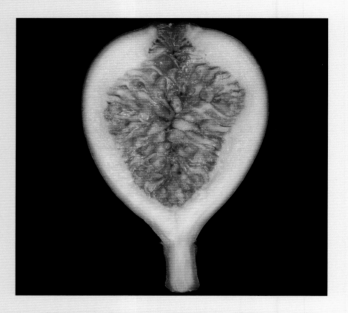

below: *Ficus sansibarica* subsp. *sansibarica* (Moraceae) – knobbly fig; native to south-east Africa where this species is classified as critically endangered – fruits (syconia); like many tropical figs, the knobbly fig is caulicarpous, i.e. it bears its fruits directly on the trunk where fruits bats, whose sonar orientation is weak, can find them more easily

bottom: *Ficus dammaropsis* (Moraceae) – dinner-plate fig; native to New Guinea – fruit (syconium); an unusual fig with fist-size syconia and large leaves up to 90cm long and 60cm wide; diameter of fruit c. 8cm

successfully in "male" syconia so that overall the arrangement benefits both the fig and its fig wasp. In gynodioecious *Ficus* species the functionally male trees are entirely devoted to providing a domicile for their respective fig wasp. Although the functionally male syconia are in theory capable of producing seeds, female fig wasps lay so many eggs in them that very few, if any, of the short-styled female flowers ever produce seeds. The vital task of producing seeds is left entirely to the female trees.

The common edible fig (*Ficus carica*), which has been so relished by civilizations around the Mediterranean for millennia, is also gynodioecious. However, only female trees produce edible fruits. The syconia of "male" trees – called "wild figs" or "caprifigs" – are full of fig wasps and not at all palatable. Historically, the only animals that have an appetite for caprifigs are goats, hence the name "goat fig" or "caprificus" for the "male" fig trees (Greek: *caper* = goat). Knowledge of the fact that female trees require the presence of a caprificus in order to be pollinated and set fruits was essential for fig cultivation; without pollination, the syconia fail to ripen and drop from the branches. To make matters even more complicated, the caprificus produces three generations of syconia every year; all of them were given Italian names. The summer generation, called *profichi*, ripens in June and contains two-thirds short-styled female flowers and one-third male flowers. The following *mammoni* generation matures in autumn and bears predominantly short-styled female flowers and only a few male flowers. Finally, the *mamme* generation is formed in September and ripens the following spring; it contains only short-styled female flowers in which the wasp larvae overwinter.

The fig's complicated pollination mechanism involving the fig wasp *Blastophaga psenes* was, in principle at least, already known to the Phoenicians and Egyptians. Aristotle (384–322 BC) reported his observations of fig wasps leaving caprifigs and entering the ostioles of female syconia more than two thousand years ago. Aristotle's favourite pupil and successor at the Lyceum, Theophrastus of Eresos (371-287 BC), who was to become the first botanical taxonomist in history, described further details of the complicated mechanism of fig pollination, also called "caprification." Later the Roman historian and natural philosopher Gaius Plinius Secundus, better known as Pliny the Elder (23-79 AD), in his *Natural History* discusses the practice of caprification in two passages. At the time, neither Pliny nor anyone else understood plant sex and the significance of pollination. Despite erroneously assuming that the "wild fig" is seedless because its seeds have been spontaneously transformed into "gnats," and various other misconceptions, Pliny understood that the caprificus never produces edible fruits itself but imparts this ability to cultivated (female) trees. Hence, the principle of caprification was to either plant a few "male" trees in a plantation of female trees or to place a branch or a basket with "male"

opposite: *Melaleuca araucarioides* (Myrtaceae) – no common name (literally "Araucaria-like melaleuca"); native to south-west Australia – fruits (loculicidal capsules) aggregated into a cone-like structure, which could also be classified as a capsiconum. The individual capsules develop from an inferior ovary that remains deeply embedded in the woody hypan-thium. The five triangular lobes around the rim of the hypanthium represent the persistent calyx. Close relatives of eucalypts (*Eucalyptus* spp.) where similar cone-like infructescences occur, the 236 species of *Melaleuca* are nearly all endemic to Australia; diameter of cluster 8mm

Liquidambar styraciflua (Hamamelidaceae) – sweet gum; native to North and Central America – fruit (capsiconum); the globose female inflorescence of the dioecious gum tree turns into a strange-looking compound fruit resembling a morning star. Each flower produces a small capsule that releases up to four tiny winged seeds. The spikes of the star represent the hardened persistent styles of the fruitlets; diameter 3cm

caprifigs among female trees around June when the *profichi* generation ripens. Among the hundreds of different cultivars of female fig trees that have been selected over the centuries are some that no longer require pollination. These cultivars are similar to the ones that scientists believe grew in the Jordan Valley more than eleven thousand years ago. They produce delicious parthenocarpic fruits without the presence of a caprificus.

For the many hundreds of wild figs, production of fertile seeds is vital to guarantee the survival of their species. In order to achieve this goal they rely entirely on the pollination service provided by members of the Agaonidae, the fig wasp family. In fact, the relationship between figs and fig wasps is so intimate and complex that every species of *Ficus* has its very own species of fig wasp and vice versa. Very few *Ficus* species call upon the services of a second species of wasp. Recent results of DNA sequence analyses suggest that fig pollination began at least 80-90 million years ago. Petrified fig syconia have been found in sediments in North America dating back to the Cretaceous time period (142-65 million years ago). During this long evolutionary history fig wasps increasingly adapted to the conditions in their respective host fig and figs adapted more and more to the needs of their respective wasp partner. Progressive co-adaptation has finally led to co-speciation between fig wasps and figs. This extreme specialization on just a single species-specific pollinator avoids hybridization among figs and is assumed to be the reason for the extreme species diversity in the genus *Ficus* compared with other members of the Moraceae.

Angiosperms with cones?

Praising the world's most famous compound fruits may have created the erroneous impression that all compound fruits are soft, juicy and edible. Although it is true that many compound fruits biologically behave like berries or drupes, offering an edible reward to attract animal dispersers, there are also quite a few inedible examples. If you have ever wondered why alder trees bear cones when cones should really be growing on conifers, you are about to find the answer.

Alders (*Alnus* spp.) belong to the Betulaceae family and as such have unisexual flowers. Both male and female flowers are arranged in inflorescences, called catkins, that are specifically adapted to wind pollination. The male catkins wither away once they have shed their pollen, but the female catkins develop into compound fruits resembling miniature pine cones. The scales of the woody cones are produced by the hardening bracts subtending the female flowers whose ovaries turn into minute winged achenes. The globose female flower heads of the dioecious gum tree (*Liquidambar styraciflua*, Altingiaceae) from America develop into spherical woody cones that resemble little morning stars. Within this curiously looking compound fruit, each of the 30-40 female flowers produces a small, pointed capsule 4-5mm wide (the spikes of the star). Each capsule opens into two halves releasing up to four tiny, winged seeds 0.8mm long. The spikes of the star are the hardened persistent styles of the fruitlets.

By far the most impressive cones of any angiosperm belong to the Australian honey-suckles (*Banksia* spp., Proteaceae). Their spectacular brush-like inflorescences, whose profusely secreted nectar is relished by the Aborigines as a sweet treat, turn into massive cone-like infructescences. What appear to be the scales of a cone are in fact follicles tightly arranged along a solid wooden axis and embedded in a compact mass of bracts and *bracteoles*. The two-winged seeds borne inside each follicle are shed only after a bush fire, an adaptation to the fire-prone habitats of the Australian drylands where many *Banksia* species make their home. The heat of the fire, increased by the dead flowers that act as tinder, triggers a remarkable opening mechanism in the follicles. Inside the thick, dead tissue of the follicle wall, which protects the precious seeds against both animal predators and fire, tension builds up between the superimposed layers of woody fibres as the follicles ripen and dry out. The tension is counteracted by a resin gluing the fibres in place. Only when extreme heat destroys the resin is the tension between the different layers of fibres released allowing the follicles to split open and each release two winged seeds. Without a bush fire, the fruits can remain on the plant for years with their seeds intact. Year by year another crop is added creating a canopy seed bank – life insurance against fire for the species.

Banksias and other Proteaceae such as *Petrophile*, also from Australia, and the South African *Aulax*, *Protea* and *Leucadendron*, are not the only plants pursuing the strategy of an *aerial seed bank* in woody cone-like fruits on their branches. *Serotiny*, as this phenomenon is called, has evolved several times independently in other, only distantly related plant families occurring in habitats with frequent wild-fires. Australia harbours a wide variety of examples, including several species of eucalypts (e.g. *Eucalyptus petila*), melaleucas (e.g. *Melaleuca araucarioides*, *M. huttensis*), and members of the she-oak family (Casuarinaceae). The Casuarinaceae are a group of around a hundred predominantly Australian species of trees and shrubs with drooping, evergreen twigs that bear a strange resemblance to horsetails (*Equisetum* spp.). Weirder even, from a distance she-oaks look like conifers although they belong to the dicotyledonous order Fagales, which also includes oaks and beeches. To add to the deception, they bear rather striking cone-shaped fruits. Like the cones of the Proteaceae, those of the Casuarinaceae are formed by entire inflorescences, albeit in a slightly more complicated manner. The Casuarinaceae produce their wind-pollinated flowers in separate inflorescences, either on the same individual (monoecious) or on separate male and female plants (dioecious). Female flowers lack a perianth and consist of

opposite: *Banskia menziesii* (Proteaceae) – firewood banksia; native to Western Australia – fruit (folliconum); only a few of the many flowers that produce a showy, brush-like inflorescence, produce fertile follicles. As indicated by the rugged appearance of the fruit in the photograph, the "cones" of the firewood banksia can remain on the plant for years; their follicles open only after they have been exposed to fire; fruit c. 8cm long

Banksia candolleana (Proteaceae) – propeller banksia, native to Western Australia – fruit (folliconum); members of the Australian genus *Banksia* produce impressive cone-like compound fruits that resemble those of conifers. What appear to be cone scales are follicles tightly arranged along a solid wooden axis and embedded in a compact mass of bracts and bracteoles. The follicles often only open to shed a pair of winged seeds after the fruits have been exposed to fire, an adaptation to the fire-prone habitats of the Australian drylands, which many *Banksia* species call their home. The curious propeller-like shape of the fruit of *Banksia candolleana* is due to the fact that of the many flowers that form the 4cm high, golden yellow, brush-like inflorescence, only three produce follicles in the particular arrangement shown in the photograph; diameter c. 10cm

nothing but an ovary. Within the tightly packed inflorescence, the ovary is subtended by a bract and flanked on either side by a bracteole. As the ovary develops into a tiny, single-seeded samara, it remains tightly enclosed within the enlarging bracts and bracteoles, which eventually become woody and form a cone. The symmetrical arrangement is revealed by its intriguing surface pattern. Often it is only after exposure to fire that the bracts and bracteoles separate and move apart to release the samaras to the wind.

Finally, serotiny is found in genuinely gymnospermous cones such as those of the North American jack pine (*Pinus banksiana*) and many other *Pinus* species. It is expressed mainly in habitats that are exposed to fairly frequent wild fires where the time gap between fires does not exceed the lifespan of the canopy seed bank. Interestingly, the bearing of seeds in relatively massive, hard structures to protect them from predation and fire for a long time is a common parallel adaptation among serotinous plants and explains the superficial resemblance between cones of the angiospermous banksias and the gymnospermous pines.

With the exploration of compound fruits we have reached the pinnacle of the angiosperms' inventiveness when it comes to "packaging" their seeds. The superficial resemblance of the seed-bearing organs of some angiosperms and certain gymnosperms takes us back to our original question, "What is a fruit?"

CARPOLOGICAL TROUBLEMAKERS

If our foray into the wondrous world of carpology has proved anything so far it is that classifying fruits according to their structure is a very difficult task. For nearly every carefully crafted scientific definition the angiosperms have produced a puzzling range of exceptions with which they have managed to foil all attempts to create a classification system of their fruits that is both logical and practical.

Botanists attempting to tackle the overwhelming diversity of fruit types thrown at them by nature have ambitiously created more than 150 different technical fruit names over the past two centuries. Unfortunately, in doing so they had little regard for previously established terms, resulting in the creation of many confusing synonyms. They also repeatedly modified definitions of established terms. All this was done using a diverse range of criteria indiscriminately, irrespective of whether they were morphological, developmental, histological, ecological or physiological in nature. Made worse by the still unsettled concept of fruit, the result was a chaotic plethora of inconsistent carpological terms that by the mid-nineteenth century many botanists found frustrating. Among the fiercest critics was the prominent German biologist Matthias Jakob Schleiden (1804–81), co-founder of

opposite: *Allocasuarina tesselata* (Casuarinaceae) – she-oak; native to Western Australia – microscopic detail of fruit (trymoconum), revealing the symmetrical structure and arrangement of the individual fruitlets. The bracts release their tiny samaras only after exposure to fire

Allocasuarina tesselata (Casuarinaceae) – she-oak; native to Western Australia – fruit (trymoconum); she-oaks look like conifers and even their fruits bear a strange resemblance to conifer cones. But the Casuarinaceae are dicotyledonous angiosperms related to the members of the birch family (Betulaceae). They are wind pollinated and produce separate male and female inflorescences. The female flowers lack a perianth and consist of nothing but an ovary, which is sub-tended by a bract and flanked on either side by a bracteole. The ovary develops into a tiny, single-seeded samara, enclosed within the bracts and bracteoles that eventually form a woody, cone-shaped compound fruit; 4.5cm long

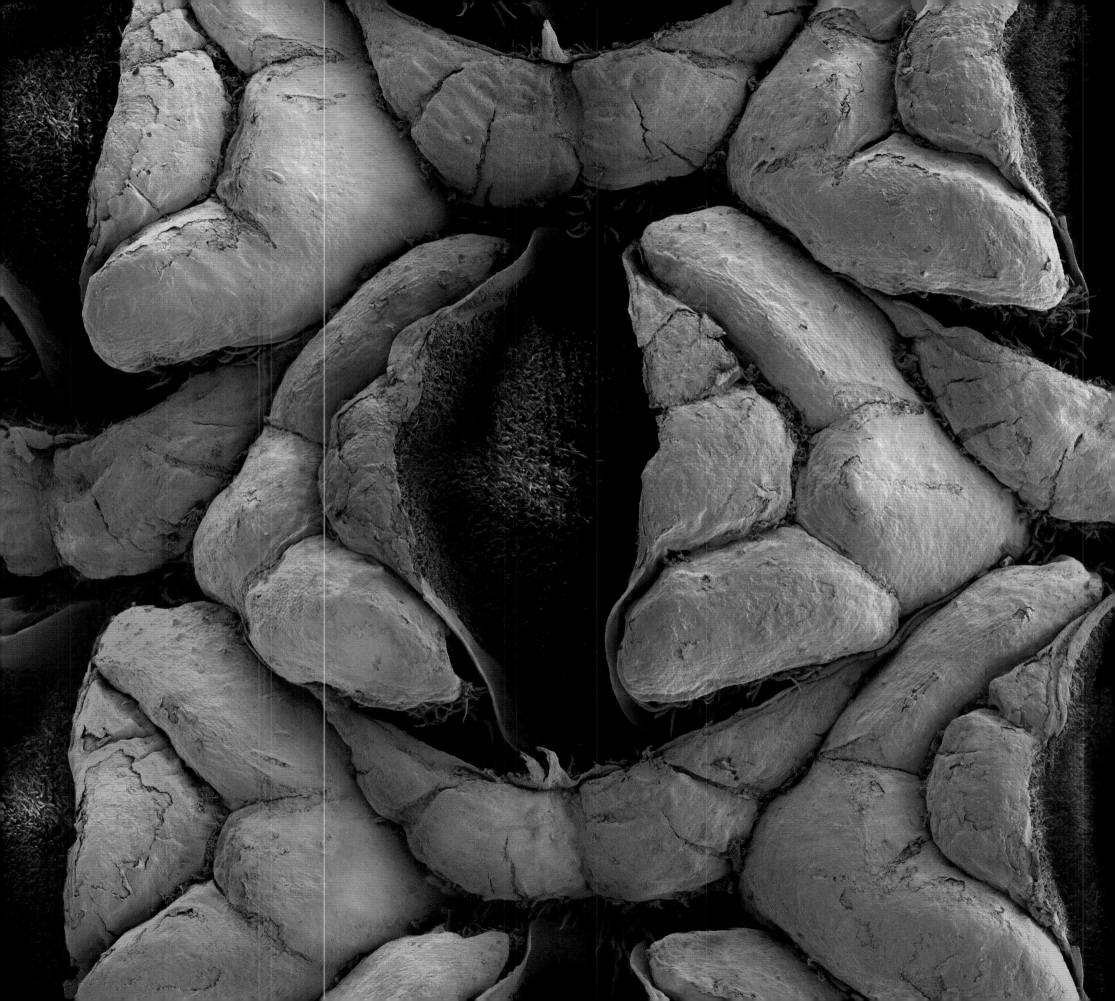

the *cell theory*. In 1849 Schleiden wrote: "Nowhere has purely diagrammatic comprehension been so prevalent as in the theory of the fruit; nowhere have botanists, starting from the language of common life and merely multiplying the words, taken so little pains to define with scientific strictness; and hence nowhere does terminology so vacillate among all definitions as in fruit ... in short, the confusion is indescribable ... Here I will merely remark ... that in the general treatment of the theory of fruit they have been playing an unaccountably frivolous game with the reader and student.*"*

Until recently the last comprehensive reviews of fruit terminology were those of Bischoff (1833), Dumortier (1835), and Lindley (1832, 1848). Hence, even today, little has changed. Confused by the tangled mass of terms and left without a standardized fruit classification system, botanists have invented their own simpler solutions in a desperate attempt to undo the Gordian knot of fruit classification. Applying Procrustean means they force-fitted the few definitions that sufficed to address the most common fruit types found in the impoverished flora of the temperate regions (especially of Europe) by creating internally contradictory names such as "dry drupes," "drupaceous nuts," "dehiscent berries" and "indehiscent capsules" for exotic misfits. The biggest troublemakers of all, anthocarpous and compound fruits, demanded even more drastic disciplinary measures. Most botanists do not even consider them genuine fruits.

Bogus fruits and how to debunk them

Surprisingly, most twenty-first-century botanical authors still subscribe either to Tournefort's or Gaertner's definition of a fruit as a mature flower or a mature ovary, respectively. But this still leaves anthocarpous and compound fruits to be sorted out. Their day of reckoning came not too long after Gaertner published his ideas on fruits. The first to address the "problem" was Carl Ludwig Willdenow (1765-1812), botanist and former director of the Berlin botanical garden, who in the third edition of his *Grundriß der Kräuterkunde* (1802) branded any fruit that consists of more than just a mature ovary a "fructus spurius" ("spurious fruit"). Later he addressed individual fraudsters directly (Willdenow 1811), denouncing the strawberry, whose fleshy part is formed by the receptacle rather than the ovary wall, as a "false berry" ("bacca spuria"). Willdenow was later backed by others, among them the French carpologist Nicaise Auguste Desvaux who called them "pseudocarpien" (Desvaux 1813), which translates into English as "false fruit." Other authors referred to anthocarpous fruits as "accessory fruits," a term still widely used today. In 1868, the well-known German botanist Julius von Sachs (1832-97) joined the alliance and debunked all fruits that include parts other than the ovary (anthocarpous fruits), and

Broussonetia papyrifera (Moraceae) – paper mulberry; native to east Asia – fruit (sorosus); formed by an entire inflorescence rather than just a single flower, the paper mulberry is a "carpological troublemaker" just like its relative the black mulberry (*Morus nigra*). The globose female inflorescence of the dioecious paper mulberry produces a compound fruit in which the individual flowers grow into loosely arranged, brightly coloured, succulent fruitlets. Each fruitlet consists of the enlarged fleshy perigone that encloses a tiny drupe. Although the fleshy proportion is small, the fruits are edible and have a pleasant taste. The fibrous inner bark of the paper mulberry tree is used to make cloth and paper, hence the name; diameter of fruit 1.5cm

those that are formed by more than one gynoecium (compound fruits) as "bogus fruits" ("Scheinfrüchte"). Many adopted this view, among them such eminent botanists as the American Asa Gray (1810-88). Nevertheless, in the nineteenth century the carpological universe was still very much in flux and many of the early carpologists harboured more liberal ideas. Among these libertines was the great Augustin Pyramus de Candolle (1778-1841), one of the most prolific botanists of all time. Whereas his orthodox contemporaries chose to reject anthocarpous and compound fruits and considered gymnosperms too primitive to be included in carpological discussions, de Candolle (1813) put forward one of the first systematic fruit classifications wherein a fruit could be the product of either one or several flowers, which included the gymnosperms. Many others, including John Lindley (1848), adopted the same broad-minded vision. However, in the twentieth century the concept of fruit became more restrictive when applied to angiosperms. This development was probably influenced by Gaertner's original definition of a pericarpium as the product of a mature pistil, which later became the basis for the definition of fruit in modern textbooks. The fact that carpology became an increasingly neglected discipline may also have played a role. In any case, most modern authors adopted a narrow fruit-concept, teaching students the seventeenth-century notion that a fruit is a mature ovary or, at most, a mature gynoecium including any associated tissues (for example, Jackson 1928; Raven et al. 1999; Mauseth 2003).

SO WHAT *IS* A FRUIT?

Finally, relief came in 1994 when Richard Spjut published his *Systematic Treatment of Fruit Types*, the most recent and most significant carpological classification of the twentieth century. Spjut's attempt to bring some order into the 300 year-old chaos of fruit classification was admirable in itself, but he also provided a precise scientific definition of the term "fruit" that effectively reinstated de Candolle's liberating definition, allowing botanists for the first time in more than a hundred years to address pineapples, jakfruits and figs, as well as the cones of conifers, cycads and *Welwitschia* simply as "fruits." In Spjut's (re)-enlightening classification system, gymnosperms and angiosperms are treated as equal, and Procrustean terms such as "pseudocarp," "false fruit" and "accessory fruit" – once again – become superfluous. They are replaced by the more appropriate expressions "anthocarpous fruit" and "compound fruit." The present authors wholeheartedly embrace Spjut's broad concept of fruit as well as most of his proposed fruit types, although in a few cases, for the sake of the non-specialist reader, we have admitted a more liberal terminology, especially with respect to the usage of the terms "nut" and "nutlet."

Fragaria x *ananassa* (Rosaceae) – garden strawberry, only known in cultivation; fruit (glandetum); famed for its delicate flavour and high vitamin content, the garden strawberry is one of the most popular fruits with an annual world production of more than 2.5 million tons. Its fleshy, edible part is not produced by the gynoecium but by the swollen axis of the flower, which is why the strawberry is considered an anthocarpous fruit. The gynoecium itself consists of many small separate carpels that turn into tiny achenes, visible as the brown granules embedded in the surface of the red fleshy part; 3cm long

The biological function of fruits and seeds

After such a bewildering overview, it is time to recall the reason for the astonishing diversity of fruits. Their startlingly complex biology is a direct consequence of their vital dispersal function, which has exposed them to high adaptive pressures during the course of evolution. Contrary to animals, which can move around and actively seek a suitable place to live, plants are rooted in the ground and so tied to one place. The chance to travel comes only once in their lifetime: when they are still tiny embryos safely tucked away inside the seed. Whether a seed is shed with the fruit, as in drupes and nuts, or released from capsules and left to its own devices, it must leave the parent plant to start a new life somewhere else. The reasons that seed dispersal plays such an important role are straightforward.

Generally it would not be advantageous for a seed to germinate in its place of origin, where it would have to compete for space, light, water and nutrients with the parent plant and its siblings; and it would probably also encounter other unfavourable conditions and hazards such as predators and diseases attracted by the parent plant. Travelling also offers the opportunity to reach and colonize new sites, expanding the range of the species. In the end, the survival of not only the individual but of the entire species depends on the seed reaching a suitable place for germination and establishment. Seeds are the spermatophytes' only means of sexual reproduction, so their dispersal plays a crucial role in evolution. Like the transport of pollen between individuals, seed dispersal is an important factor that helps prevent inbreeding by promoting gene exchange between different populations. Just as in humans, sexual reproduction in plants reshuffles the genes and creates new individuals with unique and sometimes better character combinations, thus providing the raw material for evolution. It is therefore no exaggeration to say that the survival of a species rests on the shoulders of both the seed and the fruit. With the seed phase offering such a vital window of opportunity, over millions of years spermatophytes have developed an astounding diversity of dispersal strategies. Depending on their individual lifestyle and habitat, plants surrender their seeds to the power of the elements, catapult them away, bury them in the ground or – most importantly – manipulate animals into carrying them.

DISPERSAL – THE MANY WAYS TO GET AROUND

The nature of the dispersal unit, called the *diaspore*, varies depending on the type of fruit. The diaspores of dehiscent fruits are simply the seeds themselves. In indehiscent fruits the nature of the diaspore itself varies. It can consist of one or more seeds enclosed in fruitlets (multiple fruits), fruit fragments (schizocarpic fruits), entire fruits (simple fruits) or whole infructescences (compound fruits). The most extreme diaspores are tumbleweeds in which

opposite: *Cynoglossum nervosum* (Boraginaceae) – hairy hound's tongue; native to Pakistan and India – fruit (micro-basarium) subtended by the persistent calyx; the Boraginaceae are characterized by a syncarpous gynoecium of two joined carpels divided into four single-seeded compartments. When ripe, the gynoecium breaks into four single-seeded fruitlets (mericarps). The fruitlets bear recurved hooks, an effective strategy to attach themselves to animals; diameter 1.1cm

Eryngium creticum (Apiaceae) – Crete eryngo; native to south-east Europe, western Asia and Egypt – fruitlet; the calyx of the flower persists to form two or three sharply pointed, stiff wings that appear to be able to facilitate both wind and animal dispersal (epizoochory); fruitlet 8.8mm long

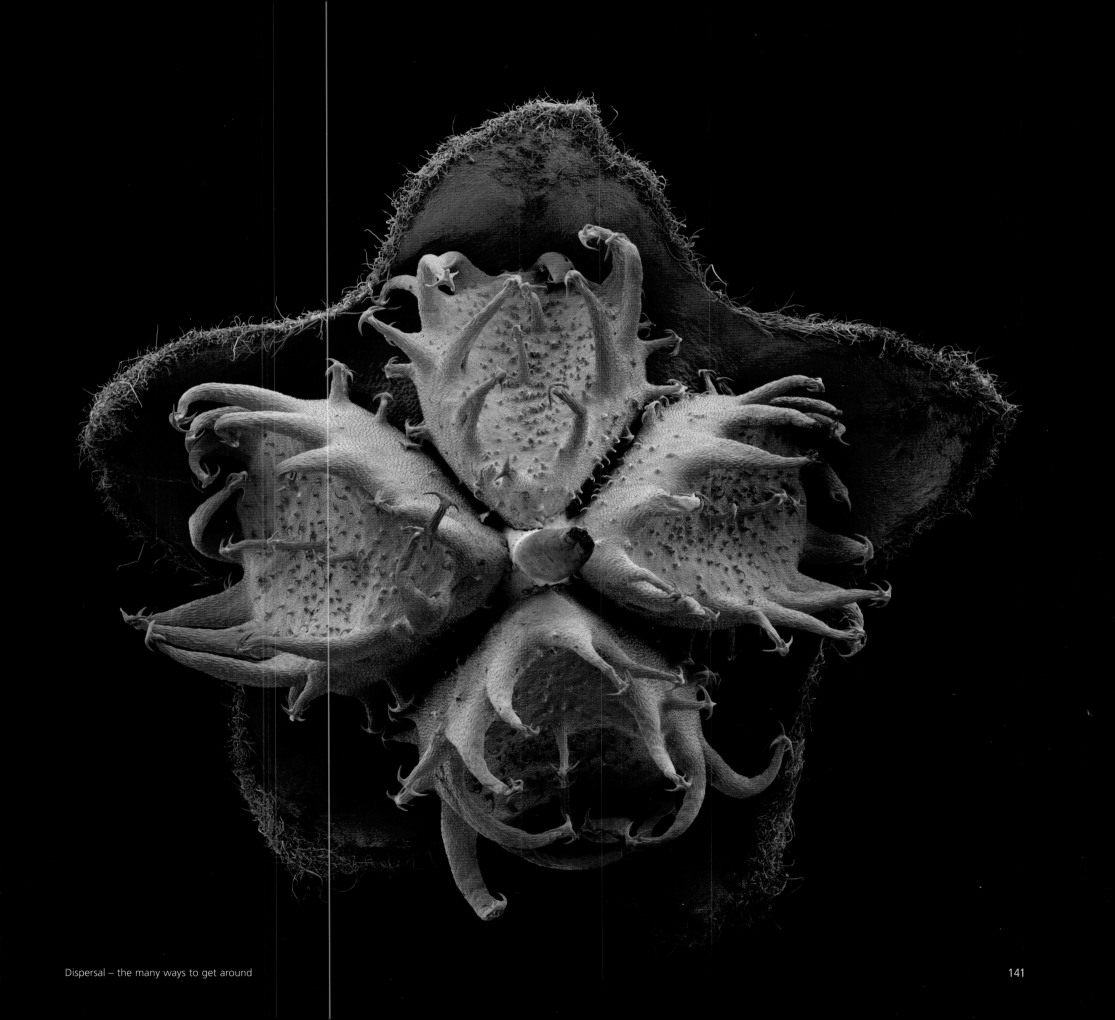

opposite: *Salsola kali* (Amaranthaceae) – Russian thistle; native to Europe but widely naturalized in America and else-where – a typical tumbleweed, shown here in Death Valley

below: *Piscidia grandifolia* var. *gentryi* (Leguminosae) – native to Mexico – fruits (samaras). The four lateral wings cause the samara to spin in the wind; c. 3cm long

bottom: *Tipuana tipu* (Leguminosae) – tipu tree; native to South America – fruits (samaras); the unilateral wing of the 3-seeded fruit causes a helicopter-like flight; c. 5cm long

the entire plant becomes the diaspore. When the seeds are ripe, the plant dies and curls up into a globular snarl of branches as it dries out. Eventually, the dead plant breaks off above the ground and is blown around by the wind. Plants demonstrating this behaviour are parti-cularly abundant in prairie and steppe environments where the wind can drive them across the ground for miles. Their seeds are scattered as they roll. Many typical tumbleweeds belong to the Amaranth family (Amaranthaceae), such as the Russian thistle (*Salsola kali*), the tumble pigweed (*Amaranthus albus*) and bugseeds (*Corispermum* spp.). A rather special tumbleweed is found in steppes and semi-deserts from Morocco to Southern Iran. *Anastatica hierochuntica*, a member of the mustard family (Brassicaceae), is famous for its reversible hygroscopic movement and often sold by florists as a curiosity called Rose of Jericho. Although anchored by a tough taproot, the plants can sometimes be torn out of the ground by the wind and behave like tumbleweed. In moist conditions, the dead plant flattens by uncurling its branches, anchors itself to the ground and sheds its seeds. As in so many other semi-desert plants, the hygroscopic movement of *Anastatica hierochuntica* is clearly an adaptation to the dry conditions of its habitat, ensuring that its seeds are dispersed only when water is available for germination.

Tumbleweeds are successful in their own way, but most plants disperse their seeds by more subtle strategies that do not require the sacrifice of the entire individual. To those able to read the sign language of nature, diaspores reveal their dispersal strategies through certain adaptations, if at various levels of encryption. Some strategies are simple, straightforward and easily spotted even by the uninitiated observer.

Wind dispersal

Among the most obvious adaptations of diaspores are those that facilitate wind dispersal (*anemochory*). The diversity of anemochorous diaspores is enormous and is found throughout seed plants, including both gymnosperms and angiosperms. Wings, hairs, feathers, parachutes and even balloon-like air chambers – anything that increases aerodynamic properties or buoyancy – all facilitate transport by air. Such specialized organs can be present in seeds in dehiscent fruits, or in the fruits themselves if they are indehiscent. Often the structures are aesthetically pleasing, resembling masterpieces of engineering. It is therefore no surprise that the dispersal strategies of fruits and seeds have long captivated the imagination of scientists and the general public alike. Despite the fact that the wind is a highly unreliable and unpredictable agent to be entrusted with one's offspring, morphological adaptations assisting wind dispersal are common among seed plants. Thin membranous aerofoils are very efficient, as proved by the great variety of winged diaspores that have evolved

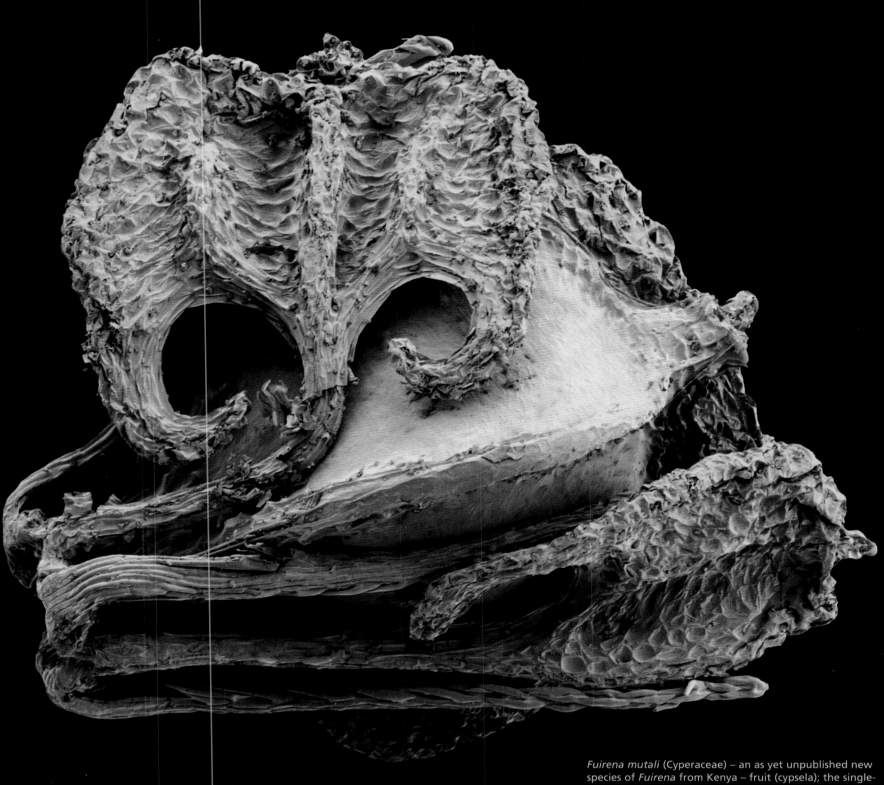

opposite: *Dasylirion texanum* (Ruscaceae) – Texas sotol; native to Texas and northern Mexico (Coahuila, Chihuahua) – fruit (achene); despite the fact that the winged fruit of the Texas sotol is clearly adapted to wind dispersal, it fails to qualify as a samara because the wings are shorter than the seeded portion of the fruit; 7.2mm long

Fuirena mutali (Cyperaceae) – an as yet unpublished new species of *Fuirena* from Kenya – fruit (cypsela); the single-seeded, achene-like ovary is surrounded by the modified perianth, which consists of six segments, three of which are narrow, awn like, and covered with barbed hooks. The other three are curiously shaped, leaf-like, with their bottom margins forming incurved hooks. Like many other plants, it has evolved a double dispersal strategy; 0.95mm long

independently. This can happen even within the same family, as shown by the various types of winged camaras present in the Leguminosae (one lateral wing in *Nissolia suffruticosa*, *Pterolobium stellatum* and *Tipuana tipu*, a peripheral wing in *Dalbergia monetaria* and *Pterocarpus* spp., four lateral wings in *Piscidia* spp. and *Tetrapterocarpon geayi*).

Wings

Among gymnosperms, it is the conifers that often possess unilaterally winged seeds that are wind dispersed, for example pines (*Pinus* spp.), firs (*Abies* spp.) and spruces (*Picea* spp.). However, the simple, uniform model of old-fashioned conifers is put to shame by the sheer diversity of winged diaspores in the angiosperms, whose winged nuts (samaras) can bear one or more wings in various arrangements. Many samaras possess just one unilateral wing that causes a helicopter-like flight as the diaspore rotates around its centre of gravity (i.e. the thickened, seed-bearing part of the fruit). This model is best known in the familiar fruitlets of maples (*Acer* spp., Sapindaceae). Obeying the laws of aerodynamics, very similar looking diaspores, including both fruits and seeds, have evolved in a variety of families. Unilaterally-winged samaras occur in the legume family (e.g. *Luetzelburgia auriculata*, *Neoapaloxylon* spp. and *Tipuana tipu*), the she-oak family (e.g. *Allocasuarina tesselata*, Casuarinaceae), in ashes (*Fraxinus* spp., Oleaceae) and the tulip tree (*Liriodendron tulipifera*, Magnoliaceae), to name just a few. Astonishingly similar in shape but representing seeds rather than fruits are the anemochorous diaspores of many Meliaceae (mahogany family; e.g. *Entandrophragma caudatum*, *Swietenia mahogani*), some Celastraceae (bittersweet family; e.g. *Hippocratea* spp.) and Proteaceae (e.g. *Alloxylon* spp.).

Monoplanes

Two opposite horizontal wings are typical of the seeds of the trumpet creeper family (Bignoniaceae), which includes the North American trumpet creeper (*Campsis radicans*) and the Central American monkey pod (*Pithecoctenium crucigerum*). The aerodynamics of this model set the seeds into a rotational movement around their longitudinal axis, or alternatively afford them a smooth, gliding flight, especially in still air. Although they are common in the Bignoniaceae, the world record for the largest, most familiar gliding seed is held by a member of the gourd family (Cucurbitaceae). The ultra-light seeds of *Alsomitra macrocarpa* (Cucurbitaceae), a liana from the jungles of south-east Asia, boast an incredible wingspan of 12–15cm but weigh just 0.2g. Their formidable gliding properties inspired the Austrian aviation pioneer Ignaz "Igo" Etrich (1879-1967) to design the "Etrich Taube" ("Etrich Dove") in 1910, a monoplane later adapted and used for reconnaissance missions during World War I.

opposite: *Artedia squamata* (Apiaceae) – crown flower, endemic to Cyprus and the eastern Mediterranean – fruit (samarium); the fruit splits into two single-seeded fruitlets, one of which is shown here; 1cm long

below: *Alsomitra macrocarpa* (Cucurbitaceae) – climbing gourd from the rainforests of south-east Asia – seed with a wingspan of 12-15cm but weighing just 0.2g

bottom: *Hippocratea parvifolia* (Celastraceae) – native to southern Africa – schizocarpic fruit (coccarium), produced by a syncarpous gynoecium of three carpels that separates into three flat cocci, one of which is shown split in half to reveal the winged seeds; coccum 5.3cm long

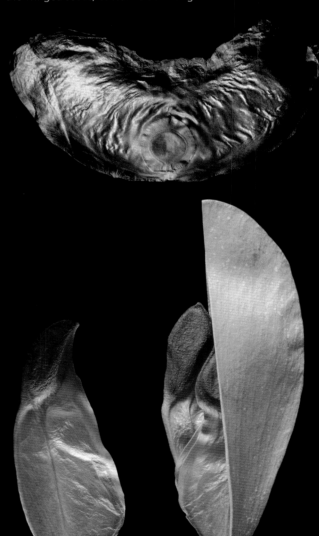

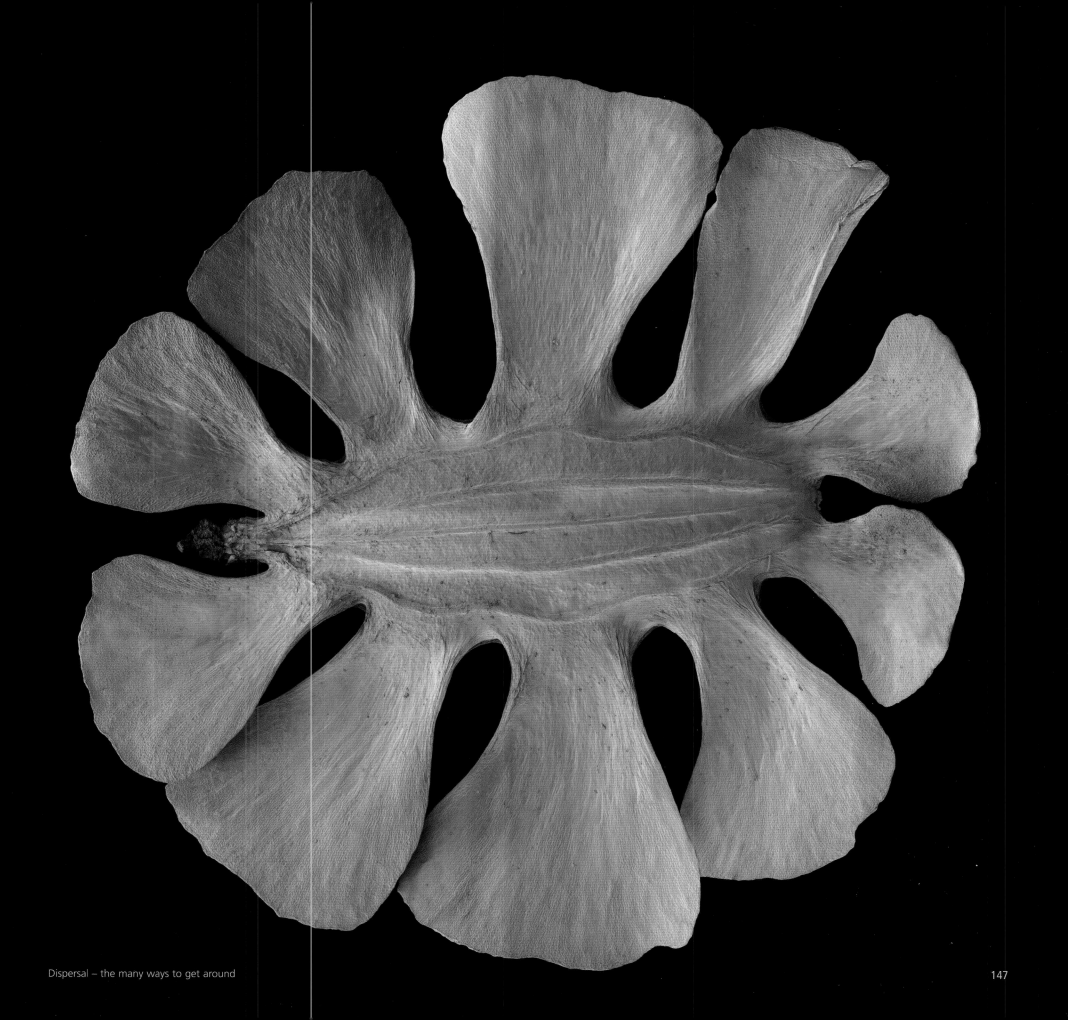

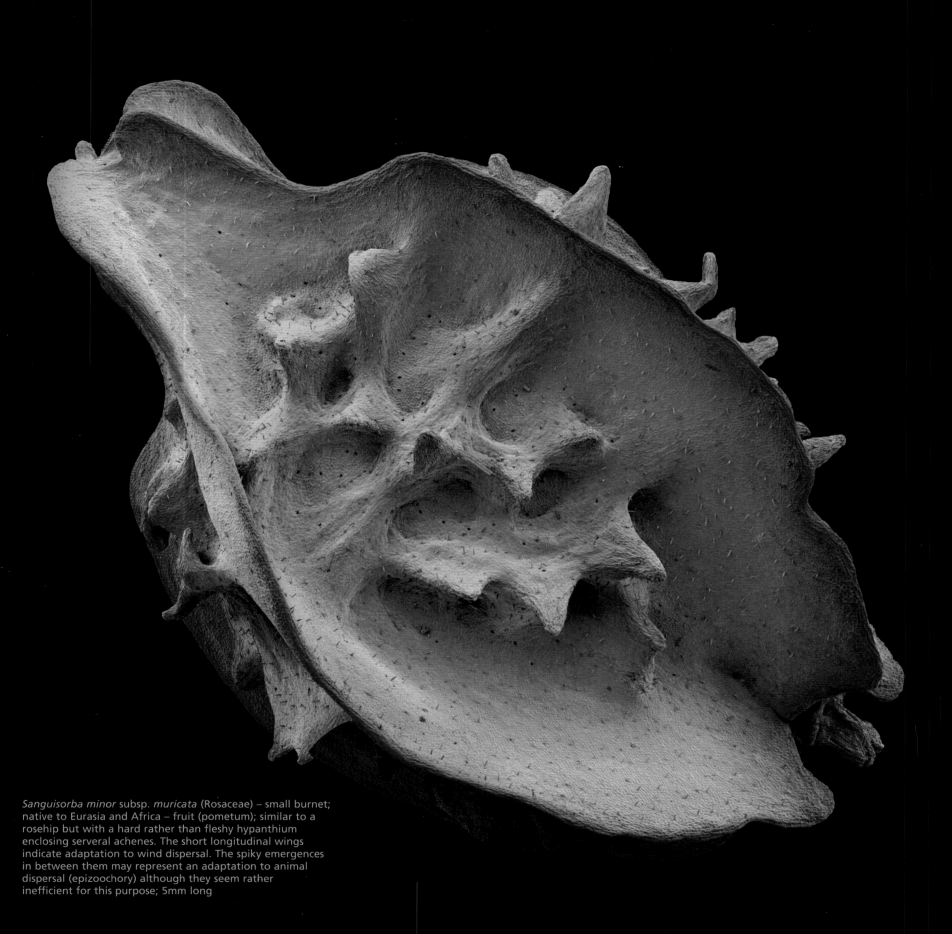

Sanguisorba minor subsp. *muricata* (Rosaceae) – small burnet; native to Eurasia and Africa – fruit (pometum); similar to a rosehip but with a hard rather than fleshy hypanthium enclosing serveral achenes. The short longitudinal wings indicate adaptation to wind dispersal. The spiky emergences in between them may represent an adaptation to animal dispersal (epizoochory) although they seem rather inefficient for this purpose; 5mm long

below: *Tristellateia africana* (Malpighiaceae) – helicopter
fruit; native to Kenya and Tanzania – fruits (pseudosamaras);
the five sepals forming the calyx of the flower persist to form
a set of wings around the seeded portion; diameter c. 2.5cm

bottom: *Paliurus spina-christi* (Rhamnaceae) – Christ's thorn;
native to the Mediterranean and western Asia – fruit
(samara) with peripheral wing. Legend has it that the
crown of thorns placed on Christ's head was made of this
tree's ferociously spiny branches; diameter 2-3.5cm

Diaspores with more than two wings can have them arranged like the spokes of a wheel around the periphery of their flattened body. This model is present in the carrot family (Apiaceae; e.g. *Artedia squamata*, *Eryngium paniculatum*) and the helicopter-like fruits of *Tristellateia africana* (Malpighiaceae), a woody climber from tropical east Africa.

Flying discs

In many Bignoniaceae two opposite wings expand into a wafer-thin peripheral wing that encircles the embryo-bearing part of the seed, as in the blue jacaranda (*Jacaranda mimosifolia*) from Argentina, the African flame tree (*Spathodea campanulata*), and the Brazilian *Zeyheria montana*. The disc-shape is a common adaptation of wind-dispersed diaspores. In addition to those belonging to the trumpet-creeper family, flattened, paper-thin seeds are also found in the Liliaceae (*Cardiocrinum giganteum*, *Fritillaria meleagris*, *Tulipa* spp.), Iridaceae (e.g. *Iris pseudacorus*), Caryophyllaceae (e.g. *Spergularia media*) and Plantaginaceae (e.g. *Nemesia* spp.). Discoid wind-dispersed fruits are also common, for example in the Leguminosae (e.g. *Pterocarpus* spp.), Rhamnaceae (e.g. *Paliurus spina-christi*, Christ's thorn), Rutaceae (e.g. *Ptelea trifoliata*, hop tree) and Ulmaceae (*Ulmus* spp., elm).

Spinning cylinders

Diaspores with multiple wings parallel to the longitudinal axis of a chunky, rather than flattened body perform a spinning motion, especially if the wind is strong. This model occurs in the fruits of the small burnet (*Sanguisorba minor*, Rosaceae) but these tiny four-winged nuts give only a temperate foretaste of what is possible in a tropical climate. Although members of the Leguminosae family have only a single carpel per flower they are endlessly versatile and have also developed various anemochorous fruit types, including four-winged samaras such as those of the Jamaica dogwood (*Piscidia erythrina*). With a diameter of up to 8cm, the spinning cylinders of the southern African raasblaar (*Combretum zeyheri*, Combretaceae) are large although not nearly as impressive as those of *Cavanillesia hylogeiton* (Malvaceae), a huge rainforest tree from the upper Amazon. With five large, paper-thin, semicircular wings radiating at right angles from the elongated seed-bearing core, the fruits of *Cavanillesia hylogeiton* can be up to 18cm in diameter and some 15cm long despite the fact that they weigh just 10g.

Shuttlecocks

Anthocarpous fruits in which parts of the perianth (sepals or petals) persist and enlarge to form a set of apical wings are particularly interesting and beautiful. The result is a

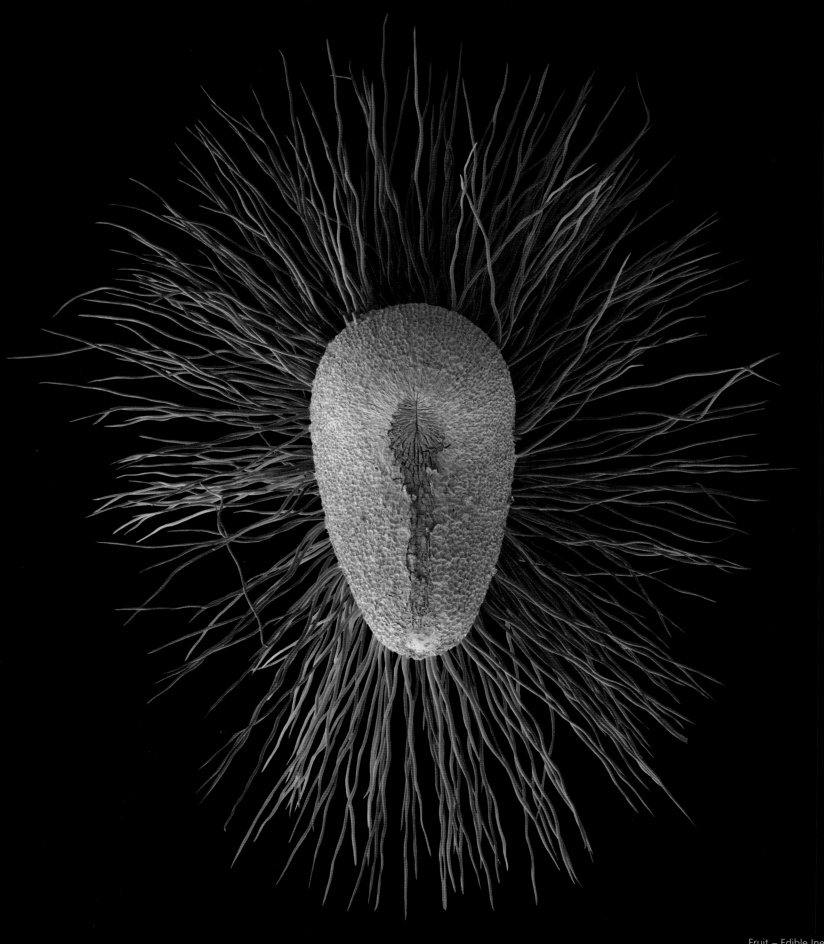

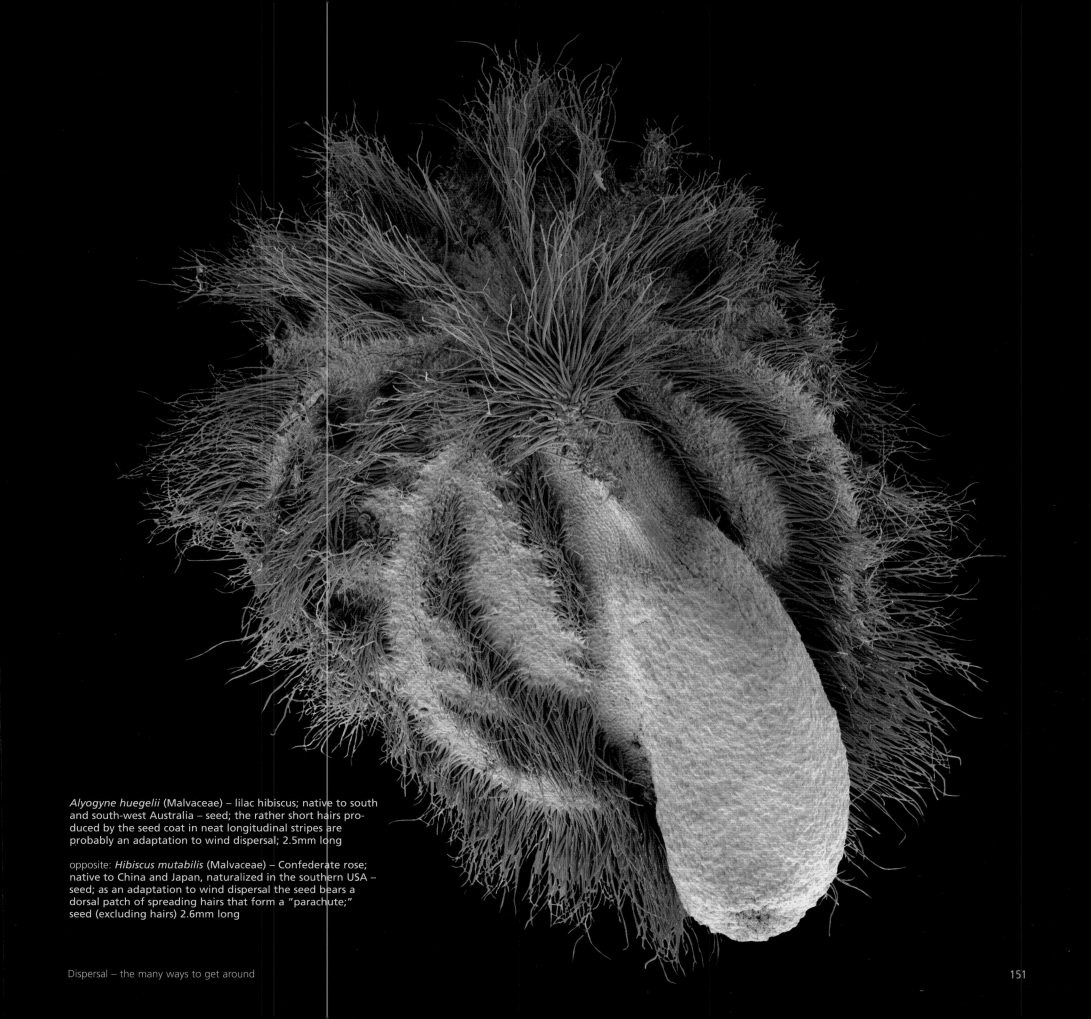

Alyogyne huegelii (Malvaceae) – lilac hibiscus; native to south
and south-west Australia – seed; the rather short hairs produced by the seed coat in neat longitudinal stripes are
probably an adaptation to wind dispersal; 2.5mm long

opposite: *Hibiscus mutabilis* (Malvaceae) – Confederate rose;
native to China and Japan, naturalized in the southern USA –
seed; as an adaptation to wind dispersal the seed bears a
dorsal patch of spreading hairs that form a "parachute;"
seed (excluding hairs) 2.6mm long

pseudosamara very similar to a shuttlecock. The most striking belong to the Dipterocarpaceae – tall rainforest trees of the south-east Asian rainforests. Their single-seeded nuts remain in the persistent calyx whose sepals grow into a set of 2-5 apical wings that can be more than 25cm long (e.g. *Dipterocarpus grandiflorus*). The wings of the very similar looking whirling nut (*Gyrocarpus americanus*, Hernandiaceae) are formed by two opposite tepals. Smaller shuttlecocks are found in the tropical genus *Homalium* (Salicaceae) and *Dicellostyles*, a peculiar genus of the Mallow family from Sri Lanka with just one species, *D. jujubifolia*. The tiniest shuttlecocks of all are the cypselas of the sunflower family (Asteraceae) whose modified calyces can turn into wings (*Galinsoga brachystephana, Xanthisma texanum*), feathers (e.g *Leucochrysum molle*) or parachutes (e.g. dandelion, *Taraxacum officinale*).

In the multiple fruits of certain Ranunculaceae (e.g. *Clematis vitalba*, traveller's joy; *Pulsatilla vulgaris*; pasque flower) and Rosaceae (e.g. *Dryas octopetala*, mountain avens), the style of the individual nutlets becomes a long, feathery appendage.

Woolly travellers

A simple but common adaptation to wind dispersal is the development of hairs in various arrangements. Diaspores that are evenly covered with hairs are relatively rare. The seeds of *Gossypium hirsutum* (Malvaceae), whose long white hairs provide us with cotton, are the best-known example. Other members of the mallow family have their seed hairs arranged in longitudinal stripes (e.g. *Alyogyne huegelii*), in a dorsal patch of spreading hairs forming a parachute (e.g. *Hibiscus mutabilis*), or in a single peripheral fringe of long hairs reminiscent of an Iroquois hairstyle (e.g. *Hibiscus syriacus*). Small woolly achenes are found in many Proteaceae in South Africa and Australia (e.g. *Isopogon* spp., *Leucadendron* spp.).

More often, hairs are arranged in localized tufts as in the seeds of many Apocynaceae (e.g. *Asclepias physocarpa, Nerium oleander*) and willowherbs (*Epilobium* spp., Onagraceae). As long as the seeds are inside the fruit their stiff hairs are folded tightly together but as soon as the fruits open, the seeds unfurl their parachutes.

Love-in-a-puff and other balloon travellers

As an alternative strategy to going airborne, diaspores develop large air spaces that help reduce their specific weight and increase their air resistance. As the ovary turns into the fruit, it can become hugely inflated, creating a thin, often transparent bladder or balloon around the seeds. Balloon fruits are dispersed whole and the seeds liberated as the fragile fruit wall disintegrates when it is blown over the ground by the wind. Fruits of this type occur in the lesser honeyflower (*Melianthus minor*, Melianthaceae), the American bladdernut (*Staphylea*

below: *Sutherlandia frutescens* (Leguminosae) – balloon pea; native to South Africa – fruit (camara); the balloon pea achieves wind dispersal by producing lightweight balloon-like fruits with an inflated, paper-thin pericarp; c. 4-5cm long

bottom: *Gossypium hirsutum* 'Bravo' (Malvaceae) – upland cotton; domesticated cultivar of the wild form, which is native to Mexico – fruit (loculicidal capsule); the hairy seeds are an adaptation to wind dispersal. About 90% of the world's cotton is produced from the seeds of *Gossypium hirsutum*; diameter of fruit 5cm

trifolia, Staphyleaceae), the sapindaceous golden rain tree (*Koelreuteria paniculata*), and love–in–a–puff (*Cardiospermum halicababum*). Their boundless inventiveness has led the Leguminosae to develop the concept of balloon fruits, as proved by the bladder senna (*Colutea arborescens*) from the Mediterranean, and the South African balloon pea (*Sutherlandia frutescens*).

Anemoballism

Wind can effect the dispersal of seeds indirectly by causing dehiscent fruits to sway so that they scatter or eject their seeds. This form of indirect wind dispersal is called *anemoballism* and involves many herbaceous plants with poricidal and denticidal capsules on long, flexible stalks. Poppy (*Papaver* spp., Papaveraceae) capsules function like pepper pots as they sway in the wind, releasing large numbers of tiny seeds. The circle of narrow openings through which the seeds leave the capsule is well protected from the rain by the protruding rim of the platform that carries the remains of the papillose stigma. Pinks (e.g. *Petrorhagia nanteuilii*), campions (e.g. *Silene* spp.), carnations (*Dianthus* spp.) and primroses (*Primula* spp.) pursue the same strategy but their denticidal capsules open with tiny teeth at the apex, leaving just a narrow opening for the seeds to escape. In the curious *foraminicidal capsules* of snapdragons (*Antirrhinum* spp., Plantaginaceae) three irregular apical pores break open with tiny recurving valves. Interestingly, in the lesser snapdragon (*Antirrhinum orontium*) the long style persists and develops into a stiff projecting rod, perhaps an adaptation to passing animals, which are able to shake them even more effectively than the wind. Although the seeds of many anemoballists display highly ornate surface patterns, they do not usually have any distinct anatomical modifications that would assist further dispersal. However, their small size allows them to travel considerable distances if they are accidentally ingested by browsing livestock or stick to the muddy feet of animals.

Water dispersal

Water facilitates dispersal in a variety of ways. The air bladders of balloon fruits and the high surface/weight ratio of many small wind-dispersed diaspores coincidentally also afford good floatability in water. Plumed fruits and seeds may stay afloat thanks to the surface tension of the water. Nevertheless, water dispersal of otherwise anemochorous diaspores is a haphazard event. Specific adaptations to water dispersal (*hydrochory*) are to be expected in aquatic plants, marsh and bog plants, and those living close to water. The most important property of hydrochorous diaspores is, of course, floatability, which is often enhanced by a water-repellent surface. Impermeability to water also inhibits premature germination of the seed and provides protection against salt water in sea–dispersed diaspores.

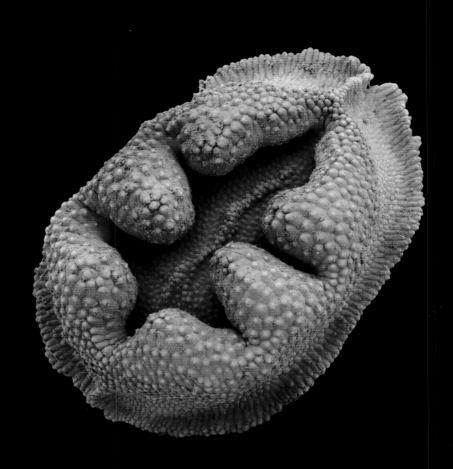

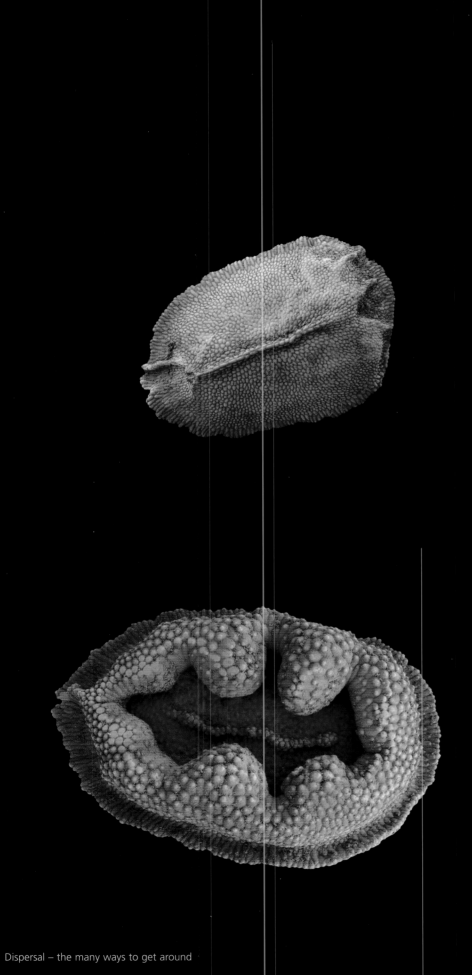

Antirrhinum orontium (Plantaginaceae) – lesser snapdragon; native to Europe – fruit (foraminicidal capsule) and seeds; the apical pores through which the seeds leave the capsule are formed by irregular ruptures of the pericarp. Dispersal is achieved as the capsule sways in the wind or is moved by passing animals, thus shaking out the seeds. The stiff, spike-like remnant of the style may assist with the latter; fruit 7mm long; seeds 1.1mm long

Buoyancy is most commonly increased by enclosed air spaces and waterproof corky tissues. Such corky floating tissues are present in the diaspores of wetland plants from families such as Cyperaceae (e.g. *Scirpus maritimus*, seaside bulrush), Iridaceae (e.g. *Iris pseudacorus*, yellow flag) and Alismataceae (*Sagittaria sagittifolia*, arrowhead). Similar adaptations are easily recognizable in the fruitlets of members of the carrot family (Apiaceae) that live in wetland habitats, such as water parsnip (*Sium suave*), cowbane, (*Cicuta virosa*), and fine-leafed water-dropwort (*Oenanthe aquatica*). The most prominent corky protuberances of any member of the carrot family are found in *Rumia crithmifolia*, the only species of this monotypic genus from the Crimea. *Rumia crithmifolia* grows on dry hillsides, which suggests that the swollen contorted ridges of the fruits are an adaptation to wind dispersal and, secondarily perhaps, rainwash. Featherlight but rather large and rounded, the fruits are easily blown across the ground. Other Apiaceae from dry habitats in the Mediterranean and western Asia possess similarly inflated wind-dispersed fruits (e.g. *Cachrys alpina*).

Tropical islands and coastal areas are rich in plants with fruits that can travel in the salty water of the sea. Among tropical fruits adapted to sea dispersal are drupes with stones that bear a thick, spongy or cork-like floating tissue around the actual seed. Fruits of this type can be dispersed by either animals or the sea, as demonstrated by the sea almond (*Terminalia catappa*, Combretaceae), whose seaworthy fruits are eaten by bats. More common are large indehiscent fruits that structurally resemble drupes but have a mesocarp consisting of a hard, fibrous-spongy, seawater-resistant floating tissue. Fruits of this type are found in palms such as the nipa palm and the coconut. The nipa palm (*Nypa fruticans*) is very common in mangrove swamps and tidal estuaries around the Indian Ocean and the Pacific. Its large, football-sized, compound fruits resemble the cone of a cycad. When ripe, they break up into typical obovoid, angular fruitlets. The single seed inside each fruitlet germinates before it is dispersed, the emerging pointed shoot assisting in the detachment. With their hard outer epicarp and underlying fibrous-spongy mesocarp, the fruitlets of the nipa palm are well adapted to seawater. But no fruit proves the success of this model better than the coconut (*Cocos nucifera*, Arecaceae), which further increases its buoyancy by adding an air bubble inside the large endosperm cavity. Coconuts are perfectly adapted to seawater dispersal, can endure months of travelling on ocean currents, and travel up to 5,000 kilometres, which explains the presence of coconut palms all over the tropics. When a coconut finally lands on a beach it germinates slowly, once the salt accumulated in the husk on its journey has been washed off by the rain. On dry, free-draining sea sand, the liquid endosperm inside the coconut, commonly called "coconut water," provides a vital moisture reserve until the roots of the seedling reach fresh groundwater.

below: *Terminalia catappa* (Combretaceae) – sea almond; widely cultivated in the tropics, probably from Indomalesia originally – fruits (drupes) and inflorescence. The spongy, sea-dispersed endocarps are common flotsam on tropical beaches; fruit 7cm long

bottom: *Nypa fruticans* (Arecaceae) – nipa palm; native from southern Asia to northern Australia – fruit; the large, football-sized compound fruits break up into obovoid angular fruitlets (nuculania). A technical term for this type of compound fruit is yet to be created

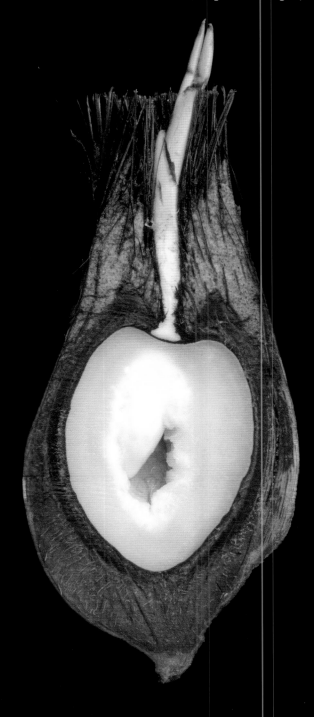

Nypa fruticans (Arecaceae) – nipa palm; longitudinal section through an individual fruitlet (nuculanium). With its spongy fibrous mesocarp and hard endocarp, the seawater-dispersed fruitlet resembles a miniature coconut; length 12cm long

Dispersal by raindrops

Just as the wind can indirectly effect seed dispersal in anemoballists, the kinetic energy of dripping rainwater or dew drops can be exploited to ensure the expulsion of seeds from their fruits. Dispersal by dripping water, *ombrohydrochory*, can be achieved in two ways.

A *splash-rain* dispersal strategy has evolved in the stone-plant family (Aizoaceae). Many Aizoaceae (e.g. *Lithops* spp., living stones) with dehiscent fruits have their seeds flushed out of their capsules by falling raindrops. Whereas most capsules dehisce as they dry out, those of the Aizoaceae repeatedly open upon contact with water and close again as they dry. This rare *hygrochastic* opening mechanism, which is an adaptation to the dry climate of southern Africa where most members of the stone-plant family are at home, ensures that the seeds are dispersed only when enough water is available for germination.

Still rare, but more common than hygrochastic capsules, are springboard mechanisms where the force of the falling drops hitting the fruit forces the diaspores to jump out. This mechanism, called *rain ballism*, is best expressed in members of the mint family (Lamiaceae), in which the persistent calyx forms a cup that holds the four nutlets (mericarps) typically produced by the ovary. In lamiaceous rain ballists such as basil (*Ocimum basilicum*), selfheals (*Prunella* spp.) and skullcaps (*Scutellaria* spp.) the upper lip of the five-lobed calyx is enlarged into a spoon-shaped springboard to enhance the chances of catching falling raindrops.

In our temperate climate, both rain and wind ballistic dispersal mechanisms are deployed to disperse the seeds throughout the winter.

Plants that do it for themselves

In view of the sophisticated strategies that facilitate wind and water dispersal, the possibility that plants disperse their seeds themselves may not sound at all advanced. Quite the contrary. Self-dispersal or *autochory* involves highly complex mechanisms in which plants rely on their own means to eject their seeds. Such *ballistic dispersal* is achieved by explosive, dehiscent capsules or other contraptions that provoke a sudden release of energy.

The explosive opening mechanisms of fruits can be triggered either by passive (hygroscopic) movements of dead tissues as they dry out, or by active movements caused by high hydraulic pressure in living cells.

Hygroscopic tension

Capsules, follicles and other dehiscent fruits usually dehisce gradually along pre-formed lines as the dead or dying pericarp shrinks. In explosively dehiscent fruits, stronger sutures can prevent the slow, continual release of energy. As a result, mechanical tension builds up

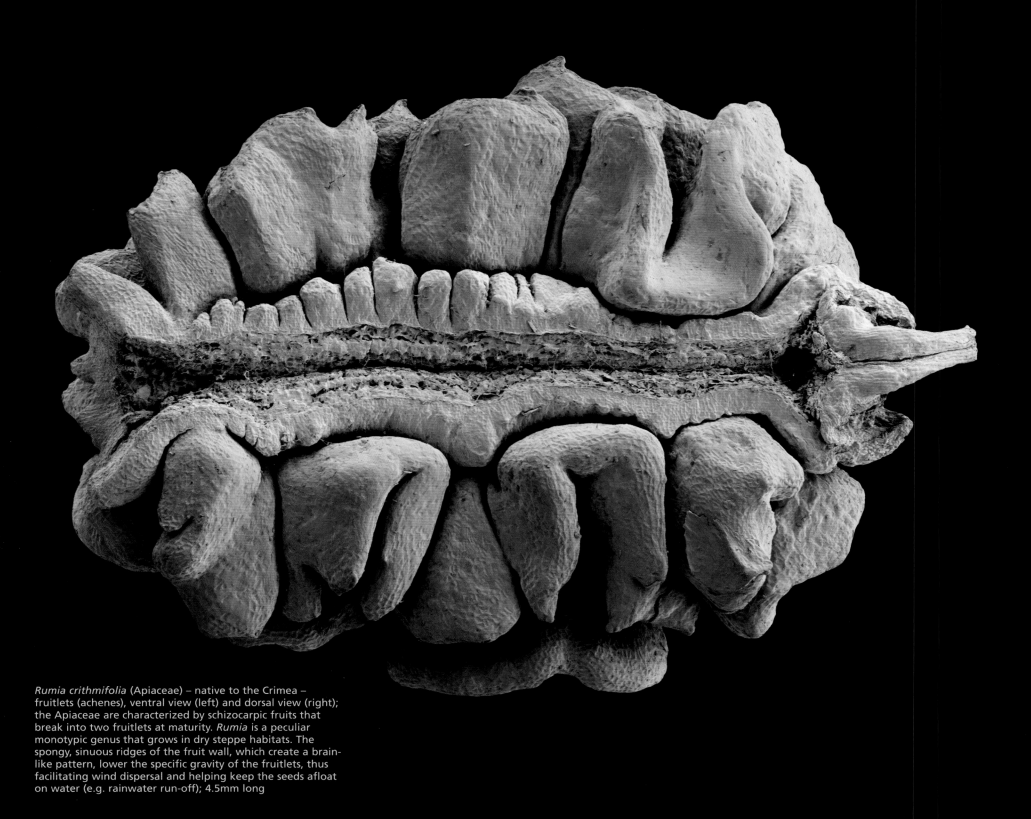

Rumia crithmifolia (Apiaceae) – native to the Crimea –
fruitlets (achenes), ventral view (left) and dorsal view (right);
the Apiaceae are characterized by schizocarpic fruits that
break into two fruitlets at maturity. *Rumia* is a peculiar
monotypic genus that grows in dry steppe habitats. The
spongy, sinuous ridges of the fruit wall, which create a brain-
like pattern, lower the specific gravity of the fruitlets, thus
facilitating wind dispersal and helping keep the seeds afloat
on water (e.g. rainwater run-off); 4.5mm long

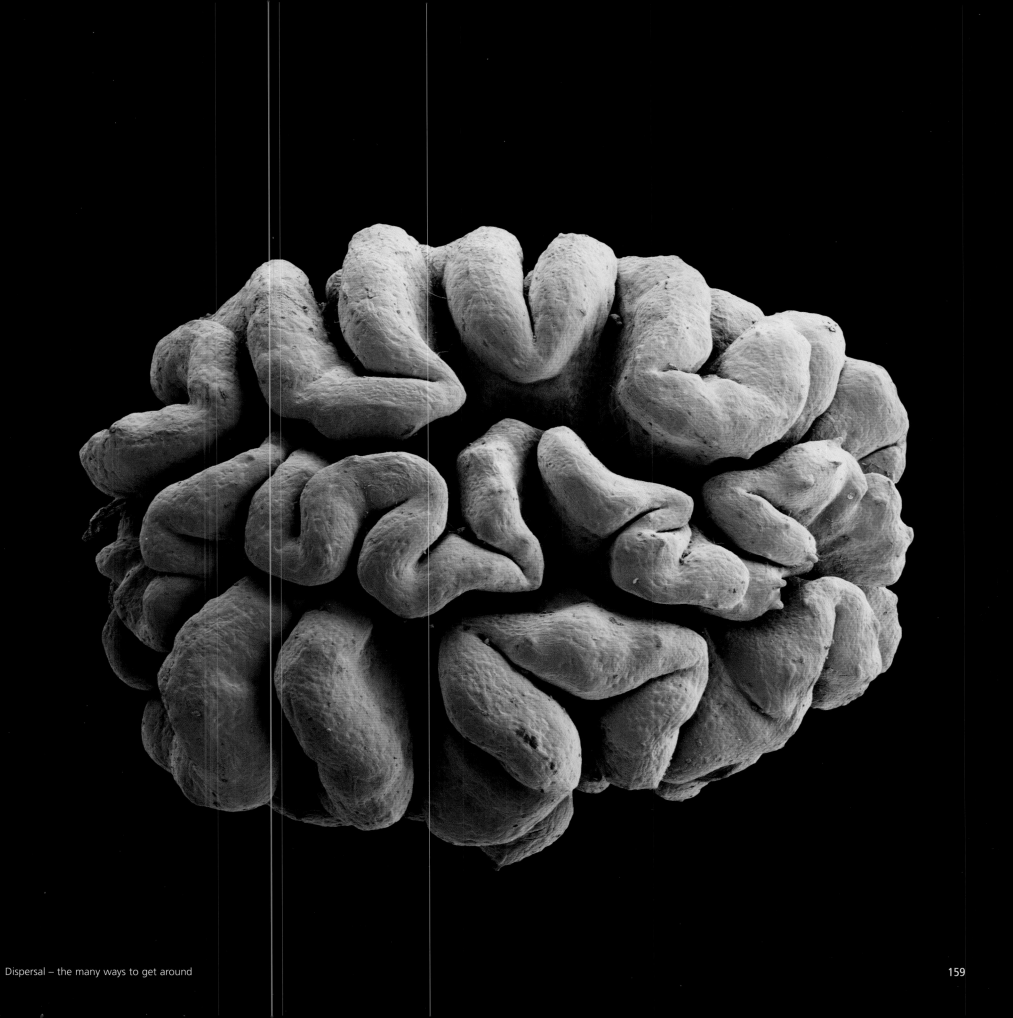

in the tissues of the fruit wall as they lose water, an effect that is enhanced by thick-walled fibrous cells with intersecting orientations in adjoining layers. The cross orientation of the fibres creates contradictory tractive forces in adjacent pericarp layers until the fruit shatters violently into pieces, which usually correspond to entire or half carpels.

The legumes of the bean family (Leguminosae) often explode. In the process the two halves of the single carpel twist in opposite directions, separate, and eject their seeds. There are many familiar examples that never fail to attract the attention of curious children, most notably lupins (*Lupinus* spp.), Chinese wisteria (*Wisteria sinensis*), common broom (*Cytisus scoparius*), gorse (*Ulex europaeus*) and sweet peas (*Lathyrus odoratus*). The distance over which the seeds are dispersed is usually quite short, a maximum of a couple of metres. As always, we need to look in the tropics for superlatives. *Tetraberlinia morelina*, an African legume tree from the rainforests of west Gabon and south-west Cameroon, holds the world record for ballistic dispersal. Assisted by its great height, the tree can shoot its seeds over a distance of up to 60 metres.

A contender for second place is the sandbox tree (*Hura crepitans*) a member of the Euphorbiaceae, a large family characterized by explosively dehiscent capsules. The mandarin-sized fruits of *Hura crepitans*, which resemble a miniature pumpkin, erupt with great force, ejecting the seeds up to 14m (some sources claim up to 45m). In our temperate climate we find only miniature versions of exploding euphorbiaceous capsules in herbaceous members of the family such as the petty spurge (*Euphorbia peplus*), sun spurge (*Euphorbia helioscopia*), dog's mercury (*Mercurialis perennis*) and annual mercury (*Mercurialis annua*).

A very interesting example of ballistic dispersal is found in *Esenbeckia macrantha*, a Mexican tree from the citrus family (Rutaceae). Although a relative of citrus trees, *Esenbeckia macrantha* has rather different fruits. The knobbly, greyish-brown capsules open slowly but the parchment-thin, hard layer of the endocarp continues to envelop each individual seed for a while longer. Eventually, in the hot Mexican sun the endocarp pockets suddenly pop open and eject the seeds, which are subsequently dispersed by ground-dwelling birds.

A specialized endocarp is also responsible for ballistic dispersal in the witch hazel family (Hamamelidaceae). Like *Esenbeckia macrantha*, witch hazel (*Hamamelis* spp.) and winter hazel (*Corylopsis* spp.) have capsular fruits (*coccaria*) that open slowly. Once they are open, further desiccation causes the hard endocarp layer to change its shape, acting like a vice on the single seed in each of the two loculi. Rising pressure pushes the hard, smooth seed to a point where it suddenly slips out of its straitjacket in a ballistic trajectory. The principle of exerting lateral pressure is also to be found in the fruits of certain pansies. The three-carpellate capsules of the European field pansy (*Viola arvensis*), dog violet (*Viola canina*), and common blue violet (*Viola sororia*) open slowly as the carpels split along their midribs and fold down their margins

opposite: *Corylopsis sinensis* var. *calvescens* (Hamamelidaceae) – Chinese winter hazel; native to China – capsular fruit (coccarium) displaying loculicidal, septicidal and septifragal dehiscence at the same time; in the witch-hazel family, members of the subfamily Hamamelidoideae are characterized by a syncarpous gynoecium of two carpels, each of which produces one seed. The capsular fruit opens slowly. As the extremely hard endocarp dries out it changes its shape, gripping the seed in each locule like tongs. Eventually, the hard, smooth, spindle-shaped seed is forcibly expelled; diameter of fruit 7mm

Viola sororia (Violaceae) – common blue violet; native to eastern North America – fruit (loculicidal capsule); the three-carpellate capsule opens slowly as the carpels split along the midrib and fold down their margins to expose the seeds. As the carpels dry out the margins bend inwards and trap the seeds like a vice. Eventually, the pressure is so high that the seeds are explosively expelled; diameter 2.5cm

to expose the seeds. As the boat-shaped carpels dry out, their margins bend inwards and grip the seeds like a vice. Eventually, the pressure is so high that the seeds are expelled explosively.

Hydraulic pressure

In addition to capsules with dead tissue that perform hygroscopic movements, there are fleshy fruits that actively build up hydraulic pressure in living tissue until their fruits detonate. Classic examples of fleshy fruits that explode at the slightest tremor when fully ripe are those of the touch-me-not (*Impatiens* spp., Balsaminaceae), the squirting cucumber (*Ecballium elaterium*) from the Mediterranean and its American relative, the seed-spitting gourd (*Cyclanthera brachystachya*), both in the gourd family (Cucurbitaceae).

The segments of the fusiform capsules of the touch-me-not curl up instantly and hurl the seeds in all directions. At the breaking point the fruits are so sensitive to the touch that anything from a passing animal, raindrops, wind, and even seeds launched by a neighbouring fruit can trigger an explosion. The gherkin-sized fruits of the squirting cucumber apply a different strategy. They squeeze their seeds and a good supply of a lubricating watery liquid through a narrow opening at the base that is formed when the fruit stalk pops out like a champagne cork. The mechanism of the seed-spitting gourd is different again. The pressure accumulated inside the fruit wall triggers an explosion that rips the fruit apart in such a way that, like a slingshot, it catapults the seeds several metres.

Another effective catapult mechanism involves the exertion of lateral pressure once again. A relative of the fig, the compound fruits of *Dorstenia contrajerva* (Moraceae) from tropical America morphologically represent an "open fig." Lots of tiny drupes are inset in the surface of the plate-like infructescence. The fleshy outer layer of the drupes is thin at the apex and thicker towards the base, surrounding the miniature stones like a pair of pliers. Below the stone lies a swelling tissue that causes the jaws of the pliers to press together. Eventually the skin of the drupe ruptures at the apex and the stone shoots up to 4m through the air like a cherry-stone flicked between finger and thumb.

ANIMAL DISPERSAL

Abiotic dispersal mechanisms are advantageous in certain habitats and so suit the lifestyles of many plants. After all, in the temperate deciduous forests of North America about 35 per cent of all woody plants produce anemochorous diaspores. Despite its obvious success, dispersal by wind and water is both wasteful and unpredictable. The strength, direction and frequency of air and water currents are variable and unreliable. When scattered randomly, most seeds inevitably end up in locations that are unsuitable for germination and go to

below: *Ecballium elaterium* (Cucurbitaceae) – squirting cucumber; native to the Mediterranean – fruit (pyxidium); as the fruits ripen, the tissue inside builds up enormous hydraulic pressure until the slightest touch or movement makes the stalk pop out like a stopper; fruit c. 3-4cm long

bottom: *Cyclanthera brachystachya* (Cucurbitaceae) – seed-spitting gourd; native to Central and South America – fruit (foraminicidal capsule); hydraulic pressure forces the fruit to burst on the weaker convex side, making it hurl out its seeds like a slingshot; fruit c. 3-4cm long

below: *Ginkgo biloba* (Ginkgoaceae) – ginkgo, maidenhair tree; native to China – what look like yellow drupes are naked seeds borne in pairs on a joint stalk. A true living fossil, the ginkgo is the last surviving species of the genus that dates back 270 million years; diameter of fruit c. 2.5cm

bottom: *Taxus baccata* (Taxaceae) – English yew; native to Europe and the Mediterranean – fruit (arillocarpium); yew "berries" consist of a single seed surrounded by a cup-shaped fleshy appendage, called an aril, the only part of the plant that is not poisonous; diameter of fruit c. 1cm

waste. Apart from being equally random, ballistic dispersal achieves only short dispersal distances. In contrast, adaptation to biotic dispersal agents in the form of animals eliminates many elements of uncertainty associated with abiotic dispersal agents and offers much more efficient options. Animal movements are less haphazard than wind and water, so fewer seeds are necessary to achieve sufficient dispersal to suitable establishment sites.

The advantages of animal-mediated dispersal are reflected in the fact that 50 per cent of gymnosperms (*Ephedra, Gnetum, Ginkgo,* a few conifers and all cycads) are animal dispersed, either offering fleshy seeds (cycads, *Ginkgo*), edible seed appendages (*Ephedra*, conifers such as *Podocarpus* spp. and yews) or the seeds themselves (e.g. scatter-hoarder dispersed *Pinus* spp.). Even though the evil-smelling seeds of the ancient *Ginkgo biloba* find no takers among living animals, they must once have been part of someone's diet. Their fleshy seed coat (sarcotesta), and the fact that they drop to the ground where they emit a pungent vomit-like smell suggests that carrion-eating, ground-dwelling dinosaurs may once have been the *Ginkgo*'s natural dispersers. Ginkgos are true living fossils that have been around for more than 250 million years. During the middle Jurassic and Cretaceous periods, about 175–65 million years ago, several species of ginkgo were widespread throughout the ancient continent of Laurasia (today's North America and Eurasia). Their phenomenal age entitles them to some anachronistic behaviour.

When it comes to animal dispersal strategies the angiosperms are much more up to date. In terms of inventiveness, they outperform the gymnosperms in every way, proving that no other dispersal mode offers more opportunities than *zoochory* (animal dispersal). Over many millions of years the angiosperms have developed and perfected a prodigious spectrum of strategies that enable them to travel with animals, either as sticky hitchhikers attached to skin or fur, or by tricking animals into carrying them in their mouth or gut.

Becoming attached

Hitching a ride on an animal is a very cost-effective way of travelling. This mode of dispersal, called *epizoochory*, may not even require specific adaptations. Small diaspores without any organs to assist a particular dispersal mechanism often travel as stowaways in mud sticking to the feet or feathers of waterfowl. This simple, effective means of travel caught the attention of Charles Darwin, who collected and sowed such diaspores. Similar chance dispersal can be mediated by any animal or human covered in mud.

Many low-growing plants that are able to position them within reach of furry passers-by possess diaspores specifically modified to attach themselves to animals. Unlike other types of animal dispersal, adhesive diaspores do not offer nutritional attractants, which means that

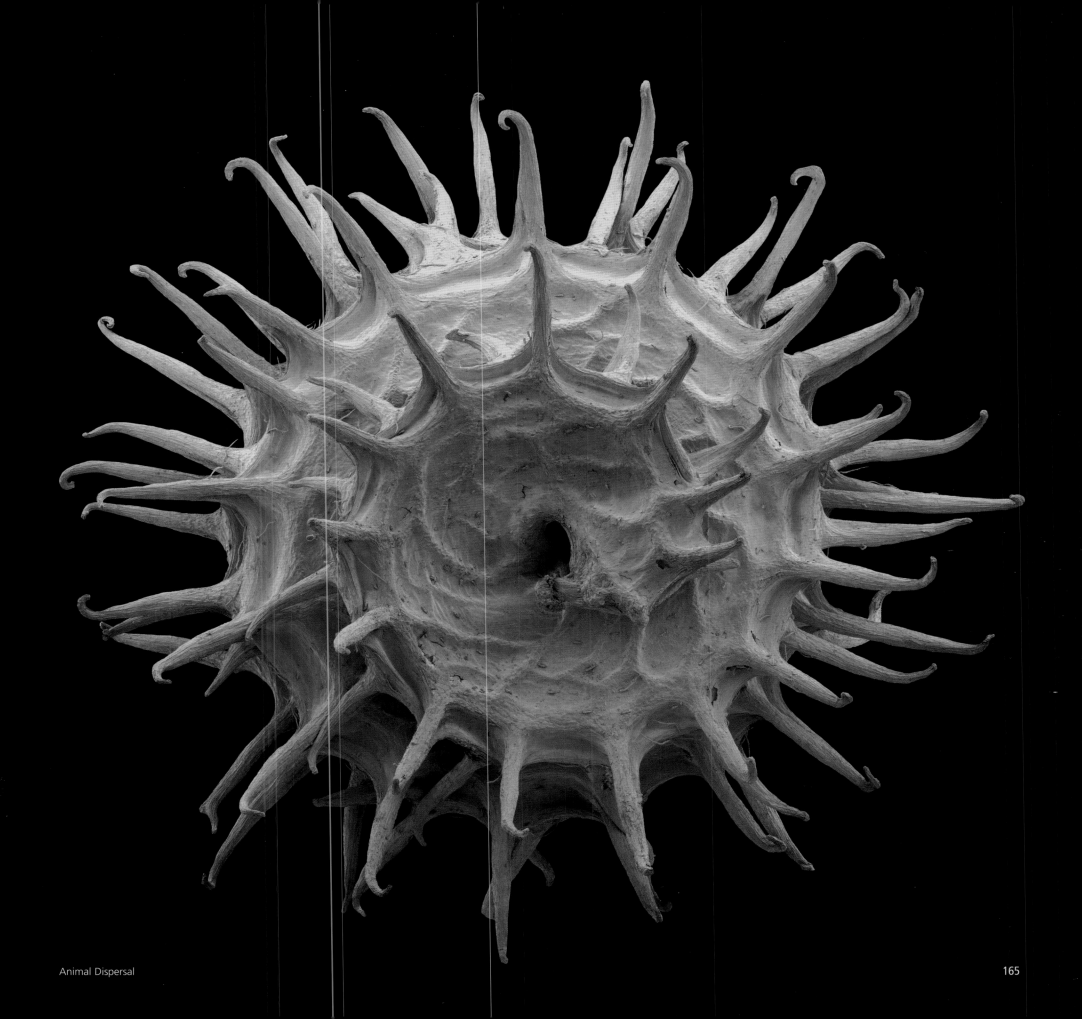

dispersal happens by chance when an animal unwittingly picks them up. In other words, adhesive fruits hitchhike rather than pay their fare – under any circumstances a cost-effective, albeit slightly unreliable, way of travelling. In addition to being physiologically cheap, epizoochory has another great advantage. Unlike fleshy diaspores, the dispersal distance of adhesive diaspores is not limited by factors like gut-retention times. Most sticky hitchhikers drop off by themselves, but if they do not, they have the potential to travel long distances before they are groomed away or the animal moults or dies.

The typical adaptations indicating epizoochorous dispersal are diaspores covered in hooks, barbs, spines or sticky substances. Examples are to be found on socks and trousers after a country walk in late summer or autumn. Among the most frequent and tenacious burrs in our temperate climate are the nutlets of stickywilly (*Galium aparine*, Rubiaceae), hound's-tongues (*Cynoglossum* spp., Boraginaceae), wild carrot (*Daucus carota*, Apiaceae) and stickseeds (*Hackelia* spp., Boraginaceae), as well as the anthocarpous fruits (pometums) of the sticklewort (*Agrimonia eupatoria*, Rosaceae) and the much larger burrs of the burdock (*Arctium lappa*, Asteraceae). The underlying principle of their adherence is simple and consists of small hooks that readily become entangled with the fur of mammals or alternatively with the tiny loops of thread in the fabric of clothes. It was the microscopic structure of these diaspores that inspired the Swiss electrical engineer George de Mestral in the 1950s to develop the hook and loop fastener that we nowadays all know under the name Velcro® (based on the French *velour* = velvet and *crochet* = hook).

Velcro-like hooks are present in the diaspores of many other plants, including the small nutlets of some buttercups (e.g. *Ranunculus arvensis*, Ranunculaceae), wood avens (*Geum urbanum*, Rosaceae), sanicles (*Sanicula* spp., Apiaceae), the curious heart-shaped fruits of the Pima rhatany (*Krameria erecta*, Krameriaceae) and even some members of the bean family (Leguminosae). Burclover (*Medicago polymorpha*) and other species of the genus *Medicago*, for example, curl up their indehiscent pods (camaras) into spiny balls. However, the gentle hook-and-loop principle is not the only way diaspores attach themselves to animals. Plants have also developed some rather sadistic means of ensuring the dispersal of their seeds.

The story of the sadistic *Tribulus*

Ferociously spiny diaspores that bite into flesh are found in a variety of unrelated families. One example from the warmer parts of Europe, Africa and Asia is puncture vine (*Tribulus terrestris*, Zygophyllaceae). To some the plant is better known as devil's thorn or caltrop, because of the diabolical insidiousness of its diaspores. As the schizocarpic fruits of the puncture vine mature they split into five indehiscent nutlets. Each nutlet is armed with two

page 164: *Agrimonia eupatoria* (Rosaceae) – common agrimony, cockleburr; native to the Old World – fruit (pometum); similar to a rosehip (although hard rather than fleshy) the fruit of the common agrimony represents a hypanthium into which several achene-like ovaries are embedded (not visible). The hooked spines around the upper margin of the ribbed hypanthium are a very efficient dispersal aid, readily catching on to animal fur or clothes; 7.5mm long

page 165: *Medicago polymorpha* (Leguminosae) – burclover; native to Eurasia and North Africa – fruit (camara); like the fruits of all other members of the bean family, the camara of burclover develops from a single carpel. Typical of the genus *Medicago*, the carpel is coiled into a spiral of 4-6 turns. With its spherical shape and hooked spines the fruit is well adapted to attach itself to furry or feathery animals; diameter (including spines) 9.5mm

Galium aparine (Rubiaceae) – stickywilly, native to Eurasia and America; opposite: fruit (achenarium); the fruit is formed by two joined carpels that break into two separate achenes when ripe. The small branch on the right bears two developing fruits with their ovaries lobed but still entire, and a flower bud whose tiny inferior ovary is crowned by the closed tetramerous perianth. Densely covered in tiny hooks, the achenes of the stickywilly are one of the most tenacious epizoochorously dispersed diaspores; mature achene 5mm long

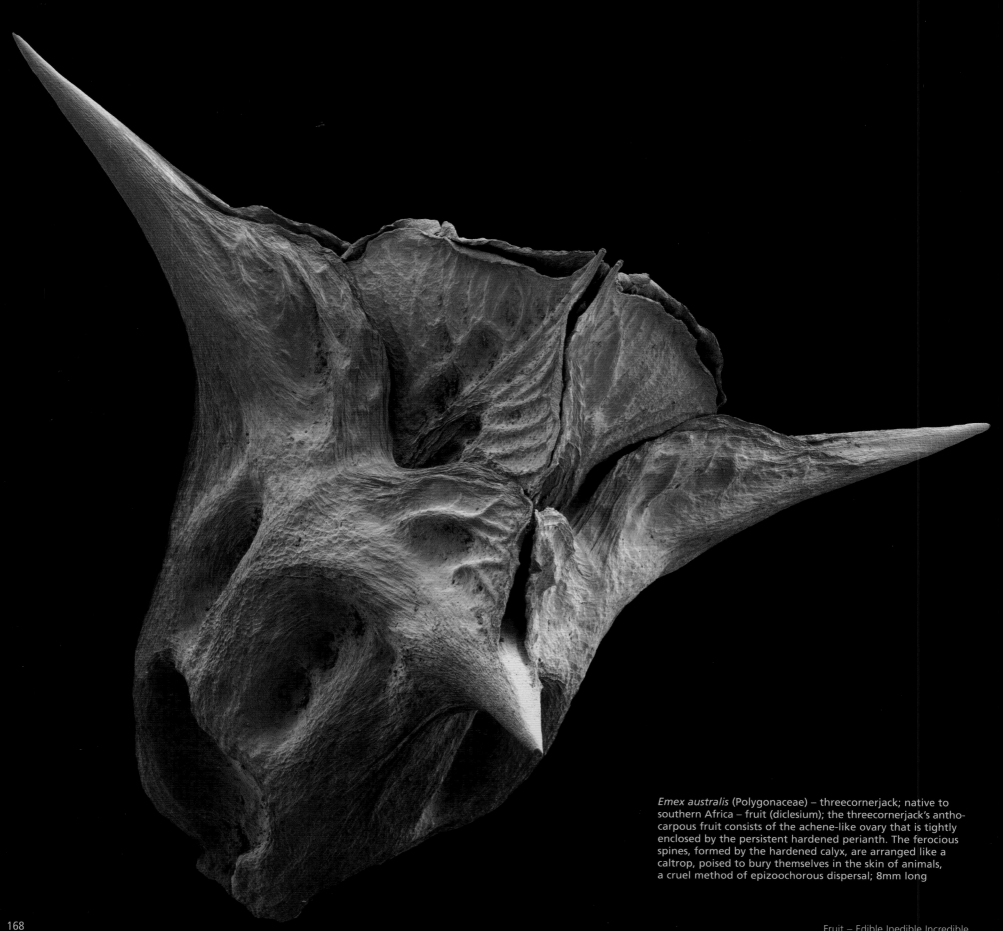

Emex australis (Polygonaceae) – threecornerjack; native to
southern Africa – fruit (diclesium); the threecornerjack's anthocarpous fruit consists of the achene-like ovary that is tightly
enclosed by the persistent hardened perianth. The ferocious
spines, formed by the hardened calyx, are arranged like a
caltrop, poised to bury themselves in the skin of animals,
a cruel method of epizoochorous dispersal; 8mm long

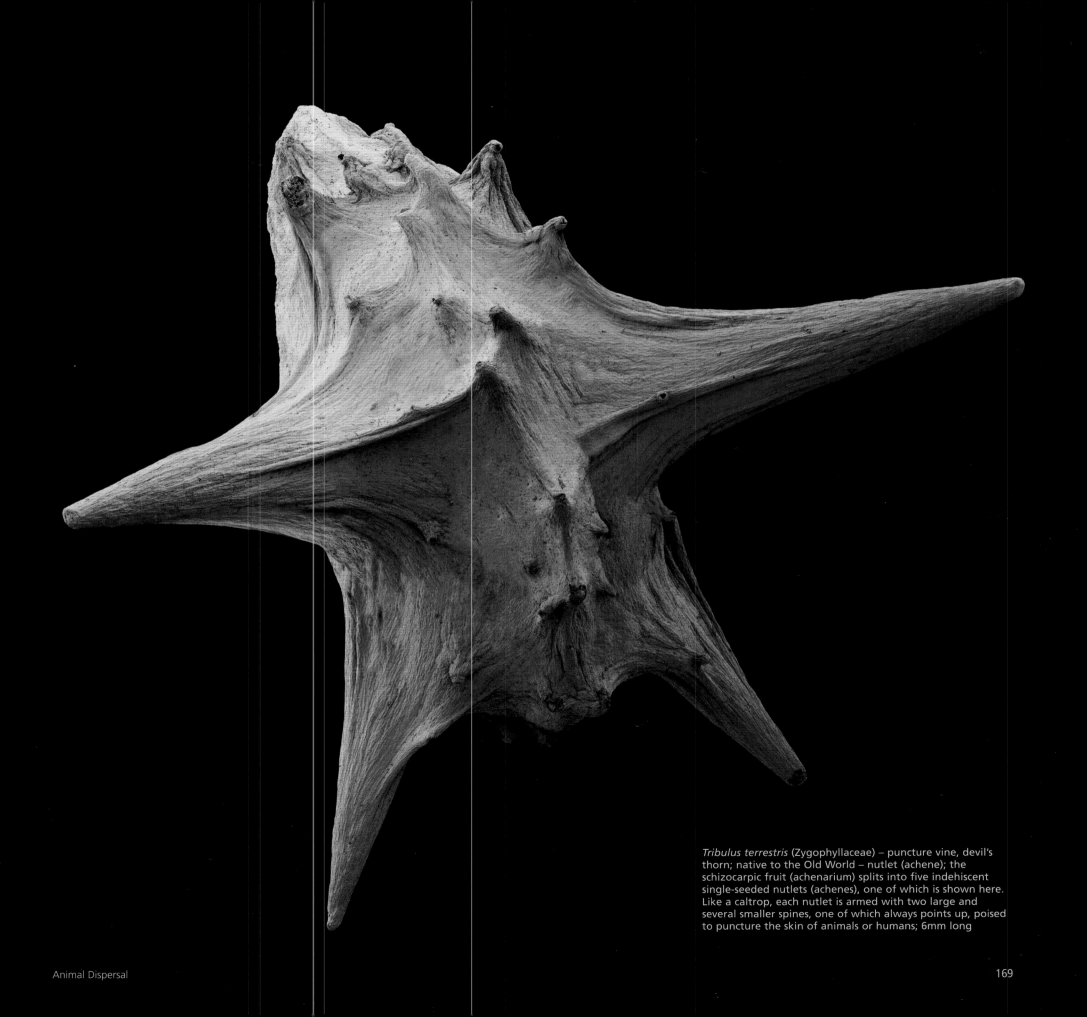

Tribulus terrestris (Zygophyllaceae) – puncture vine, devil's thorn; native to the Old World – nutlet (achene); the schizocarpic fruit (achenarium) splits into five indehiscent single-seeded nutlets (achenes), one of which is shown here. Like a caltrop, each nutlet is armed with two large and several smaller spines, one of which always points up, poised to puncture the skin of animals or humans; 6mm long

large and several smaller spines. Whichever way up they land, some of their spines point upwards like some medieval caltrop poised to penetrate animal skin or human feet. In the Hungarian plains these prickly hitchhikers cause sheep farmers considerable problems by inflicting festering wounds that make it difficult for their animals to walk. Obviously a successful concept, the caltrop model has been perfected by the threecornerjack (*Emex australis*), an annual herb from the knotweed family (Polygonaceae). Originally from southern Africa, it has spread throughout the warmer regions of our planet, where it has become a serious weed. The threecornerjack's fruit is of the anthocarpous type and correctly addressed as a *diclesium*. It consists of an achene-like ovary, which remains tightly enclosed by the persistent, hardened perianth. The hardened calyx is shaped to form three large straight spines that are arranged into a perfect caltrop poised to cripple animals and injure humans.

In the claws of the devil

To encounter the largest, most notorious burrs, the so-called devil's claws, we must visit the tropical and subtropical semi-deserts, savannahs and grasslands of America, Africa and Madagascar. The New World devil's claws belong to the genus *Proboscidea* (esp. *Proboscidea louisianica*) and their smaller relative *Martynia annua*, both members of the unicorn family (Martyniaceae). In South America their carnivorous brethren from the genus *Ibicella* produce similar devil's claws, or unicorn fruits as they are sometimes called (e.g. *Ibicella lutea*). Whilst immature, the green fruits look harmless. Their true intentions are revealed only when the ripe fruits shed their outer fleshy cover to expose highly ornate endocarps. At their tip the endocarps possess a beak, which eventually splits down the middle to produce a pair of sharply pointed, recurved spurs. Once the ferocious spurs are unfolded the fruit is ready to cling to fur or hoof, or even bore into skin.

Old World devil's claws belong to the sesame family (Pedaliaceae); not surprisingly they are close relatives of the Martyniaceae. The burrs of the Malagasy genus *Uncarina* look like miniature sea mines with long, sharply barbed spines. But in terms of cruelty, nothing can beat the fruits of the grapple plant, *Harpagophytum procumbens* from southern Africa. The tardily dehiscent woody capsules bear numerous thick, sharply-pointed, barbed hooks that inflict serious wounds on anyone who is unfortunate enough to step on them.

How to catch a bird

Glue is seldom developed as an alternative to hooks in epizoochorous diaspores. The most notorious plants applying this rare strategy are several species of the genus *Pisonia* (e.g. *P. brunoniana*, *P. umbellifera*, Nyctaginaceae), aptly called birdcatcher trees. *Pisonia brunoniana* is

below: *Martynia annua* (Martyniaceae) – small-fruited devil's claw; native to America – fruit (capsule); structurally and functionally resembles the golden devil's claw but smaller; fruit 2.5-3cm long

bottom: *Arctium lappa* (Asteraceae) – greater burdock; native to temperate Eurasia – inflorescence; typical of Asteraceae, the flowers are crowded in head-like inflorescences (capitula). At the fruiting stage, the hooked, needle-like tips of the bracts attach the entire capitulum, fruits (achenes) inside, to animal fur or human clothes; diameter c. 3cm

Fruit – Edible Inedible Incredible

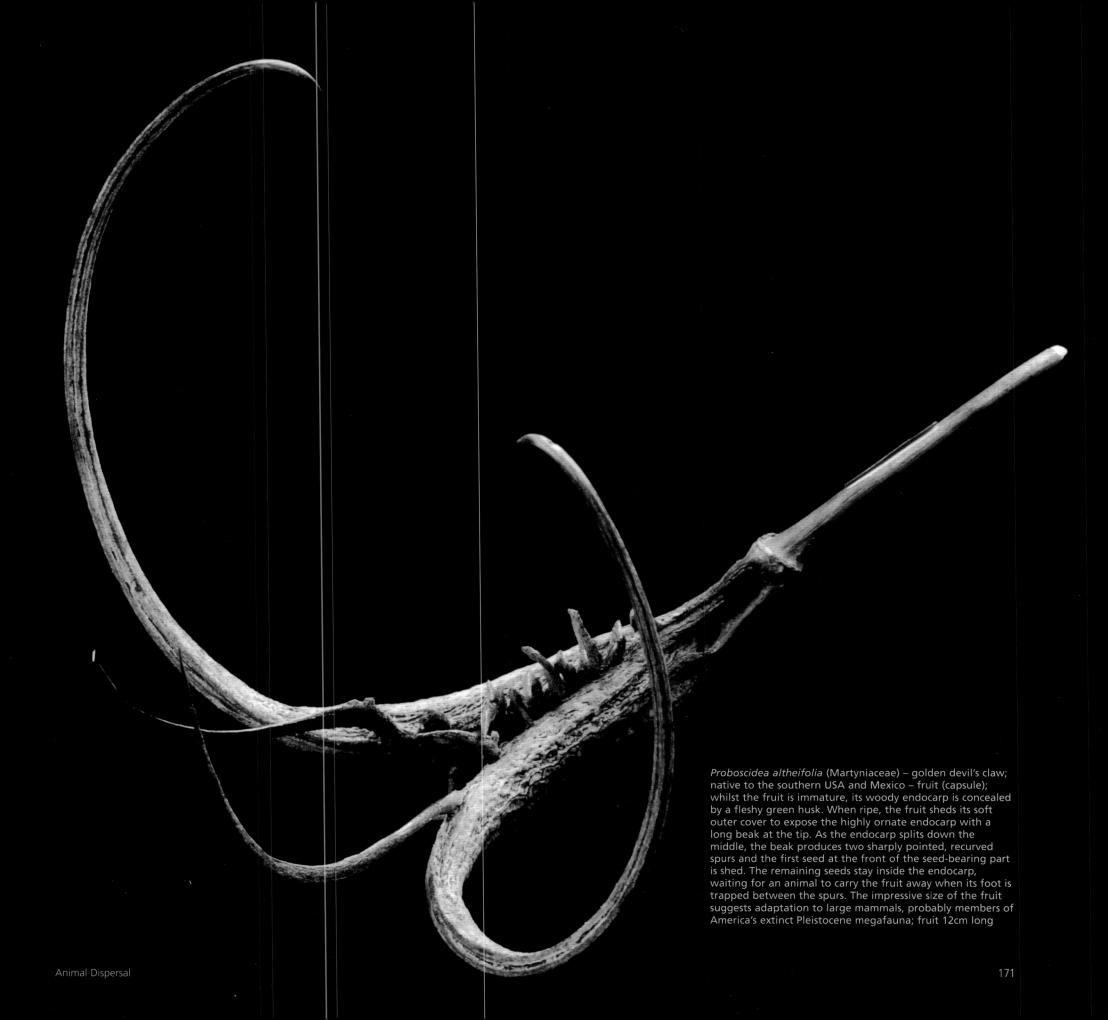

Proboscidea altheifolia (Martyniaceae) – golden devil's claw; native to the southern USA and Mexico – fruit (capsule); whilst the fruit is immature, its woody endocarp is concealed by a fleshy green husk. When ripe, the fruit sheds its soft outer cover to expose the highly ornate endocarp with a long beak at the tip. As the endocarp splits down the middle, the beak produces two sharply pointed, recurved spurs and the first seed at the front of the seed-bearing part is shed. The remaining seeds stay inside the endocarp, waiting for an animal to carry the fruit away when its foot is trapped between the spurs. The impressive size of the fruit suggests adaptation to large mammals, probably members of America's extinct Pleistocene megafauna; fruit 12cm long

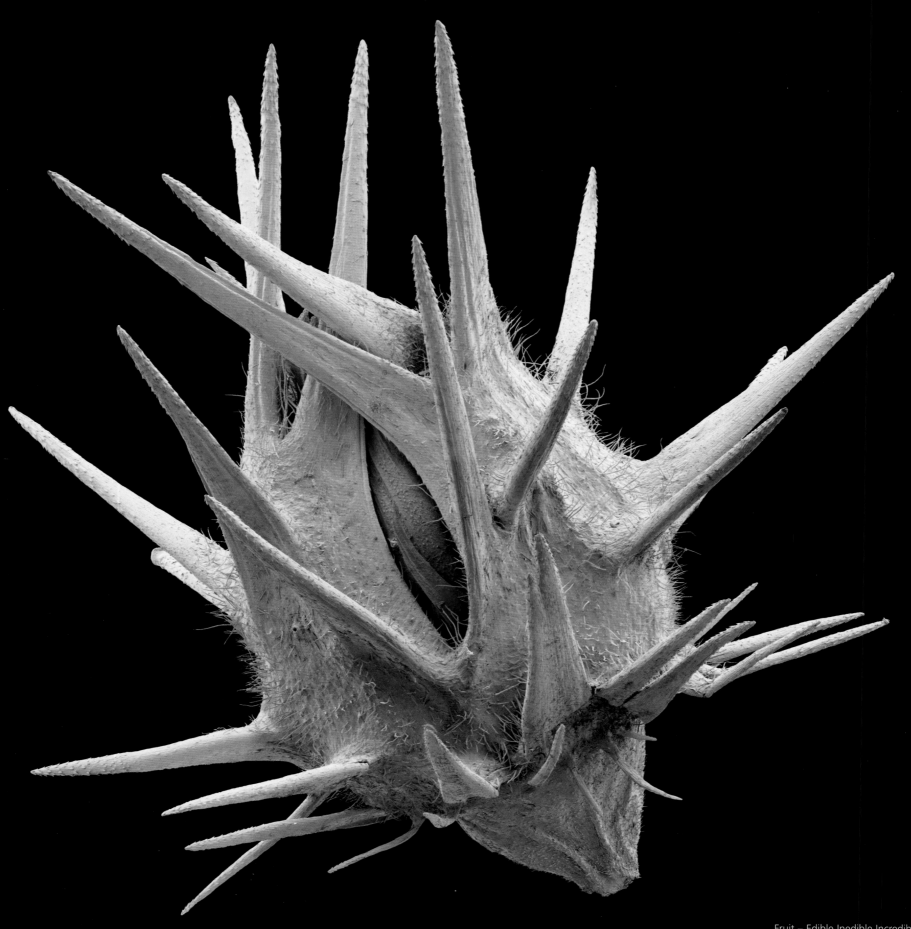

native to New Zealand, Norfolk Island, Lord Howe Island and Hawai'i, whereas the very similar *P. umbellifera* is more widespread throughout the tropical Indo-Pacific region. Their elongated fruits are wrapped in a persistent calyx that secretes a viscid substance along five longitudinal ridges. Like flypaper, the sticky strips trap many insects and these in turn attract birds; they also become ensnarled by the extremely sticky fruits. As the birds attempt to free themselves they become covered in more and more fruits. Apparently the targeted dispersers of the fruits – larger seabirds such as petrels, gannets and shearwaters – usually escape, somewhat disabled but laden with fruits. Alas, smaller birds often become completely immobilized and die.

Dispersal by scatter-hoarders

The majority of zoochorous plants have developed mutually beneficial partnerships with animals, rather than simply exploiting their ability to travel. The animals receive an edible reward in exchange for their dispersal services. The simplest means of compensation, which does not require any special adaptations, is to surrender part of the seed crop to scatter-hoarding animals. Squirrels and other rodents bury acorns (*Quercus* spp.), beechnuts (*Fagus* spp.), sweet chestnuts (*Castanea* spp.) and walnuts (*Juglans* spp.) underground to last them through the winter when other food sources are scarce. Birds such as bluejays (*Cyanocitta cristata*) also cache acorns and beechnuts in cracks and crevices of tree trunks or in loose soil. Dispersal by jays is assumed to be the reason for the rapid post-Pleistocene northward spread of oaks and other species.

Among gymnosperms there are about twenty species of pines (*Pinus* spp., Pinaceae) that are dispersed by scatter-hoarding jays and nutcrackers. Most other pines are wind dispersed although some benefit from secondary dispersal by scatter-hoarders. Several species native to the semi-arid forests of western North America, among them the torrey pine (*Pinus torreyana*) and the digger pine (*P. sabiniana*), possess large, heavy seeds with wings that are obviously too small to provide an effective means of wind dispersal. The seeds drop almost straight to the ground where they are collected by rodents and jays, which scatter-hoard them in the soil. Pines are known to have shifted several times back and forth between wind and animal dispersal during their evolutionary history. Therefore, it is unclear whether the sugar and the digger pines are progressing towards perfecting the scatter-hoarder dispersal syndrome and losing their wings entirely, or whether current selection pressures are directing them towards improving their aerodynamic properties by enlarging their wings. As an interesting adaptation to dispersal by scatter-hoarders, the wings of the seeds of the digger pine and others detach very easily along a predetermined breaking line, as if they are

opposite: *Cenchrus spinifex* (Poaceae) – coastal sandbur; native to America – fruit (anthecosum); the burr-like fruit of the coastal sandbur has a complex structure. It develops from a fertile spikelet that is surrounded by an involucre of sterile spikelets. In the ripe fruit the sterile spikelets form a hard, spiny burr assisting in animal dispersal (epizoochory); 9.5mm long (including spines)

Pisonia brunoniana (Nyctaginaceae) – bird catcher tree; native to Australia, New Zealand and other Pacific islands – fruits (diclesia); the elongated mature ovary is wrapped in the persistent, tubular calyx, which secretes a viscid substance along five longitudinal ridges. The glue traps insects and these in turn attract unsuspicious birds, which also become ensnarled by the extremely sticky fruits. Larger birds such as petrels and gannets usually manage to escape laden with fruits, just as the plant intended. Unfortunately, smaller birds often become completely immobilized and die in the tree; fruits c. 3cm long

trying to make the animals' job of collecting and burying the seeds easier. Given enough time, the wings drop off by themselves.

The strategy of sacrificing part of the seed crop to scatter-hoarding animals relies on the fact that not all cached seeds will be eaten. Because of the forgetfulness or premature death of their collector, a good number of seeds always survive unharmed, planted in suitable establishment sites well away from the shadow of the parent tree.

Oaks, beeches, and other long-lived plants such as bamboos produce large crops of fruits once every few years in an unpredictable pattern. The phenomenon of synchronous production of bumper crops within a population has been called "masting." Ecologists have paid much attention to mast fruiting during the last twenty years and have come up with numerous hypotheses to explain the phenomenon. One possible reason is linked to the fact that many well-known mast-fruiting species are wind pollinated. Mast seeding could therefore be a consequence of mast flowering, which increases the chances of successful wind pollination in mast-flowering years. A second hypothesis suggests that plants can predict which years will be the most promising for the establishment of seedlings. This can apply to fire-prone habitats, for example in Australia where the mast flowering of grass trees (*Xanthorrhoea* spp., Xanthorrhoeaceae) is triggered by fire. The burning of the surrounding vegetation reduces competition and the ash of burnt plant matter increases the availability of nutrients. However, serotinous dicotyledons and conifers storing seeds in armoured fruits that release their seeds after a fire can exploit this opportunity much more quickly.

Another well-received theory suggests that occasional large reproductive efforts are more economic in terms of seed production and survival than regular smaller ones. Mast fruiting produces more seeds than the available populations of predators (e.g. scatter-hoarders, seed-eating insects) are able to consume, ensuring that in mast years larger numbers of seeds escape unscathed. The great variation of seed production from year to year can have a strong effect on the plant populations involved and on the populations of the animals that eat their seeds. In between mast years predators starve and their numbers fall until they are overwhelmed by food during the next bumper crop. Although this surprise tactic sounds plausible, its success depends on the response of the predator. Specialist predators can be hit hard but numbers of generalists who gorge on the nutritious seeds only when available (e.g. acorn-eating deer and pigs) remain the same. However, even specialists are sometimes able to outwit the cunning ruse of mast seeding.

New Zealand's kakapo (*Strigops habroptilus*), the world's heaviest (up to 3.5kg) and only flightless parrot has long been known to breed only in years with an overabundance of food supplies. One event that incites the kakapo to lay eggs is the mast fruiting of the rimu tree

(*Dacrydium cupressinum*), a conifer from the family Podocarpaceae. Although kakapos are generally herbivorous and live on a variety of plant matter, they are particularly fond of the fruits of the rimu tree. In fact, during mast years of the rimu, which occur every 2–5 years, the nocturnal birds feed exclusively on rimu fruits. Another example where animals have adapted to preempt masting events was reported in the prestigious journal *Science* in 2006. Scientists have discovered that both the American red squirrel (*Tamiasciurus hudsonicus*) and the Eurasian red squirrel (*Sciurus vulgaris*) have found a way to predict the mast seeding of their food suppliers (e.g. spruces, *Picea* spp.). In anticipation of the extra provision the squirrels produce a second litter. The clue for the squirrels could come earlier in the year with the profusion of flowers or pollen cones, which they also consume.

The strategy of mast seeding could help plants that rely on scatter-hoarders for the dispersal of their seeds to reduce the loss of offspring even though sacrificing some seeds is part of the strategy. A far more common method to remunerate animals for their dispersal efforts is the provision of separate edible rewards that divert attention from the seeds.

Dispersal by ants

On close examination, seeds of many plants, especially in dry habitats, turn out to carry a small yellowish-white oily nodule. In 1906 the Swiss biologist Rutger Sernander described the strategy that lies behind these strange appendices and called it *myrmecochory*. (Greek *myrmex* = ant + *choreo* = to disperse). Sernander observed that seeds bearing such an "oil body," or *elaiosome* as he called it in Greek, were irresistible to ants, which avidly collect the seeds and carry them to their nest. What triggers the stereotypical seed-carrying behaviour is the presence of ricinolic acid in the elaiosome. Obviously the result of millions of years of co-adaptation, myrmecochorous plants have evolved to produce the same unsaturated fatty acid in the tissue of their elaiosomes as that found in the secretions of the ants' larvae. Once workers have hauled the seeds into the nest, they dismantle the nutritious appendage but leave unharmed the rest of the seed, which is well protected by a hard seed coat. The tissue of the elaiosome, rich in fatty oil, sugars, proteins and vitamins, is not consumed by the ants but used to feed their larvae. Once the fatty nodule has been removed, the seeds become waste and are discarded on a nest midden, which can be either underground or above ground. The organic substrate of such refuse piles is rich in nutrients and offers better conditions for seedling growth and establishment than the surrounding soil.

Myrmecochorous diaspores occur in over eighty plant families in which they have evolved many times independently. Adaptation to ant-dispersal is common among herbaceous plants in temperate deciduous forests in Europe and North America. In dry habitats

below: *Strigops habroptilus* (Psittacidae) – kakapo; endemic to New Zealand – the world's heaviest (up to 3.5kg) and only flightless parrot. It breeds only in years of mast fruiting

bottom: *Dacrydium cupressinum* (Podocarpaceae) – rimu; endemic to New Zealand – fruits (epispermatia); the fruit consists of one or two seeds subtended by a brightly coloured, fleshy swelling that attracts animals (especially birds) for dispersal; fleshy part 5-10mm long, seed 4mm long

Fruit – Edible Inedible Incredible

prone to frequent fires, such as the Australian heathlands and the fynbos in the Cape region of South Africa, myrmecochory plays an even more important role. Storage in an underground formicary greatly increases the chance of escaping destruction by fire as well as of being consumed by seed-eaters (*granivores*) such as rodents. Obviously, ant-dispersed seeds have to be small to match the physical strength of their dispersers. Because of their multiple evolutionary origins, the organs that participate in the formation of elaiosomes vary greatly. Most of the time they are genuine seed appendages (arils), derived from different parts of the seed coat or the *funiculus*. In the spurge family (Euphorbiaceae) and milkwort family (Polygalaceae) the elaiosomes are formed by an outgrowth of the seed coat around the micropyle. In the seeds of the greater celandine (*Chelidonium majus*, Papaveraceae) and many Aristolochiaceae such as the wild ginger (*Asarum canadense*) from North America or the European asarabacca (*Asarum europaeum*), the *raphe* develops into a swollen oily appendage.

Elaiosomes of funicular origin are very common in ant-dispersed seeds. If the seeds of the Leguminosae possess any fleshy appendages, and many of them do, they invariably represent a modified part of the funiculus. Many of them have developed funicular arils to attract ants, for example gorse (*Ulex europaea*), broom (*Cytisus scoparius*) and Australian *Acacia* species (e.g. *Acacia vittata*). The seeds of some cacti (e.g. *Aztekium*, *Blossfeldia*) and Caryophyllaceae (e.g. *Moehringia trinerva*) also possess elaiosomes of funicular origin.

As an alternative to seed appendages, many plants attract ants by offering edible swellings on their tiny indehiscent fruits and fruitlets. The achenes of Ranunculaceae, such as liverleaf (*Hepatica nobilis*), the nutlets of Lamiaceae, for example bugle (*Ajuga reptans*) and deadnettles (*Lamium album, L. maculatum, L. galeobdolon*), bear elaiosomes that are formed by their short stalks or the pericarp. Other examples of fruit-borne elaiosomes include the nutlets of certain Boraginaceae (e.g. *Myosotis sparsiflora, Pulmonaria officinalis, Symphytum officinale*) and the cypselas of some members of the sunflower family (e.g. *Carduus* spp., *Centaurea* spp., *Cirsium* spp.), which are equipped with elaiosomes that attract ants. Interestingly, in the myrmecochorous cypselas of some species of Asteraceae the pappus that usually facilitates wind dispersal is reduced so that it is unable to fulfil its original function. In the cornflower (*Centaurea cyanus*) the pappus rays are transformed into hygroscopically moving scales that enable the fruit to slowly creep on the ground as the humidity in the environment changes. Morphologically the most unusual elaiosome probably belongs to the grasses (Poaceae). In the itchgrass (*Rottboellia cochinchinensis*) and the related pitscale grass (*Hackelochloa granularis*) the individual spikelets with their caryopses are totally sunken into the shoot axis. At maturity, the shoot axis disintegrates at the nodes into single-fruited segments (internodes), exposing the elaiosome growing from the tissue of the diaphragm inside the shoot.

page 177: *Cnidoscolus* sp. (Euphorbiaceae) – seed with elaiosome attracting ants; photographed in northern Mexico – ant-dispersed diaspores typically bear a fatty nodule (elaiosome) as a nutritious attractant. As a strategy, ant-dispersal (myrmecochory) is found in over eighty plant families where it has evolved many times independently. Because of their multiple evolutionary origins, the organs that produce the elaiosome vary greatly. In the spurge family (Euphorbiaceae) the elaiosome is formed by an outgrowth of the seed coat around the micropyle; seed c. 1cm long

opposite: Ant-dispersed mericarps (nutlets) of the mint family (Lamiaceae) with an elaiosome (coloured green) formed by part of the pericarp – white: *Lamium album* (Lamiaceae) – white deadnettle; native to Eurasia; c. 3mm long, yellow: *Lamium galeobdolon* (Lamiaceae) – yellow archangel; native to Eurasia; c. 4mm long

Lamium album (Lamiaceae) – white deadnettle; native to Eurasia – plant in flower; the softly hairy leaves with their serrated margins resemble those of the stinging nettle (*Urtica dioica*, Urticaceae) but, unlike the latter, the white deadnettle does not sting, hence its name

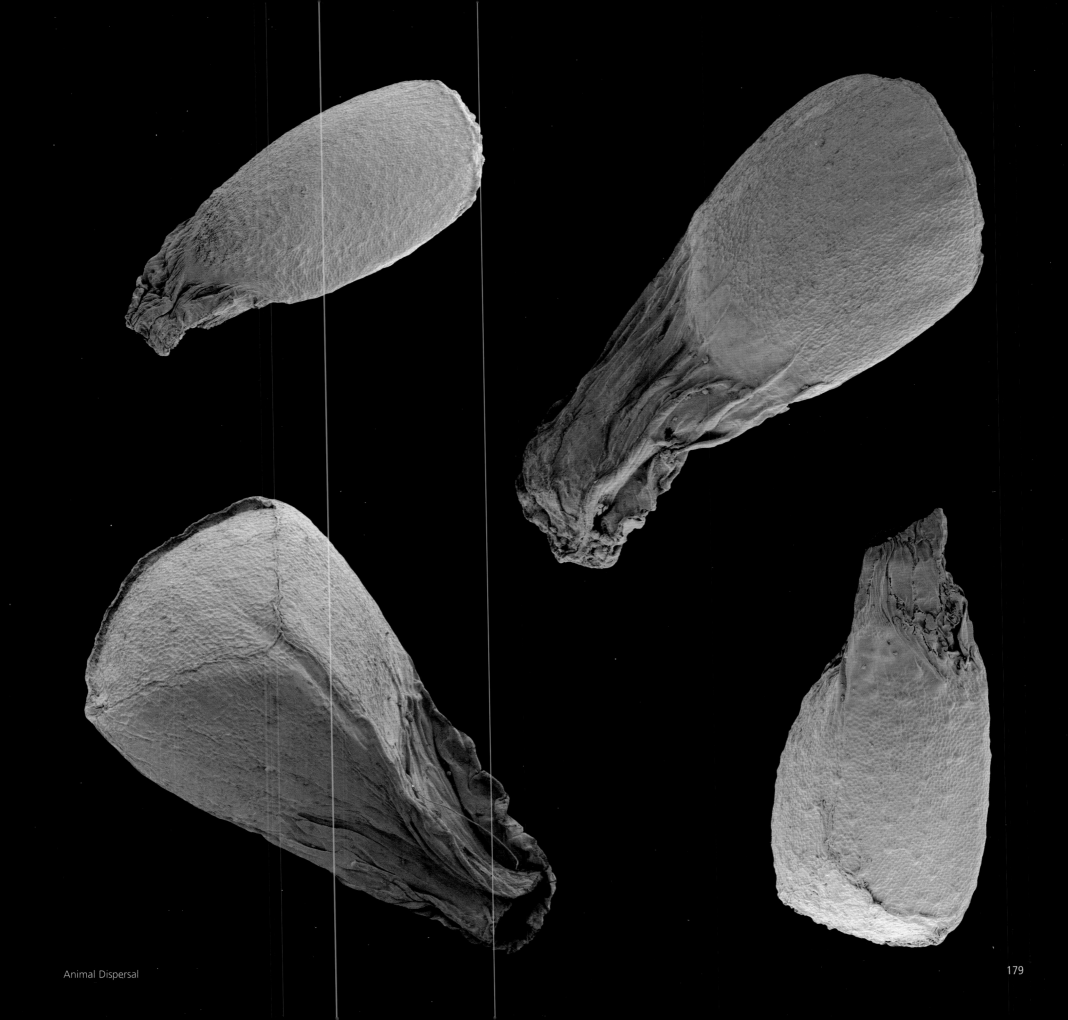

COMBINING STRATEGIES

Ant-dispersed plants often combine myrmecochory with ballistic dispersal. For example, the seeds of certain violets (*Viola* spp., Violaceae), spurges (*Euphorbia* spp., Euphorbiaceae), the squirting cucumber (*Ecballium elaterium*, Cucurbitaceae) and many shrubs and trees in the arid regions of South Africa and Australia are first ejected from the fruit and then further dispersed by ants attracted by the seeds' elaiosomes. Fleshy fruits often combine ant dispersal with bird or mammal-dispersal by offering berries whose seeds are equipped with elaiosomes that survive passage through the gut. This form of diplochory, where ants later collect the seeds from the animals' faeces is mostly found in the tropics, but also in some temperate plants such as yellow fairy bells (*Disporum lanuginosum*, Ruscaceae) from the Great Smoky Mountains. The Australian quinine bush (*Petalostigma pubescens*, Picrodendraceae) spares the ants an unsavoury task by interposing yet another dispersal step. When the quinine bush is in fruit, emus (*Dromaius novaehollandiae*) eat hardly anything other than its fleshy drupes. After passing through the emu's gut the hard endocarps reappear in the bird's droppings. As the dung pile sizzles in the Australian sun, the endocarps shrink and finally explode to propel the seeds up to 3m through the air. Exposed on the ground, the seeds' elaiosomes soon catch the attention of ants, which take them to the safety of their nest.

Other diaspores possess structures that can be interpreted as adaptations to facilitate both wind and animal dispersal. North American pine species with heavy-bodied, winged seeds are first (rather inefficiently) wind dispersed and subsequently scatter hoarded by animals (e.g. *Pinus coulteri, P. lambertiana, P. sabiniana, P. torreyana*). The suspicion that the large samaras of the wild teak (*Pterocarpus angolensis*, Leguminosae) from Africa and the Brazilian zebra wood tree (*Centrolobium robustum*) also pursue a double strategy (anemochory and epizoochory) has already been raised. Smaller suspects are found in temperate climates. The cypselas of the teasel family (Dipsacaceae) are characterized by a collared "airbag" that is formed not by the flower itself but by an outer calyx of four laterally-fused bracts surrounding the gynoecium. The actual calyx of the flower is transformed into a set of stiff awns. The cypselas of the Dipsacaceae, which are heavier and much plumper than their delicate counterparts in the sunflower family (Asteraceae), apparently also pursue a double strategy: the collared airbag assists wind dispersal whilst their stiff calyx awns can easily become entangled in the fur of a passing animal. The tiny *pseudanthecia* (achenes enclosed by joined modified bracts) of the Asian spikesedge (*Kyllinga squamulata*, Cyperaceae) and the fruitlets of certain members of the Araliaceae (e.g. *Hydrocotyle coorowensis, Trachymene ceratocarpa*) possess both wings and hooks, indicating the same combined dispersal syndrome.

below: *Petalostigma pubescens* (Picrodendraceae) – quinine bush; native to Malesia and Australia – top: seeds; below: fruit (drupe); dispersal of the seeds of the quinine bush involves three different steps. In Australia, the fruits are eaten by emus (*Dromaius novaehollandiae*). After passing through the emu's gut the hard endocarps reappear in the bird's droppings. As the endocarps dry out and shrink while the dung pile sizzles in the Australian sun, they eventually explode to propel the seeds up to 3m through the air. During the third and last phase of dispersal, the seeds' elaiosomes attract ants which take them to the safety of their nest; seeds 1.2cm long, fruit 1.5-2cm in diameter

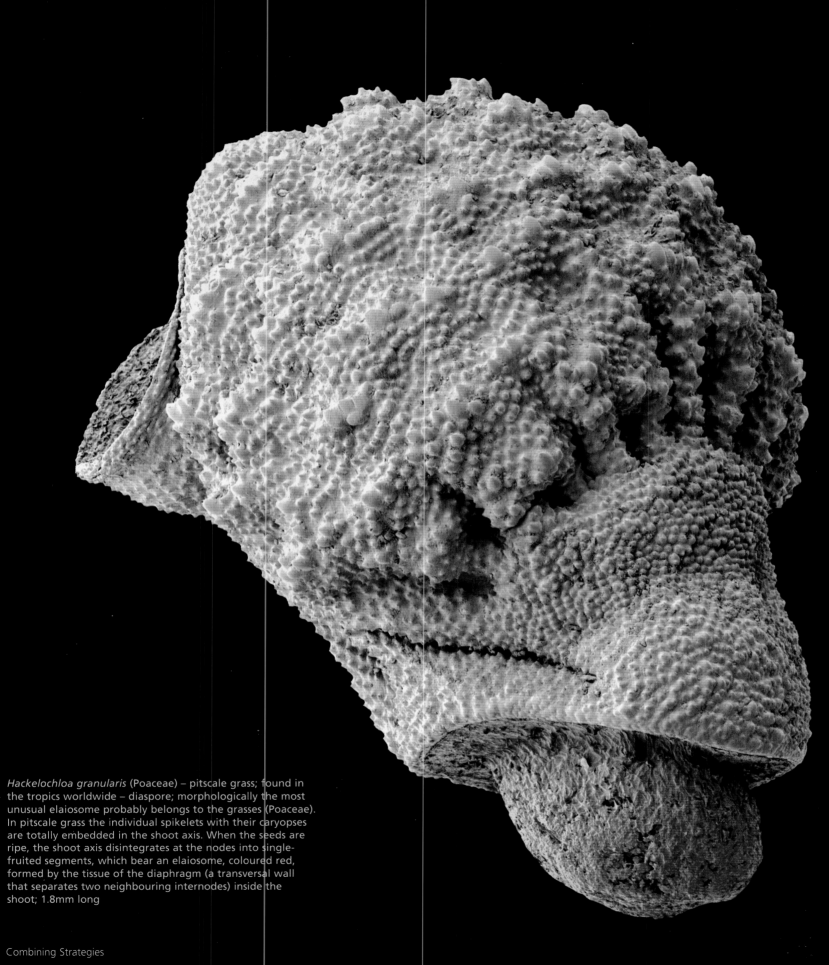

Hackelochloa granularis (Poaceae) – pitscale grass; found in
the tropics worldwide – diaspore; morphologically the most
unusual elaiosome probably belongs to the grasses (Poaceae).
In pitscale grass the individual spikelets with their caryopses
are totally embedded in the shoot axis. When the seeds are
ripe, the shoot axis disintegrates at the nodes into single-
fruited segments, which bear an elaiosome, coloured red,
formed by the tissue of the diaphragm (a transversal wall
that separates two neighbouring internodes) inside the
shoot; 1.8mm long

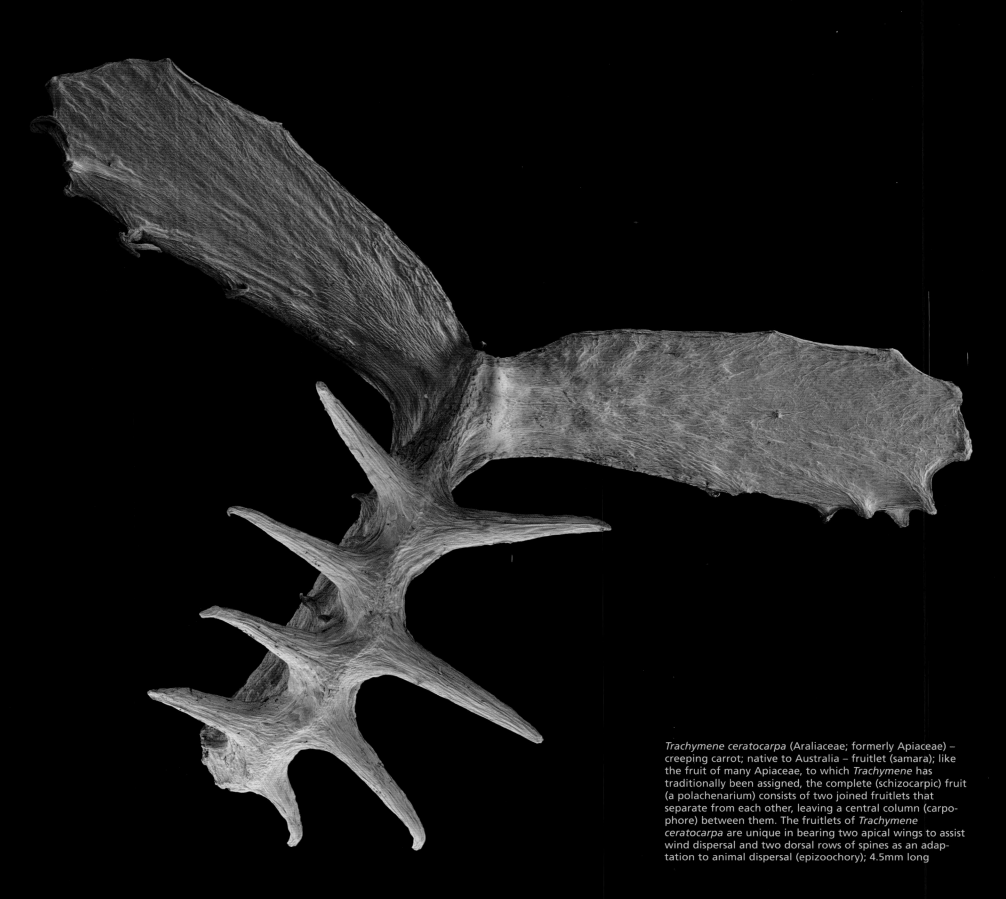

Trachymene ceratocarpa (Araliaceae; formerly Apiaceae) –
creeping carrot; native to Australia – fruitlet (samara); like
the fruit of many Apiaceae, to which *Trachymene* has
traditionally been assigned, the complete (schizocarpic) fruit
(a polachenarium) consists of two joined fruitlets that
separate from each other, leaving a central column (carpophore) between them. The fruitlets of *Trachymene
ceratocarpa* are unique in bearing two apical wings to assist
wind dispersal and two dorsal rows of spines as an adaptation to animal dispersal (epizoochory); 4.5mm long

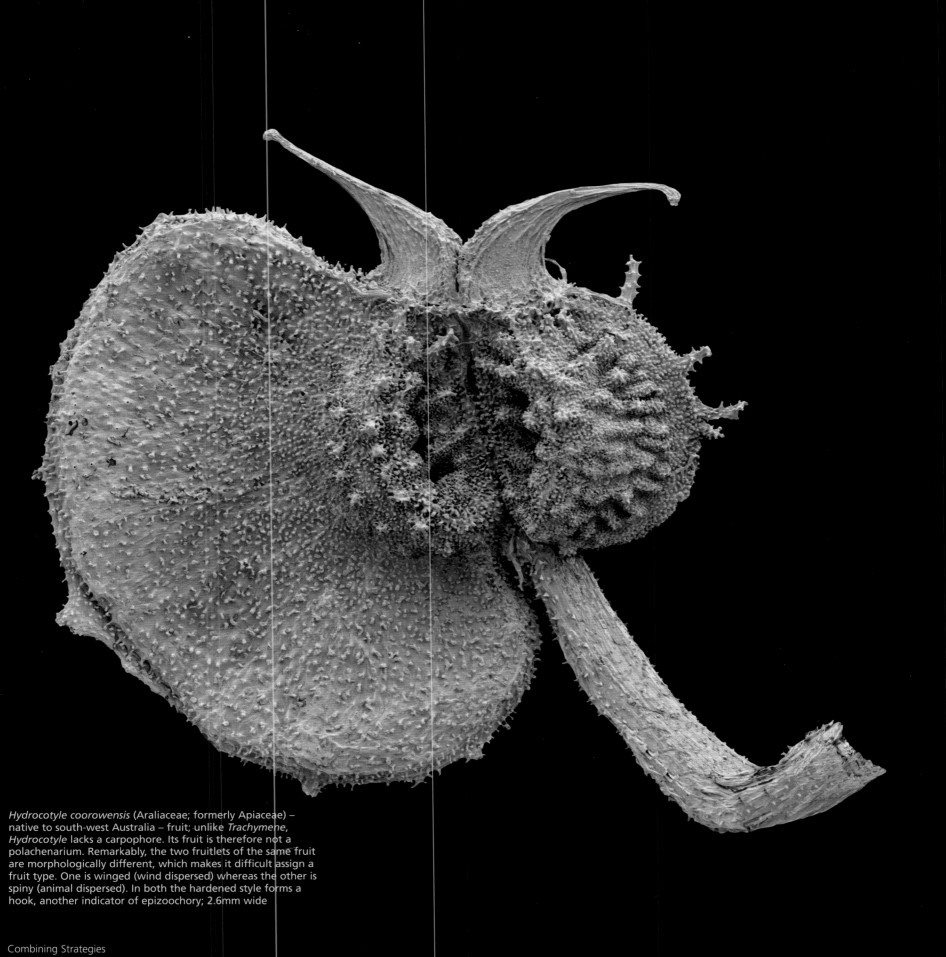

Hydrocotyle coorowensis (Araliaceae; formerly Apiaceae) –
native to south-west Australia – fruit; unlike *Trachymene*,
Hydrocotyle lacks a carpophore. Its fruit is therefore not a
polachenarium. Remarkably, the two fruitlets of the same fruit
are morphologically different, which makes it difficult assign a
fruit type. One is winged (wind dispersed) whereas the other is
spiny (animal dispersed). In both the hardened style forms a
hook, another indicator of epizoochory; 2.6mm wide

DIRECTED DISPERSAL

The phenomenon of seed dispersal by a sequence of two or more phases involving different dispersal agents is called *diplochory*. Many plants both in temperate and tropical habitats are diplochorous and have evolved to capitalize on the advantages offered by different dispersal agents. The first phase of dispersal usually allows the seeds to escape competition in the shadow of the parent plant where seedling mortality is high. During the second phase seeds are often moved to predictable locations, especially if animals are involved. Ants and scatter-hoarders such as jays and nutcrackers, for example, deposit seeds below ground where their chance of survival and establishment is much higher than above ground, especially in fire-prone habitats. By attracting dispersers with particular lifestyles who ensure that seeds are carried to sites where they have a predictably higher survival rate than in random sites, plants are able to direct the dispersal of their seeds, sometimes with just a single dispersal step. The evolutionary advantages of such directed dispersal are obvious. Until recently it was thought that plants were unable to evolve dispersal relationships with animals close enough to allow for directed dispersal; however, more and more cases are emerging that prove the opposite. Besides myrmecochory and dispersal by scatter-hoarders, the best-known example of directed dispersal involves mistletoes and frugivorous (fruit-eating) birds.

Mistletoes are parasitic plants that live on the branches of trees. Taxonomically they belong to a group of three families, the largest of which is the Loranthaceae with more than 900 species. Mistletoes are found worldwide and wherever they occur they have entered into close relationships with small frugivorous birds that feed largely on their fruits. The European mistle thrush (*Turdus viscivorus*, Turdidae) even owes its name to its great appetite for the berries of the European mistletoe (*Viscum album*, Santalaceae). The pulp of mistletoe berries contains viscin, a sticky mucilage that makes it difficult to separate the seeds from the flesh, even after they have passed through a bird's gut. When, after a meal, a bird wipes its beak or bottom against a branch to get free of the sticky seeds, the viscin ensures that the seeds stick to the branch. The branch of a host tree is the only place where mistletoe seeds can germinate. But since the birds eat hardly anything other than the berries of their mistletoe, the seeds stand a good chance of reaching their preferred establishment site.

FLESHY FRUITS

Mistletoe berries are just one of many examples of fleshy fruits that have evolved to be eaten by animals, who then carry the seeds inside their guts until they deposit them with their faeces. This strategy, which involves ingestion of the seeds by the dispersing animal, is called *endozoochory*. Although ants perform a valuable role as seed dispersers they can carry only small

opposite: *Viscum album* (Santalaceae) – European mistletoe; native to Eurasia – branch with fruits (berries); wherever mistletoes (hemi-parasitic plants in the order Santalales) occur, they have entered into close relationships with small frugivorous birds that feed largely on their fruits. The European mistle thrush (*Turdus viscivorus*, Turdidae) owes its name to its large appetite for the berries of the European mistletoe. Since the birds eat hardly anything other than the berries of their mistletoe, the seeds stand a good chance of reaching their preferred establishment site

Ribes rubrum (Grossulariaceae) – redcurrant; native to Eurasia – fruits (berries); small red fleshy fruits such as redcurrants are typically dispersed by birds

opposite: *Macrozamia riedlei* (Zamiaceae) – endemic to Western Australia; the picture shows a northern form of *Macrozamia riedlei*, regarded by some as a separate species and provisionally referred to as *Macrozamia* sp. Eneabba – despite the fact that cycads look like palms, they are gymnosperms. The fossil record proves that cycads have existed since the earliest Permian (248-290 million years ago). About 290 species of cycads have survived almost unchanged for more than 200 million years. In some areas, like here in Western Austra;a, they still constitute a prominent feature of the landscape. The fact that these "living fossils" produce seeds with fleshy seed coats suggests that offering an edible sarcotesta to attract animal dispersers is an ancient strategy

Macrozamia lucida (Zamiaceae) – native to eastern Australia (Queensland, New South Wales) – seed cone, c. 20cm long, 9cm in diameter; when ripe, the seed cones of cycads disintegrate and reveal their large, brightly coloured seeds to attract animal dispersers. In Australia, the fleshy seeds of cycads provide food for a wide variety of animals such as birds, small marsupials and fruit bats. In Western Australia, the seeds of *Macrozamia* spp. are mainly dispersed by emus who swallow the entire seeds. Grey kangaroos, brush wallabies, quokkas and quolls also act as dispersers but rarely achieve long distances since they only eat off the sarcotesta and drop the rest. In Queensland, cassowaries are responsible for the long-distance dispersal of the seeds of *Lepidozamia hopei*. During their long evolutionary history, cycads must have seen many dispersers come and go. For example, until some 50,000 years ago, Australia was home to a unique group of giant flightless birds, the heaviest of which reached half a ton in weight. At the time, these so-called mihirungs (family Dromornithidae), were probably the most important dispersers of cycad seeds on the continent

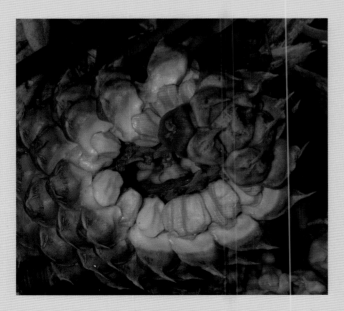

seeds for just a few metres. Other invertebrates occasionally disperse small seeds either after feeding on fruits (e.g. snails eating strawberries), by accidentally ingesting them with soil (e.g. earthworms) or when burying seed-laden balls of herbivore dung (dung beetles). Bigger animals that are able to transport much larger seeds over far greater distances offer much better options. In fact, in the history of seed plants, no other dispersal mechanism has proved more successful than endozoochorous dispersal by vertebrate animals. Today, about one third of the species in temperate deciduous forests produce fleshy fruits that are dispersed by vertebrates. The proportion of endozoochorously-dispersed species climbs to nearly 50 per cent in Mediterranean shrublands and neotropical dry forests, and reaches 70 per cent in subtropical humid forests. In tropical rainforests between 80 and 95 per cent of all plant species rely on fruit-eating vertebrates to disperse their seeds. Of these, birds and mammals form the most important groups but in tropical environments fish and reptiles also play a (minor) role.

The tremendous success of the mutually beneficial relationships that have evolved between the angiosperms and frugivorous animals has been a major contributor to the evolutionary success of this plant group. Fleshy fruits offered the angiosperms endless possibilities to engage with animals. The following chapters are devoted to this most enthralling aspect of the natural history of fruits.

The evolution of fleshy fruits

The dispersal of fruits and seeds through vertebrate animals (i.e., fish, amphibians, reptiles, birds and mammals) is a common feature of many gymnosperms and modern angiosperms. Of all dispersal strategies, endozoochory is the most efficient. Apart from affording much more reliable transport over potentially long distances, the seeds of many endozoochorously-dispersed species germinate much better after passing through the gut of an animal.

Little is known from the fossil record about how the close relationships between fruit-eating animals and plants began to evolve millions of years ago. Most likely, it all started with the evolution of land-dwelling plant-eaters (*herbivores*), a way of life that later allowed the dinosaurs to become the largest animals that ever roamed the Earth. The first seed plants evolved some 360 million years ago, during the late Devonian time period, long before the first dinosaurs existed. Only towards the end of the following Carboniferous time period (354-290 million years ago) did the first reptiles appear on land, and it was not until the Permian (290-248 million years ago) that herbivores began to diversify. These first herbivores belonged to a group of small reptiles (diapsids) that were the ancestors of the dinosaurs; their rise began during the Triassic (248-206 million years ago). The fossil record proves that seed dispersal through vertebrate animals was in place by the late Permian at the latest. It was the arrival of these early plant-eaters that led to the development of consistent dispersal relationships between animals and plants. The most common indication of animal dispersal, a fleshy outer layer or appendage, is perishable and so likely to escape fossilization. Therefore, evidence of dispersal from fossils can only ever be indirect: for example, the presence of seeds in coprolites (fossil faeces) and gut cavities of exceptionally well-preserved fossils. The oldest known fossil containing seeds belongs to a rhynchocephalian *Protorosaurus* from the late Permian. However, it turned out that the seeds belonged to a primitive conifer of the genus *Pseudovoltzia*, which was wind dispersed, suggesting that the *Protorosaurus* ingested the seeds accidentally, probably whilst browsing on the leaves of the tree.

The Permian was followed by the Mesozoic era, spanning the Triassic (248–206 million years ago), Jurassic (206–142 million years ago) and Cretaceous (142–65 million years ago). The flora of the Mesozoic was dominated by gymnosperms, many of which produced large seeds with a fleshy exterior, suggesting adaptation to animal dispersers. Most notably among them were the cycads and ginkgos, the remnants of which have survived almost unchanged as "living fossils." In their Mesozoic heyday their dispersers were almost certainly dinosaurs, among them such awesome, herbivorous creatures as the Jurassic *Brachiosaurus*, which weighed between 30,000 and 60,000 kilos, and the Cretaceous *Argentinosaurus* – the largest land animal ever found – with an estimated weight of 70,000 to 100,000 kilos. During the late Triassic, some 220 million years ago, the dinosaurs were joined by the first true mammals; and in the following Jurassic period the earliest known birds and the first angiosperms appeared. From then on, with all the major groups of animals and plants in place, plenty of opportunities existed for the establishment of biotic dispersal mechanisms in both gymnosperms and angiosperms.

But just how did plants come to produce nutritious fruits in exchange for the dispersal of their seeds? The classic assumption is that seeds and fruits came to produce edible fleshy rewards to attract animal dispersers by "co-evolution," a reciprocal process of evolutionary selection in which the dispersing animal influences the ways in which the diaspore evolves and vice versa. However, it would be teleological to assume that the fleshy seeds of the cycads and ginkgos and the nutritious fruit pulp of the angiosperms evolved from the outset to attract animal dispersers. Like a chicken and egg conundrum, fruit-eating animals could not have existed before fleshy seeds and fruits were available. A much more likely scenario is that fleshy layers were initially developed as a defence barrier against predators, such as insects and diseases caused by fungi and bacteria. Fleshy layers consist mainly of water and can be produced at much lower cost to the plant in terms of energy and materials than hard, dense tissues. Besides providing additional, if weak, physical protection, the fleshy cover most

Throughout the evolution of seed plants, defence against predators has been of prime importance. Therefore, the seeds of many Rosaceae contain poisonous cyanogenic glycosides:

below: *Prunus armeniaca* (Rosaceae) – apricot; native to northern China, cultivated in China since 2000 BC – fruit (drupe); diameter c. 3.5cm

bottom: *Malus pumila* 'Katy' (Rosaceae) – domesticated cultivar of apple; the ancestor of our modern apples came originally from Asia – fruit (pome); diameter c. 7cm

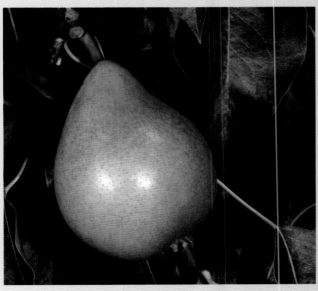

probably also retained distasteful or even poisonous chemical compounds to ward off unwanted attention, including herbivores that killed the seeds while feeding on the foliage. Modern seeds and fruits apply the same strategy, especially when immature.

It is probable that during the next steps in the evolution of fleshy seeds and fruits, vertebrate herbivores consumed the seeds and fruit and thus became "proto-frugivores." Initially a consequence of seed defence, enhanced seed dispersal and its benefits to the plant then began to induce the escalating co-adaptation between seeds or fruits and frugivores: the plants adapted their pulp so that the animals found it more palatable and nutritious; and the fruit-eaters adapted their feeding habits to consume the fleshy tissue without killing the seeds inside. The latter is still encouraged today, both by hard seed coats and endocarps encasing the seeds, as well as by bitter-tasting or even poisonous seeds in otherwise non-poisonous fruits that are harmless as long as they are not chewed. For example, cyanogenic glycosides are found in the seeds of many Rosaceae (e.g. apples, plums, apricots and bitter almonds). In yew (*Taxus baccata*, Taxaceae), highly toxic cyanogenic glycosides are present in the seed coat (and all vegetative parts of the plant) while both the fleshy aril and the seed content are free from toxins.

In this co-evolutionary scenario, pulp first developed as a defence and was later, through a transfer of function, co-opted for animal dispersal. Of course, these fleshy rewards associated with fruits did not evolve only once in the plant world. The same or similar sequences of evolutionary steps occurred many times in many different groups of plants, from the early gymnosperms of the late Devonian and Carboniferous to the angiosperms. As soon as mutually beneficial relationships began to form, increased opportunities for specialization resulted in bursts of evolution among both plants and animals. Relatively late in their evolution, in the early Tertiary (65-2 million years ago), the angiosperms took advantage of the potential to recruit birds and mammals as seed dispersers. It was probably the seren-dipitous availability of new ecological niches after the mysterious demise of terrestrial dinosaurs at the Cretaceous-Tertiary boundary that allowed them to develop new co-evolutionary relationships with these animals. The wealth of opportunities for co-adaptation triggered the rapid radiation of fruit and seed dispersal strategies, as well as a burst of adaptive radiation among birds and mammals. During this time, many frugivorous and seed-eating birds (particularly among the *passerines*) emerged. Simultaneously, many modern angiosperm genera appeared, often with fruits and seeds that closely resemble those of their living descendents.

The traditional concept of co-evolution assumes that the ongoing co-adaptation leads to a situation in which the extinction of one partner would greatly harm the other or, in extreme cases, lead to its extinction. As proved by figs and their respective fig wasps, such

exclusive co-evolution is possible between flowers and pollinators. However, new evidence suggests that similar relationships between fruits and frugivores are rare. Where fruits depend on the continued existence of a single specific disperser, the animal is most probably the last remnant of a once diverse group of dispersers. Evidence from the fossil record and recent observations suggest that the co-evolution of fleshy fruits and their dispersers has been diffuse, involving not just two species but rather classes of animals – for example, birds or mammals and plants. During diffuse co-evolution, the dispersing animal influences certain traits of the diaspore but the disperser and the diaspore are not exclusively co-adapted. In other words, conflicting ecological requirements among the species of a certain combination of dispersers consuming a given fruit blur the selective pressures. As a result, the traits of most fruits have evolved to make them attractive to more than one animal species. In return, most frugivorous (fruit-eating) animals do not live exclusively on the fruits of a single plant species, if only for the simple reason that very few plants bear fruit all year round. Thus, diffuse co-evolution creates "ecological redundancy" that has a clear evolutionary advantage for the plant species involved: the loss of one or two dispersers or even an entire group does not result in the total collapse of the plant's dispersal system.

The good, the bad and the ugly, *or* why fruits are poisonous

The hypothesis that fleshy seeds and fruits initially developed as a defence mechanism that was later converted into a bait to entice hungry animals into consuming the diaspores and subsequently (and unwittingly) disperse the seeds with their faeces seems plausible. After millions of years of evolution, the fleshy fruits of today's plants should have had enough time to perfect their seductive skills and provide us with a cornucopia of delicious treats. After all, they were designed by Nature to be eaten. Nevertheless, anyone adventurous enough to experiment with wild fleshy fruits will find that a sour taste is the rule not the exception. And for those not put off by the bad taste, a much more memorable experience lies in wait: deadly poison. We all remember the warnings of our parents not to eat the berries, pods and other fruits we encountered while playing outdoors, no matter how appetizing they looked. All too easily, young children assume that shiny red or black berries are sweet gifts from Mother Nature. Such a mistake can have fatal consequences if the berries are those of privet (*Ligustrum vulgare*, Oleaceae), deadly nightshade (*Atropa belladonna*, Solanaceae), Virginia creeper (*Parthenocissus quinquefolia*, Vitaceae), white bryony (*Bryonia dioica*, Cucurbitaceae), daphne (*Daphne mezereum*, Thymelaeaceae) or mistletoe (*Viscum album*, Santalaceae).

So why, if the co-evolution between fruits and frugivores has driven fruits to become increasingly attractive to animals, do so many wild fruits look delicious but taste so bad or

Viscum album (Santalaceae) – European mistletoe, native to Eurasia – fruits (berries); despite the fact that small birds such as the mistle thrush (*Turdus viscivorus*, Turdidae) devour the berries of the European mistletoe, they are extremely poisonous to mammals, including humans. Most poisonings of people involve children under the age of five, who mistake the harmless-looking berries for a sweet treat. Symptoms include drowsiness, nausea, stomach pain, vomiting, convulsions, and eventually coma and death. For centuries the European mistletoe has been considered either evil or divine. In Britain, the Druids saw the fruits, which appear around the time of the Winter Solstice, as a symbol of immortality. A familiar Christmas custom is that mistletoe branches are hung in houses to ward off lightning and fire. Tradition also gives young men permission to kiss girls if they meet them under a branch of mistletoe. For each kiss, a berry has to be plucked from the mistletoe branch and when all the berries are gone, the privilege comes to an end. On the scientific side, mistletoe extracts have proved effective as a natural remedy in cancer treatment

are even poisonous? It seems like an evolutionary paradox, or one of nature's evil tricks. As with all misunderstandings, it is necessary to know the whole story. The truth is that fruits still have to defend themselves against all kinds of predators. In this respect, little has changed. From the very beginning, the potentially trouble-free relationship between fruits and dispersers was spoiled. All kinds of creatures, from fruit flies to humans, learned to capitalize on the nutritious rewards provided by fruits, without dispersing their seeds. These pulp thieves eat the flesh but do not swallow the seeds, either because their gape is too small (e.g. fruit flies, racoons) or because their intelligence and dexterity allow them to distinguish and remove indigestible parts of the fruit (e.g. some parrots, monkeys and apes). By taking the reward without providing a service, pulp thieves have effectively become parasites in a previously mutualistic system.

Other thieves are less interested in sweet, sugary pulp. For example, armies of insects have become specialized in preying on the most precious parts of the fruits, the seeds. What makes seeds such worthwhile targets is the highly nutritious food reserve that provides the tiny embryo plant with energy during germination. Among the worst seed predators are beetles. True weevils – also called snout beetles for obvious reasons – form the largest group of beetles, taxonomically classified as the superfamily Curculioinidea. Their membership counts more than fifty thousand species and nearly all of them are plant predators feeding on leaves, shoots, roots, cambium, wood, flowers, fruits and seeds. In the face of such overwhelming diversity among their predators it is not surprising that there is hardly a plant that cannot be infested by at least one weevil species. Weevils are easily recognizable by their long snout, called a rostrum. The chewing mouthparts with which they bore their way into plant tissue are located at the end of it. Many species of snout beetle are serious agricultural pests. The granivores among them specialize in infesting fruits and seeds with their eggs. The females feed on young fruits and leaves, and insert their eggs into holes that they drill through the fruit wall with their long rostrum. The larvae feed on the developing seeds inside the immature fruit. Either they leave the fruit as adults, or the grubs gnaw their way out of the fruit to burrow into the soil where they pupate and metamorphose into beetles. Among the most dreaded pests are the hazelnut beetle, the grain weevil, and the boll weevil. The hazelnut beetle (*Curculio nucum*) is responsible for huge losses in hazelnut orchards in Europe and Turkey. Infestation of stored cereals, especially wheat, corn and barley, by the grain or granary weevil (*Sitophilus granarius*), which is 3-4mm long, can lead to the devastation of entire granaries. In our homes, they can be found feeding on a bag of flour or a packet of muesli. The boll weevil (*Anthonomus grandis*) feeds inside young cotton bolls and is feared by cotton farmers in the southern United States.

An unknown snout beetle of the Curculionoid family Nanophyidae, many of which feed on seeds. The illustrated specimen was found in a collection of seeds that were sent to the Millennium Seed Bank from Madagascar; 4.8mm long

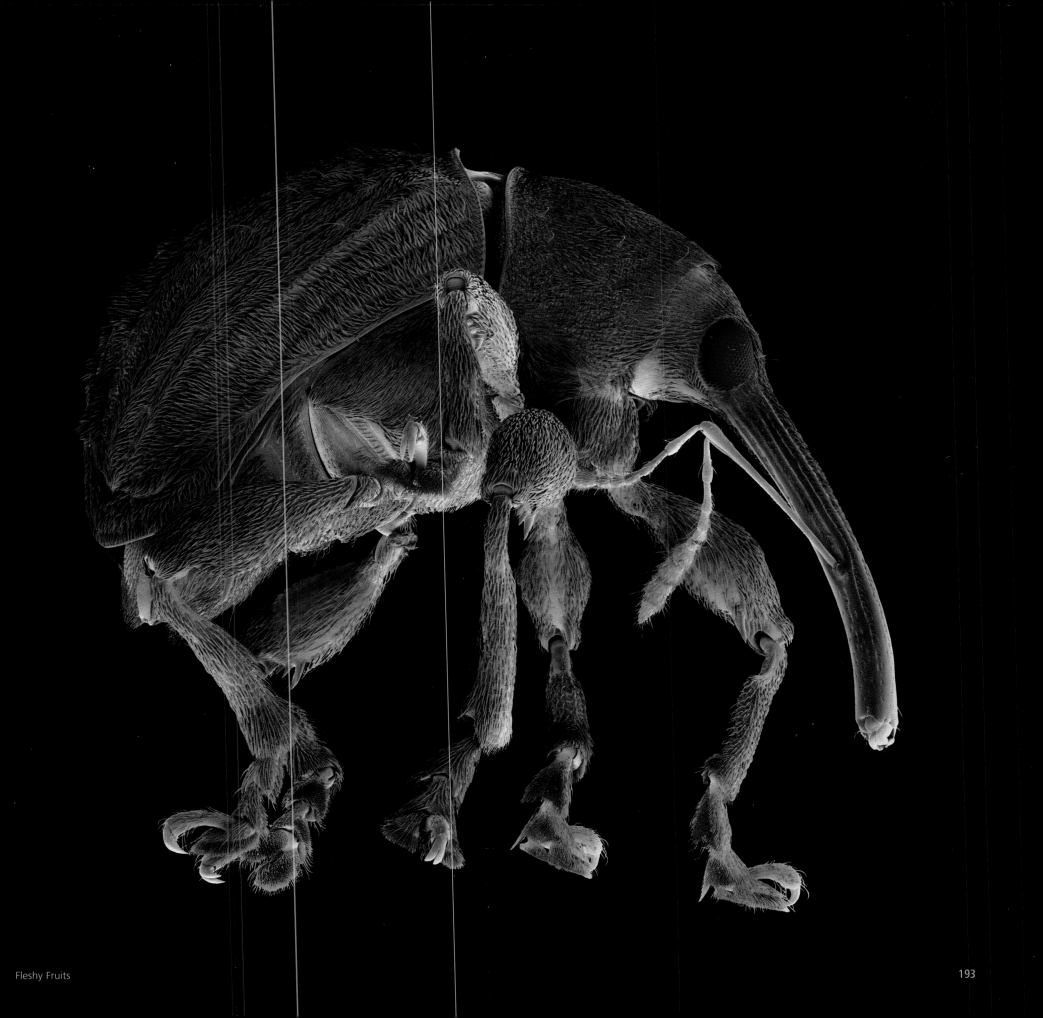

The bruchids are a group of beetles that are exclusively specialized in parasitizing seeds. They are sometimes called seed weevils but are not closely related to true weevils. They belong to a different family of beetles, the Chrysomelidae (leaf beetles), of which they constitute the subfamily Bruchinae (formerly family Bruchidae). Adult bruchids feed on the pollen and nectar of their host plants and lay their eggs on the ovaries of the flowers. Upon hatching, the young larvae burrow through the ovary wall into a developing seed and feed on its tissue. The larvae pupate inside the same seed and emerge as young adults, boring characteristic exit holes in the seed coat and fruit wall, telltale signs of bruchid infestation. The fact that there are more than thirteen hundred different species of bruchids, ranging in size from 1mm to 2cm, is proof of the success of their destructive strategy. Seed weevils are granivores (seed-eaters), many of which specialize in legumes. They can devastate entire harvests of beans (*Phaseolus vulgaris*), peas (*Pisum sativum*), chickpeas (*Cicer arietinum*), cowpeas (*Vigna unguiculata*) and other pulses in the world's poorest countries.

However, granivorous insects are not the only animals that prey on seeds. Birds and mammals also boast legions of granivores. Finches specialize in eating seeds and so do many rodents (Rodentia), the largest order of mammals, which includes mice, rats, hamsters and squirrels. Seeds form at least part of the diet of many other animals, including deer and pigs, which feed on acorns when they are available.

Apart from insects, birds and mammals hungry for seeds, Nature also harbours an endless diversity of much smaller predators prepared to devour any organic matter unable to defend itself: fungi and bacteria. With their resilient spores omnipresent in air, water and soil, these dangerous enemies can cause all kinds of diseases. A fruit riddled with fungal or microbial infection is unattractive to potential dispersers even if it is not completely destroyed. In both cases the fruit fails in its vital mission to disperse its seeds. This ancient game of attack and counter-attack has sparked off another, sinister kind of co-evolution between animals and plants, an evolutionary arms race. Whilst selection pressures drove plants to constantly upgrade their physical and chemical defences, not only in their seeds and fruits but in all parts of their bodies, predators strove to breach them by constant adaptation. For example, the seeds of legumes (Leguminosae) contain a gamut of toxic deterrents, ranging from cyanogenic glycosides, tannins and toxic amino acids to lectins (sugar-binding proteins), trypsin inhibitors (blocking protein-digesting enzymes in the intestine) and bitter-tasting alkaloids, to name but a few. Small amounts of golden chain (*Laburnum anagyroides*), lupin (*Lupinus* spp.) or crab's eye (*Abrus precatorius*) seeds can cause lethal poisoning in both animals and humans. Even pulses, although bred for centuries for human consumption, still require careful soaking, boiling, sprouting or fermenting to denature the toxins before consumption.

opposite: Mixture of edible legume seeds (Leguminosae): three different cultivars of *Phaseolus vulgaris* var. *vulgaris* – red kidney beans (dark red), pinto beans (mottled), canellini beans (white), Dutch brown beans (medium brown) – *Glycine max* (soya beans; the smallest of the selection), *Cicer arietinum* (chick pea; round and rugose). Legume seeds are one of the prime targets of seed predators such as curculioinid snout beetles and members of the Bruchidae, which is why most pulses are poisonous when raw. Heat destroys the poison and the seeds become edible after boiling. Red kidney beans are particularly notorious for their high content of the poisonous lectin phytohaemagglutinin (a sugar-binding protein) and need thorough cooking before being eaten

Abrus precatorius (Fabaceae) – crab's eye; found in all tropical regions – fruit (legume); a pantropical climber with beautiful red and black seeds, this species has been given many other common names, including coral bean, jequirity bean, paternoster bean and rosary pea. Despite their striking looks and popularity with makers of botanical jewellery, the seeds of the crab's eye are extremely poisonous. The reason these and the seeds of many other legumes, including the ones we eat, are so toxic is because during their evolution, they had to develop chemical defences to ward off seed predators, especially snout beetles and seed weevils. Many seed weevils have specialized in legume seeds and managed to adapt their physiology so as to overcome the plants' chemical defences. Seed weevils can devastate entire harvests of beans (*Phaseolus vulgaris*), peas (*Pisum sativum*), chickpeas (*Cicer arietinum*) and other pulses, causing severe problems in the world's poorest countries; diameter of seed 4mm

Fruit – Edible Inedible Incredible

This arsenal of chemical weapons discourages most potential aggressors but not bruchid beetles. During their long co-evolutionary relationship with legumes, bruchids have acquired resistance to seeds full of toxic compounds that would poison other animals. In the course of evolution, the increasingly sophisticated weaponry of the legumes required extreme specialization on the part of the bruchids to a point where today a species of beetle can cope with only a few species of legumes. Similar host-specific relationships are found between true weevils and legumes (e.g. Australian *Acacia* species). Some have even focussed on the chemically well-defended seeds of the ancient cycads. *Antliarhinus zamiae* and *A. signatus*, two straight-snouted weevils of the family Brentidae, feed exclusively on cycad seeds.

In addition to waging chemical warfare on thieving elements, fruits have to provide a worthwhile meal for *bona fide* dispersers. With so many parasites causing conflicting selective pressures, the traits of fleshy fruits are probably the result of a trade-off between being sufficiently repulsive to the bad and ugly whilst remaining attractive to the good. Therefore, the ecology of fruits and vertebrate frugivores can be understood only by taking into account the evolutionary triangle between fruiting plants, their mutualists, and their predators and parasites, including granivores. Toxic chemicals in fruits and seeds almost certainly evolved to mediate these interactions by balancing the potential cost of losing dispersers with the benefits of protecting seeds. The good news is that not only the parasites but also the mutualists co-adapted with fruits and seeds. Whether a certain chemical compound is poisonous or not depends on the species involved, a fact proved not only by the villainous bruchids. For example, many birds, the most important group of animal dispersers, are able to eat fruits that are toxic to humans and many other mammals. The berries of the European mistletoe (*Viscum album*, Santalaceae), a winter delicacy for small birds such as the mistle thrush (*Turdus viscivorus*, Turdidae), contain several small proteins that are highly toxic to mammals. The sweet-tasting, black berries of the deadly nightshade (*Atropa belladonna*) and henbane (*Hyoscyamus niger*, Solanaceae) contain a very potent mixture of tropane alkaloids (e.g. hyoscyamine, scopolamine and atropine) that interfere with acetylcholine receptors in the nervous system. Atropine in particular causes severe symptoms in humans, including sweating, vomiting, breathing difficulties, confusion, agitation, hallucinations, and finally coma and death. Atropine also has a pupil-widening effect that was already known in ancient Greece, where an extract of belladonna (Italian for "beautiful woman'") was frequently applied by women to enlarge their pupils. Widened pupils, naturally evoked by arousal, were supposed to intensify eye contact in romantic encounters. Deadly nightshade and henbane are also thought to have been among the main ingredients of the hallucinogenic brews of medieval Europe, including "flying ointments" to give witches the sensation of flight. This kind of early drug abuse held dangers other than the risk of being burnt at the stake. *Atropa belladonna* and *Hyoscyamus niger* are two of the most poisonous plants in the Western hemisphere; just three berries are enough to inflict severe poisoning, not only in children but also in domesticated animals such as cats, dogs and livestock. However, wild birds and certain mammals, including rabbits and deer, are able to eat the fruits and other parts of the plants without suffering any ill effects.

The presence of poisons that are harmless for one group of animals but toxic for others enables plants not only to ward off predators but also to select the preferred dispersers from the available repertoire of frugivores.

Enough is as good as a feast

The fact that certain birds and mammals can eat fruits that are poisonous to other animals and to humans does not necessarily mean that they are totally immune to the toxins. Observations in temperate Europe of hawthorn (*Crataegus monogyna*, Rosaceae), ivy (*Hedera helix*, Araliaceae) and holly (*Ilex aquifolium*, Aquifoliaceae) – all three poisonous to humans – have shown that different species of birds spend similar periods of time on a certain plant and eat a similarly limited number of fruits (always fewer than ten) at each feeding session. The birds' instinctive behaviour suggests that they are not entirely immune to the poisons, just less sensitive. As with everything else in life, too much of anything is not good. Even animals seem to realize this. In fact, most frugivores – birds and others – rely only partially or not at all on fruits. This is especially true in temperate regions with cyclical fruitless periods. But even if there are only a few fruits that cause acute symptoms of poisoning, others often have constipating or laxative effects. For instance, figs, prunes and the fruits of the aptly named purging buckthorn (*Rhamnus cathartica*, Rhamnaceae) have long been known for their cathartic effects and are used as natural laxatives. The presence of purgative or constipating compounds in fruits (e.g. sorbitol in Rosaceae, glucosides of emodin in Rhamnaceae) could be an adaptation to regulate seed passage rates in certain frugivores. Slowing down the digestion in loitering mammals can increase the dispersal distance, whereas rapid passage through faster moving animals, such as passerine birds, can reduce damage to the seeds caused by gizzards and enzymes.

Despite these unsavoury side effects, there are animals that specialize in eating nothing but fruit. Such *obligate frugivores* are found predominantly in the tropics where fruits are available all year round. Most of these exclusive fruit-eaters are birds (e.g. macaw parrots, fruit pigeons, toucans, hornbills, cassowaries) and mammals (e.g. flying foxes, ruffed lemurs, spider monkeys), but rarely reptiles (two species of monitor lizard). Somehow these

Atropa belladonna (Solanaceae) – deadly nightshade; native to Europe, North Africa and western Asia – despite their edible looks, sweet taste and the fact that wild birds and certain mammals, including rabbits and deer, are able to eat them without no ill effects, the black berries of the deadly nightshade are extremely poisonous to humans. Besides causing severe symptoms or even death, the main toxin, atropine, also has a pupil-widening effect that was already known in ancient Greece. Women believed they could enhance eye contact in romantic encounters by applying to their eyes an extract of belladonna (Italian for "beautiful woman")

below and opposite: Lithographic plate from Köhler, F.E. (1887) *Köhlers Medizinal-Pflanzen in naturgetreuen Abbildungen mit kurz erläuterndem Texte*, Vol. 1. Gera-Untermhaus, Leipzig, Germany

specialists have found a way of coping with the large quantities of toxins in their one-sided diet. Although little is known about how obligate frugivores detoxify harmful substances in their bodies, certain animals are known to practise self-medication. For example, South American macaw parrots gather almost every morning on the banks of the Manu River in Peru to eat soil from the riverbank. But they do not eat just any old dirt. They carefully choose one particular layer exposed along the riverbank that consists of very fine clay. Clay has the ability to deactivate toxins such as tannins and alkaloids that are present in fruits, leaves, stems and roots. It can also absorb pathogenic bacteria and viruses. The beneficial effects of clay were known to many native peoples in Europe, Asia, Africa and America, where geophagy (dirt-eating) has been practised by humans for centuries if not millennia. For example, American Indians added clay to neutralize bitter tannins in acorns and toxic alkaloids in wild species of potatoes to make them edible. Even today, China clay, also known as kaolin, is sold in pharmacies as a natural cure for diarrhoea. For the macaw parrots who live on the bounty of fruits and seeds provided by the rainforest, their daily portion of clay is a remedy that mitigates the potential side effects of their fruity diet. But not only fruit-eaters practise geophagy as a way of self-medication. The defensive chemicals present in fruits are also found throughout the rest of the plant's body, so herbivores face the same digestive challenges as frugivores. It is therefore not surprising that sheep, cattle, elks, giraffes, zebras and elephants, as well as chimpanzees and gorillas, have been observed eating generous helpings of earth, especially when sick. Other examples of animal self-medication include the instinctive consumption of rough-haired leaves by geese, dogs, bears and gorillas as herbal scours to dislodge intestinal parasites.

Young and dangerous

A fruit's chemical defences are at no time more important than during its development. As long as a fruit is not fully mature, its seeds are not yet viable. Their unfinished seed coats are soft and vulnerable, the embryo is underdeveloped, and the endosperm has not yet been filled with energy-rich fat, protein or starch. Therefore, the premature harvest of its fruits, even by legitimate dispersers, is as much against a species' survival strategy as an attack by predators and diseases. This explains why unripe fruits make themselves as unattractive as possible by staying hard, green and firmly attached to the parent plant, while at the same time maintaining their chemical defences at a very high level. Fruits that are poisonous when ripe tend to be even more toxic when still green, for example, elderberries (*Sambucus nigra*, Adoxaceae), the fruits of many species of nightshade (*Solanum* spp., Solanaceae), and numerous members of the gourd family (Cucurbitaceae). The immature fruits, as well as the rest of the plant of the

Hyoscyamus niger (Solanaceae) – henbane; native to temperate Eurasia – henbane and its sister, the deadly nightshade, are two of the most poisonous plants in the Western hemisphere; three berries are enough to inflict severe poisoning, not only in children but also in domesticated animals such as cats, dogs and livestock. Most symptoms of poisoning are unpleasant, but the very potent mixture of tropane alkaloids (e.g. hyoscyamine, scopolamine and atropine) also has hallucinogenic effects. This effect was known to the "witches" of the Middle Ages. Allegedly, deadly nightshade and henbane were the main ingredients of their hallucinogenic brews, including the infamous "flying ointments," which afforded those who indulged in them the sensation of flight. Not surprisingly, when tried for witchcraft, the accused admitted that they were indeed able to fly and therefore ended up being burnt at the stake

African balsam pear (*Momordica balsamina*, Cucurbitaceae) have in the past been used as one of the ingredients of arrow poison by tribes in the Benué region of Nigeria.

There is also the possibility that fruits that are poisonous during their development become perfectly edible when fully mature, as proved by passionfruits (*Passiflora edulis*, Passifloraceae), many solanaceous berries (e.g. Chinese lantern, *Physalis alkekengi*; cape gooseberry, *Physalis peruviana*) and the North American mayapple (*Podophyllum peltatum*, Berberidaceae), whose immature fruits can be fatal if eaten. Most famous for being dangerous when young is the ackee. Scientifically named *Blighia sapida* in honour of Captain William Bligh who presented the first plants to Kew Gardens in London in 1793, the ackee belongs to the soapberry family (Sapindaceae), which also includes the lychee (*Litchi chinensis* subsp. *chinensis*) and rambutan (*Nephelium lappaceum*). Although originally at home in West Africa where it was called anke or akye-fufuo, the ackee somehow reached Jamaica where the edible parts of its fruit became a vital ingredient in the island's national dish of ackee and saltfish. Some sources credit Captain Bligh with introducing the ackee to Jamaica in February 1793 when he delivered 1,200 precious breadfruit plants and other potentially valuable crops from the Old World. However, historical records indicate that Thomas Clarke, Jamaica's official botanist and supervisor of the botanical garden at the time, received the ackee in 1778, probably on board a slave ship. Whichever way the ackee crossed the Atlantic, it has gained great popularity in the West Indies, despite the fact that its unripe fruits are highly poisonous. Technically, the beautiful, pear-sized, bright red to yellow–orange fruits of *Blighia sapida* are loculicidal capsules. When ripe they dehisce to expose three shiny black seeds that are partly covered in a soft, yellowish aril. The texture of the seed appendage has been likened to scrambled eggs or cooked brains, hence its common name, "vegetable brain." The tasty aril is the only edible part of the fruit and is full of nutritious, energy-rich oil. It is therefore not surprising that birds, the ackee's natural dispersers, find the arils irresistible and compete with humans who want to enjoy ackee and saltfish. There is a strong temptation to beat the birds by harvesting the fruits before they split open but the consequences can be lethal. As long as the fruit is immature and closed, even the arils contain the same defensive chemical that renders the rest of the fruit, including the seeds, inedible. The toxicity is caused by an unusual non-proteinogenic amino acid that induces severe hypoglycaemia in animals and humans, hence its name hypoglycin A. Between 1880 and 1955, before the toxic principle of the ackee was known, a mysterious illness emerged in Jamaica that killed approximately 5,000 people. Symptoms of the Jamaican Vomiting Sickness or Toxic Hypoglycaemic Syndrome included violent bouts of vomiting, abdominal pain, aciduria, hypoglycaemia, coma and, in the most serious cases, death. Although the

below: *Solanum luteum* subsp. *luteum* (Solanaceae) – yellow nightshade; native to the Mediterranean – fruits (berries); typical of bird-dispersed fruits, as the berries ripen, they change colour from green to red; diameter c. 2cm

bottom: *Blighia sapida* (Sapindaceae) – ackee; native to West Africa – fruit (loculicidal capsule); upon opening, the leathery, pear-shaped capsules reveal three large black seeds that bear a soft, yellowish aril, the only edible part of the otherwise poisonous fruit. The orange-red pericarp, the yellowish aril and black seeds create a colourful display typical of bird-dispersed fruits; fruit c. 8-10cm long

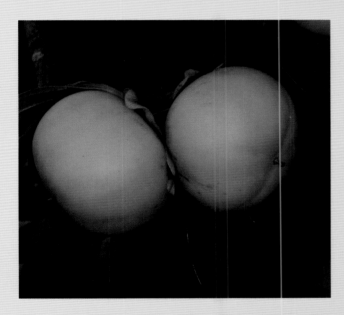

ackee was suspected of being toxic as early as the nineteenth century, the fact that it was eaten by so many people with no ill effects made it difficult for researchers to identify the delicious fruit as the cause of the illness. It is assumed that the consumption of unripe ackees continues to be the reason for many unexplained deaths in small children in West Africa.

Fortunately, our most popular fruits can be enjoyed without risking life-threatening side effects. But whether poisonous or not, eating unripe fruit is generally not pleasant. Even if potent toxins are absent, a chemical defence remains in immature fruits in the form of a bad taste. High concentrations of fruit acids, tannins and lignins create a sour, bitter, astringent tang, sufficient to deter anybody with a potential interest in the fruit. Wild species, including the ancestors of many of our domestic fruits retain relatively high levels of these compounds even in ripe fruits, which is the reason that they do not generally have a particularly nice taste. The main chemical compounds that render fruit flesh and other plant parts unpalatable are condensed tannins. Tannins are large, complex polyphenols that bind and precipitate proteins, an attribute that has long been exploited by tanners to turn animal hides into leather. In the mouth, tannins interact with salivary mucoproteins, causing astringency and the typical, furry taste of unripe fruit. In the gut they inhibit digestion by reducing the availability of protein. Although bred to please our palates, experience tells us that even apples, pears, bananas, plums, peaches, persimmons, grapes and many more of our favourite fruits contain high levels of condensed tannin when unripe. During the ripening process, tannins are inactivated by forming large polymers or complexes with pectins. However, toning down astringency is just one of many complicated changes a young fruit has to undergo in order to become soft, sweet and juicy – in other words, a delicious treat.

Climacteric fruits

During maturation and ripening, fleshy fruits undergo many visible and invisible changes that ultimately render them both palatable and attractive to potential dispersers. These changes either take place as a continuous, gradual process, or unfold swiftly as the result of a hormonal signal. Hormone-controlled or *climacteric* fruits (Latin: *climactericus* = a dangerous or critical period in life) enter a phase of rapid ripening during which their rate of respiration (oxygen consumption) and their temperature increase. The trigger for climacteric ripening comes from a burst of ethylene produced by the fruit itself. Ethylene is structurally a very simple gaseous hydrocarbon ($H_2C=CH_2$), which acts as a plant hormone. Once the signal is received a whole range of new enzymes are produced bringing about a multitude of changes. Tannins, organic acids and other chemical deterrents such as cyanogenic glycosides and alkaloids rapidly decrease in concentration as they are broken down or otherwise inactivated. Amylase enzymes convert starch into osmotically more active sugar, thereby making the dry, mealy flesh sweet and juicy. Depending on the kind of fruit, there is an increase not only in sugar content, but also in the level of other nutritious reward substances such as proteins and lipids (in avocados, for example). As the pulp becomes more edible, pectinase enzymes soften the flesh by hydrolysing pectin, the substance that glues the cells together into a firm tissue. The same enzymes induce weakening of a particular layer of cells in the fruit stalk. This "abscission zone" marks the spot where the stalk eventually separates from the plant, if only by its own weight. Other ripening enzymes create the characteristic aroma of the pulp by preparing a cocktail of fruit esters through the chemical combination of organic acids and alcohols. Ethylene also triggers enzymes that mediate the breakdown of chlorophyll and the synthesis of new pigments. As a result, the peel changes from green to the typical colour of the mature fruit.

One bad apple spoils the barrel

Apples, tomatoes, and especially bananas are climacteric fruits that produce large quantities of ethylene. The ethylene they emit is a gaseous hormone that can even be used to ripen other climacteric fruits. This is useful because most fruit sold by supermarkets is unripe. If a ripening banana or apple is added to a bag of unripe avocados, peaches, pears, apricots, mangoes, papayas, figs, guavas, or melons, they will be ready to eat within a few days. Artificially produced ethylene is used on an industrial scale to synchronize the post-harvest ripening of tomatoes, bananas and pears, all harvested and shipped while immature because they are highly pressure sensitive and perishable when ripe. Timing of the ethylene treatment is crucial. Once climacteric fruits have started the ripening process, there is little that can be done to slow it down as proved by the old English proverb "one bad apple spoils the barrel." Apples used to be stored in barrels in cool cellars to provide the family with fresh fruit during the winter. It was important to exclude all damaged fruit because plant tissues also produce ethylene if they are damaged or diseased. One wormy or rotten apple can cause premature ripening of all the other apples in the barrel and thus spoil the whole batch. However, not all fruits are sensitive to ethylene. Non-climacteric fruits such as cherries, grapes, strawberries and most citrus fruits produce very small quantities of ethylene. Although they gradually undergo the same changes during ripening as climacteric fruits they remain unaffected by the hormone.

Dispersal syndromes, the sign language of fruits

The ripening processes that turn fleshy fruits from hard, green, floury, tart, odourless lumps into brightly coloured, soft, sweet, juicy, fragrant morsels are essentially an advertising

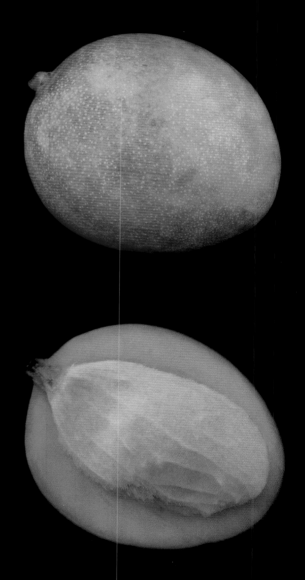

Mangifera indica (Anacardiaceae) – mango; only known in cultivation, its precise origin is uncertain but assumed to be somewhere between India and the Malay Peninsula – fruit (drupe); fruit (top) and longitudinal section of fruit (bottom); mangoes have been cultivated for millennia and rank among the most popular fruits in the world. Among animals, fruit bats are fond of mangoes and seem to be the main dispersers. However, the large size of the hairy stone, which is difficult to separate from the pulp (a ruse to encourage swallowing), usually suggests adaptation to large-bodied dispersers. Elephants, for example, relish mangoes. Their digestive system lets the kernel pass unharmed and removes the pulp which would otherwise attract bacteria and fungi that could harm the germinating embryo; fruit c. 10cm long

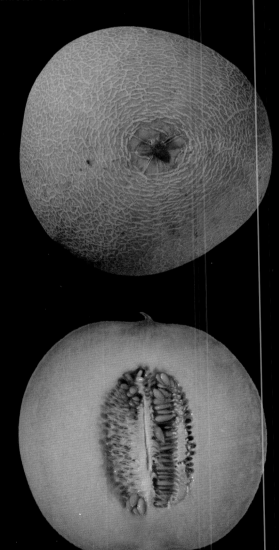

Cucumis melo subsp. *melo* var. *cantalupensis* 'Galia' (Cucurbitaceae) – Galia muskmelon – fruit (pepo); like mangoes, melons are climacteric fruits that enter a phase of rapid and unstoppable ripening, triggered by the gaseous phytohormone ethylene. The ripening of "green" mangoes and melons can be accelerated by keeping them together in a bag with a ripening banana, also a climacteric, ethylene-producing fruit. Although the Galia muskmelon is a cultivar, bred to produce large fruits, there are also big wild melons. Assuming that fruits generally grow to a size that does not exceed their co-adapted dispersers' gape size, the natural dispersers of melons may be large mammals, most of which became extinct at the end of the last Ice Age; diameter c. 16cm

strategy aimed at recruiting legitimate dispersers. For an advertising campaign to be successful, its message must be communicated in such a way that it reaches the target customer group. The customers of plants with fleshy fruits are animal dispersers. We should not forget that the diffuse nature of selective pressures prevented the evolution of tight co-adaptive relationships between fruits and animals with the result that the sign language used by fleshy fruits to catch the attention of potential customers is not highly specific because it evolved to appeal to a broad audience. Nevertheless, over millions of years of co-adaptation, frugivorous animal dispersers (predominantly birds and mammals) have strongly influenced the traits of fruits. Quantitative observations have shown that the two characters that distinguish bird-dispersed (*ornithochorous*) from mammal-dispersed (*mammaliochorous*) fruits are size and nutrient content. As a general rule, ornithochorous species have smaller fruits than mammaliochorous species. The larger a fruit, the more mammals are attracted as preferred dispersers, while at the same time birds are increasingly less effective dispersers, if only because of their smaller gape-size and lack of teeth. The continuous spectrum of fruit sizes means that intermediate-sized fruits often draw on a mixed suite of vertebrate dispersers, including both birds and mammals. Because of the strong correlation across species between fruit size and the dry mass of pulp the total energy reward per fruit increases from bird- to mammal-dispersed species. This observation – though generally valid – is slightly biased by the fact that bird-dispersed fruits tend to contain more lipids (e.g. olives, many palm fruits) than mixed- and mammal-dispersed fruits, and therefore have a higher energy content relative to their weight.

This is about as far as scientists from all relevant biological disciplines are prepared to agree on the extent to which a specific disperser group has influenced the evolution of the traits of fleshy fruits. The general notion is that there are simply too many other factors involved that outweigh the influence of *bona fide* dispersers on the survival of a species. Adaptive responses limited by genetic constraints, inconsistent selection pressures over space and time, and considerable uncertainty as to the outcome of the interactions themselves because of unpredictable post-dispersal events (e.g. secondary dispersal, seed predators) are considered the main obstacles to more tightly co-evolved relationships between fruits and mutualistic frugivores. However, in the 1970s and 1980s, when seed dispersal ecology began to develop into a scientific discipline, ecologists ventured to take visual, olfactorial and other unquantifiable traits into account in their attempts to find adaptive explanations for the enormous diversity of fleshy fruits and frugivore behaviours. Based on the hypothesis that co-adaptation between fruits and their preferred mutualists would lead to suites of morphological, physiological, biochemical and behavioural characters, ecologists defined "specialized" and "generalized" dispersal syndromes. Despite recent criticism of the theoretical foundation underlying the concept of dispersal syndromes, many ecological studies confirm that certain fruit traits such as colour, size and protection are clearly associated with either bird or mammal dispersal. Whether quantifiable or not, colour, texture and aroma usually provide the best clues as to whether birds or mammals are the preferred dispersers of a particular fruit.

The bird-dispersal syndrome

Volant birds are by far the most important dispersers of seeds. The ability to fly affords them excellent mobility and enables them to transport seeds in their gut quickly and over long distances. In terms of their effectiveness as seed dispersers, birds are probably only rivalled by fruit bats. Mostly diurnal creatures with excellent colour vision but a poor sense of smell, birds rely on their eyes rather than on their noses to locate food that they are unable to chew because of their lack of teeth. Only a few groups of birds, including parrots, crows and New World blackbirds (family Icteridae), are capable of breaking up tough-skinned fruits with their feet and beaks to reach pulp and seeds. Most other birds swallow fruits whole, which could be the reason that they are not put off by the sour or bitter taste that renders many bird-dispersed fruits unpalatable for humans. Nevertheless, birds have a clear bias towards certain fruits and prefer them when they are fully ripe. The colours of ornithochorous fruits are most likely the result of a co-adaptive response to birds' sensory preferences. Their function is to provide a conspicuous, reliable signal that indicates a safe, nutritious reward. Taking into account all general avian strengths and limitations, except those of ground-dwelling flightless specialists such as ratites (ostriches, emus, rheas and cassowaries), a predictable pattern can be formulated, termed the volant bird dispersal syndrome. When ripe, bird-dispersed fruits are typically small and have an attractive, brightly coloured but inodorous edible part enclosing bitter or poisonous seeds that are protected by hard seed coats (berries) or endocarps (drupes). Hard outer rinds are absent and the fruits remain attached to the plant until they are harvested. Naturally, not all characters of the syndrome are necessarily expressed at the same time.

Meeting the sensory capabilities of their dispersers, the most obvious signal sent out by bird-dispersed fruits is colour. The majority of ornithochorous fruits are red or black, less frequently yellow, orange, blue, white or green; or they display a mixed pattern that combines several of these colours. Although it has often been assumed that red is the colour that birds can distinguish best against a background of green leaves, there is little evidence to support this theory. Moreover, the visual sensitivity of birds extends into ultraviolet

Ramphastos toco (Ramphastidae) – toco toucan; native to tropical South America – the toco toucan is the largest member of the toucan family. *Ramphastos toco* and other toucan species live primarily on fruits which is why they play such an important role as seed dispersers in the tropical parts of Central and South America

wavelengths (320–400nm), an area of the spectrum that is inaccessible to humans and other primates, who can only perceive wavelengths between 400 and 800nm. The UV-reflectance of black and purple berries and drupes has been shown to be an important signal for birds. The same phenomenon could explain the existence of white fruits such as the snowberry (*Symphoricarpos albus*, Caprifoliaceae) that are not particularly striking to the human eye. The wax layers of glaucous fruits such as sloe (*Prunus spinosa*, Rosaceae) and blueberries (*Vaccinium corymbosum, V. myrtillus*, Ericaceae) have also been shown to be highly UV-reflective.

Colour is clearly the most important signal of ornithochorous fruits, but birds are perfectly able to spot drab-coloured fruits (e.g. common hackberry, *Celtis occidentalis*, Ulmaceae). Colour preferences in birds might, therefore, be the result of learning rather than dictated by innate visual sensitivity. The fact that many fruits are red when fully ripe while others are the same colour although they are still immature and unpalatable (e.g. blackberries) indicates that birds have to learn how to choose certain fruits. Experiments have confirmed that the colour preferences of most species of birds are inconsistent, even between individuals of the same species, and likely to change with age. For example, the instinctive bias for red fruits in young blackcaps (*Sylvia atricapilla*, Sylvidae) disappears with experience and is absent in adult birds. In the face of volatile colour preferences among birds, fruit colour is more likely to have evolved as a long-distance signal to enhance detectability. If the latter is indeed the motivation for fruit choice among birds, then contrasts between fruits and their environment play a bigger role than their colour *per se*. This still leaves open the question why red is the dominant colour among ornithochorous fruits. The reason could be a complex mixture of influences, involving the birds' innate instincts and colour sensitivity, their ability to learn by association how to interpret certain colour combinations, and enhanced contrast against foliage

How to catch the eye of a bird

In temperate regions, colour displays can be flashy but most are plain (for example, uniformly red), whereas in the tropics, many bird-dispersed fruits advertize flamboyantly with contrasting combinations of red, purple, yellow, blue and black. The mature ovary can be assisted by other organs to achieve these colour contrasts. In *Ochna* (Ochnaceae), the fleshy receptacle provides a bright red background for the black drupelets of the schizocarpic fruit (*glandarium*). A persistent red calyx accentuates the berrylets of the schizocarpic fruit (baccarium) of *Tropaeolum speciosum* (Tropaeolaceae), the black drupes of *Heisteria cauliflora* (Olacaceae) and the blue drupes of several *Clerodendrum* species (e.g. *C. indicum, C. minahassae, C. trichotomum*, Verbenaceae). The black nuts of the tropical genus

Vaccinium corymbosum (Ericaceae) – American blueberry; native to eastern North America – below: fruits (berries); right: microscopic detail of fruit skin; although blueberries are primarily ornithochorous (bird dispersed), they do not seem to display the usual bright, flashy colours typical of the bird-dispersal syndrome. However, birds may take a different perspective. Their visual sensitivity extends into ultraviolet wavelengths (320-400nm), an area of the spectrum that is inaccessible to humans. Glaucous fruits like blueberries, sloes and plums are covered in a powdery layer of wax platelets ("wax crystalloids"), which have been shown to be highly UV-reflective. Therefore, the blueberry, which appears dark blue to us, may look bright red to a bird; diameter of breathing pore (stoma) 20µm

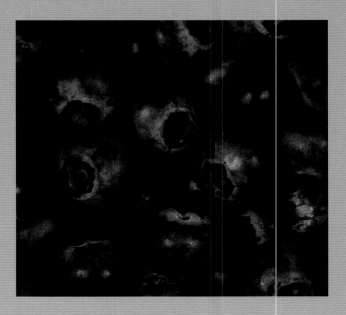

Hernandia (Hernandiaceae) are surrounded by a loose, brightly coloured, fleshy envelope formed by two or three bracteoles (modified leaves). The bracteoles can be joined, as in the Chinese lantern tree (*Hernandia nymphaeifolia*, Hernandiaceae), or free, as in the grease nut (*Hernandia bivalvis*). In the Japanese raisin tree (*Hovenia dulcis*, Rhamnaceae), the contorted swollen fruit stalks subtending the small brownish drupes turn dull red and become sweet and juicy after frost. The succulent stems are edible with a pear-like, astringent flavour. They not only enrich the visual display but also provide an edible reward for both animals and humans. By contrast, the thick, red fruit stalks that help advertize the poisonous fruits of the white baneberry (*Actaea pachypoda*) from North America are a warning signal rather than an invitation to a sweet snack. A member of the buttercup family (Ranunculaceae) with a monocarpellate gynoecium, each baneberry flower produces a single white berry. The black dots left by the stigma at the apex of the berries greatly enhance the conspicuousness of the infructescence and have earned the plant the alternative name "doll's eyes." Finally, the anthocarpous fruits of the pomegranate vine, *Palmeria scandens* (Monimiaceae), a subtropical liana from the temperate rainforests of Australia, should be mentioned for their rather drastic approach to attracting avian attention. Like an exploding rosehip, their fleshy hypanthium bursts open to reveal an eye-catching exhibition of several red drupelets against the pinkish-red background of the hypanthium wall.

These are just a few of many examples of how the angiosperms enhance the appearance of their fruits by including all kinds of accessory organs, proving once more their boundless inventiveness. But the visual feast does not end with fleshy drupes and berries. It continues with fruits that open to release their seeds.

Fleshy seeds

Dehiscent ornithochorous fruits – where the seed rather than the entire fruit is the diaspore – achieve similar contrasting effects by integrating the seeds and seed appendages into the display. Offering sarcotestal seeds (seeds with fleshy seed coats) is probably the most ancient strategy to trick vertebrate animals into becoming dispersal agents. The ancient gymnospermous cycads have adopted it as their sole dispersal strategy, attracting not only birds (for example, hornbills in Africa, emus in Australia) but also mammals (bats, rodents, monkeys), including Australian marsupials (possums, quokkas, quolls, brush wallabies, grey kangaroos).

Sarcotestal seeds have also evolved many times independently in the angiosperms. Familiar examples are magnolias (*Magnolia* spp., Magnoliaceae). Their primitive apocarpous gynoecia produce clusters of green, brown or red follicles. When ripe, the follicles split along their ventral side to reveal shiny, bright red seeds, offering their fleshy outer seed coat layers

below: *Hernandia bivalvis* (Hernandiaceae) – grease nut; native to Queensland – fruit (glans); when fully ripe, the fruit consists of a black oily "nut" (name!) surrounded by a loose, fleshy, brightly orange-red envelope formed by two or three bracteoles. The colours suggest bird dispersal; 2.5cm long

bottom: *Ochna natalitia* (Ochnaceae) – coast boxwood; native to southern Africa – fruit (glandarium); only joined at their base, the five carpels mature into separate drupelets, conspicuously presented on the enlarged red flower axis. The contrasting colour combination clearly indicates bird dispersal; diameter of fruit c. 2.5-3cm

below: *Clerodendrum trichotomum* (Verbenaceae) – harlequin glory bower; native to Japan – fruit (glans); in order to achieve a contrasting display that catches the attention of birds, the blue drupe is assisted by the enlarged bright red calyx; diameter 3.5cm

bottom: *Actaea pachypoda* (Ranunculaceae) – white baneberry, doll's eyes; native to eastern North America – fruits (berries); the white berries are advertised on red fruit stalks and marked by a black dot, giving the fruits the appearance of porcelain eyes, hence the name "doll's eyes." All parts of the plant are highly poisonous; fruit c. 1cm long

below: *Magnolia* sp. (Magnoliaceae) – photographed in Yunnan (China) – fruit (follicetum); to enhance their visibility to birds, the seeds dangle from silken threads, adding motion to the colourful display; seed c. 6-8mm long

bottom: *Palmeria scandens* (Monimiaceae), – pomegranate vine; native to eastern Australia – fruit (trymetum); the fleshy hypanthium bursts to expose several red drupelets which, in the tropical rainforests of Queensland, are dispersed by the spectacular regent bowerbird; diameter of fruit 3cm

mescal bean (*Sophora secundiflora*), a native of the south-western United States and Mexico, and the Panamanian *Ormosia cruenta*. Even more eagerly sought are the bi-coloured red and black seeds of the pantropical rosary pea (*Abrus precatorius*, also called crab's eye, jequirity bean or paternoster bean), the jumby bean (*Ormosia monosperma*) from South America and the Caribbean, and the American rosary snoutbean (*Rhynchosia precatoria*). Certain coral trees, such as the Malagasy *Erythrina madagascariensis*, also have red and black seeds.

Dangerous beauty

Despite their great beauty, the seeds of the crab's eye contain one of the strongest plant poisons known. The toxic principle, called abrin, is a lectin (glycoprotein) that attacks the ribosomes of eukaryotes, the factories of protein biosynthesis in cells. With an estimated lethal dose for humans between 0.1 and 1 microgram per kilogram of body weight, less than 0.003gm could kill a child. Fortunately, the seeds have a very hard seed coat and as long as they are not damaged, they are harmless, even if swallowed. However, the manufacture of botanical jewellery can be a dangerous occupation if holes are drilled through jequirity beans. The dust from the seeds can cause blindness on contact with the eyes, and inhalation or contact with open wounds has even worse consequences. Symptoms of poisoning appear hours or even days after contamination; they include nausea, vomiting, severe abdominal pain, diarrhoea and a burning sensation in the throat. Later, drowsiness and ulcer-like lesions of the endothelia in the mouth and oesophagus, convulsions and shock finally lead to coma and death. A simple way to detoxify the perilous seeds is to destroy the poison with heat. At temperatures above 65°C, abrin denaturates so that the seeds are edible once boiled.

Colourful appendages

As demonstrated by the ackee (*Blighia sapida*), offering edible arils (seed appendages) as attractants can be an efficient alternative to sarcotestal seeds. Arils can arise from localized growths of the seed coat, but mostly they are formed by the funicle, which becomes partly or entirely swollen and fleshy. In many small seeds, tiny pale arils have evolved as an adaptation to ant dispersal. Larger seeds such as those of the ackee can bear similar but larger appendages to attract birds. Playing to the animals' visual acuity, stark contrasts are once again of prime importance if avian dispersers are to be lured, and the arils have to play their part. One of the few north-temperate examples with spectacularly coloured fruits offering arillate seeds is the spindle tree (*Euonymus europaeus*, Celastraceae). The bright red loculicidal capsules have the shape of a bishop's mitre and open to expose three or four seeds wrapped in a deep orange aril. Once the pendulous fruits are fully open the seeds drop out and

page 209: *Iris foetidissima* (Iridaceae) – stinking iris; native to Europe and northern Africa – fruit (loculicidal capsule); offering fleshy (sarcotestal) seeds is an alternative dispersal strategy to berries, drupes and other fleshy fruits. The stinking iris displays its sarcotestal seeds on the spreading valves of its capsule, where they are clearly visible to birds thanks to the bright orange colour of their seed coat; diameter of seed c. 8mm

opposite: *Pararchidendron pruinosum* (Leguminosae) – snow wood; native to Malesia, New Guinea and eastern Australia – fruit (legume); the showy fruits of this monotypic genus open to reveal their seeds through torsion of the carpel halves. Although the colour scheme clearly indicates ornithochory, the fruits hold no edible reward for birds. Still a controversial concept, fruit mimicry, i.e. one fruit imitating the appearance of another one, may at least fool some young, inexperienced frugivorous birds into swallowing the hard seeds; fruit 8-12cm long

Harpullia pendula (Sapindaceae) – tulipwood; native to Australia (Queensland, New South Wales) – fruits (loculicidal capsule); despite its attractive appearance that suggests a fleshy, ornithochorous fruit, the tulipwood offers no edible reward for potential dispersers; diameter of seed c. 1.5cm

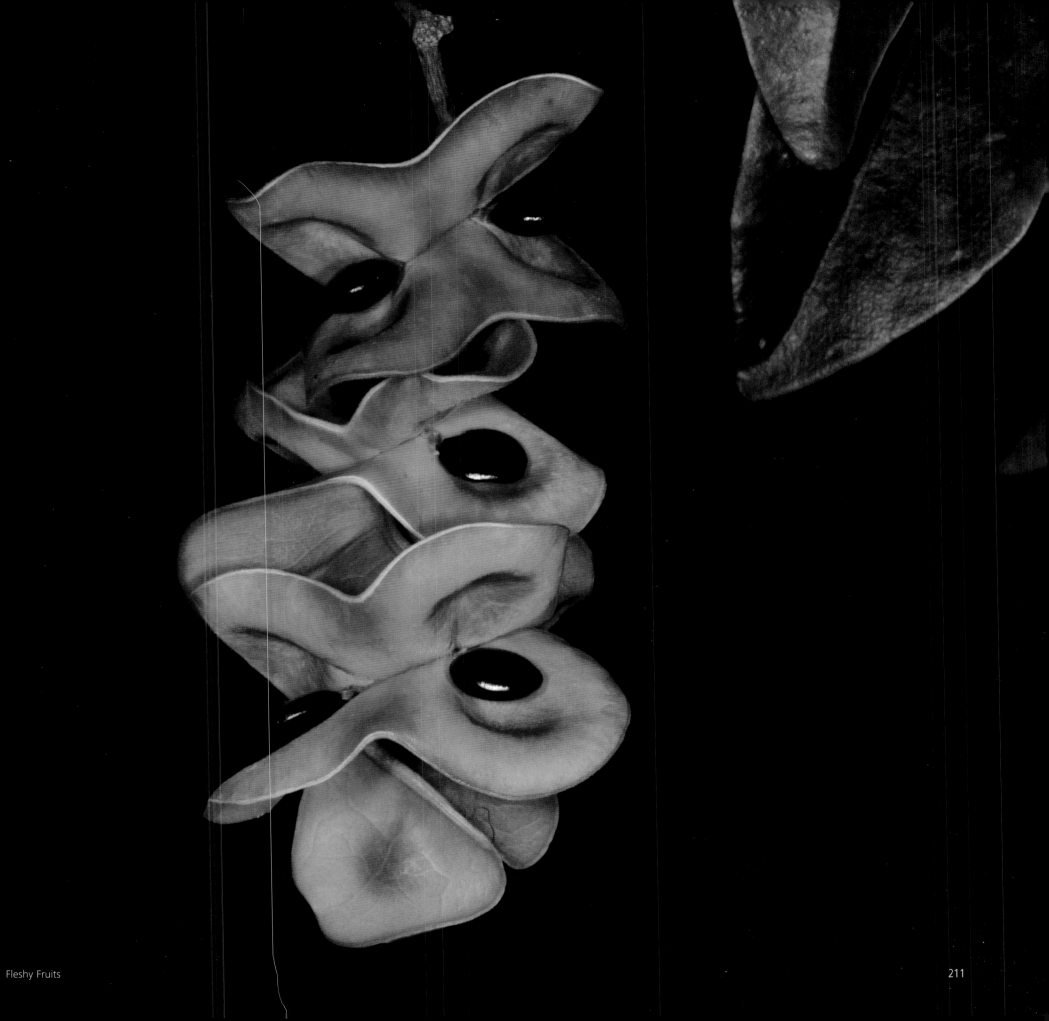

dangle from their short funicles to add movement to the display, a ruse also employed by magnolias and others. Although the fruits of the spindle tree impress observers in temperate Europe, as usual the tropics boast much larger and far more colourful specimens.

The flamboyant display of the West African ackee is a good example of a tropical advertising strategy that involves seed appendages as bird catchers. Its striking colour scheme – orange–red capsules, black shiny seeds with yellowish–white arils – is replicated by some of the ackee's relatives such as the guarana (*Paullinia cupana*), a vine from the Amazon rainforest. As the bright orange capsules split, one to three black seeds embedded in white arils emerge, looking curiously like human eyes. The seeds have gained fame for their very high caffeine content. Guarana is added to soft drinks and has become a fashionable stimulant in developed countries. Black seeds with white arils against red fruit walls would appear to be a successful dispersal concept as it has evolved independently in various other plant families, including the Annonaceae (negro pepper, *Xylopia aethiopica*, tropical Africa), Dilleniaceae (*Dillenia alata*, Malaysia to Australia; seeds almost entirely covered by the white aril) and Leguminosae (e.g. Manila tamarind, *Pithecellobium dulce*, Central America; *Swartzia auriculata*, South America).

Another typical variation of the bird–dispersal syndrome is the display of black seeds with orange or red appendages against the bright background of the fruit wall. This pattern is found in the loculicidal capsules of the bird-of-paradise flower (*Strelitzia reginae*, Strelitziaceae) from South Africa, whose seeds are adorned with a curious aril resembling a shaggy orange wig. Although involving less eccentric arils, the same syndrome is found in bird-dispersed legumes, for example in the African mahogany (*Afzelia africana*) and certain Australian wattles (*Acacia* spp.). Whilst the funicular aril of *Afzelia africana* is a substantial lump of tissue that partly covers the seed from the hilar end, the arils of the coast wattle (*Acacia cyclops*), the blackwood (*A. melanoxylon*) and the dead finish (*A. tetragonophylla*) consist of a very long, fleshy funiculus, folded in two and wrapped around the periphery of the seed. Other Australian acacias, such as the earpod wattle (*Acacia auriculiformis*) and the hickory wattle (*A. mangium*) have similar extended edible funicles but they unfold when the legumes open so that their seeds dangle in the air like those of the spindle tree and magnolias.

Some tropical and subtropical plants keep a lower public profile and camouflage their colourful arillate seeds in inconspicuous, dull-coloured capsules. The signal indicating readiness for dispersal is sent when the black seeds that contrast with their colourful arils are suddenly exposed. The greenish-brown fruits of the titoki tree (*Alectryon excelsus*) from New Zealand, another member of the soapberry family (Sapindaceae), burst open to expose a single black seed half-sunken in an intensely red, fleshy aril. The eagerness with which

opposite: *Acacia cyclops* (Fabaceae) – coastal wattle; native to south-western Australia – seed surrounded by a brightly orange-coloured aril formed by a double layer of the funiculus, which encircles the seed first in one direction and then folds back to surround it once more in the opposite direction. There are several species of *Acacia* with bird-dispersed, arillate seeds but the aril of the coastal wattle has by far the highest fat content. The fruits (legumes) are not shed but remain on the plant, displaying the seeds in a similar way to the snow wood (*Pararchidendron pruinosum*). In its native Western Australia, the energy-rich arils are an important food for birds. In South Africa, where *Acacia cyclops* has become an invasive weed, the nutritious funicles are also eaten by rats and baboons; seed 9mm long (including aril)

Carmichaelia aligera (Leguminosae) – North Island broom; native to New Zealand – fruit (type yet to be defined); carpologists have yet to coin a technical term for the curious looking fruits of the genus *Carmichaelia*. As in a craspedium, the carpel halves drop and leave a black frame. But unlike in a craspedium, the red seeds remain attached to the black frame. The display of one or two hard but brightly coloured seeds in such a conspicuous way strongly suggests ornithochory. However, the fact that the fruit holds no edible reward raises the suspicion of deceit; fruit c. 1cm long

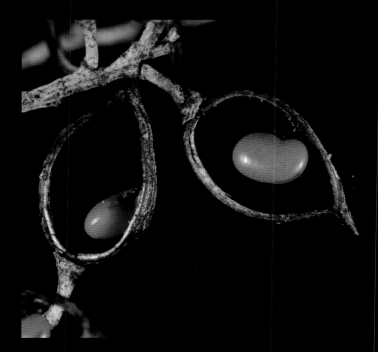

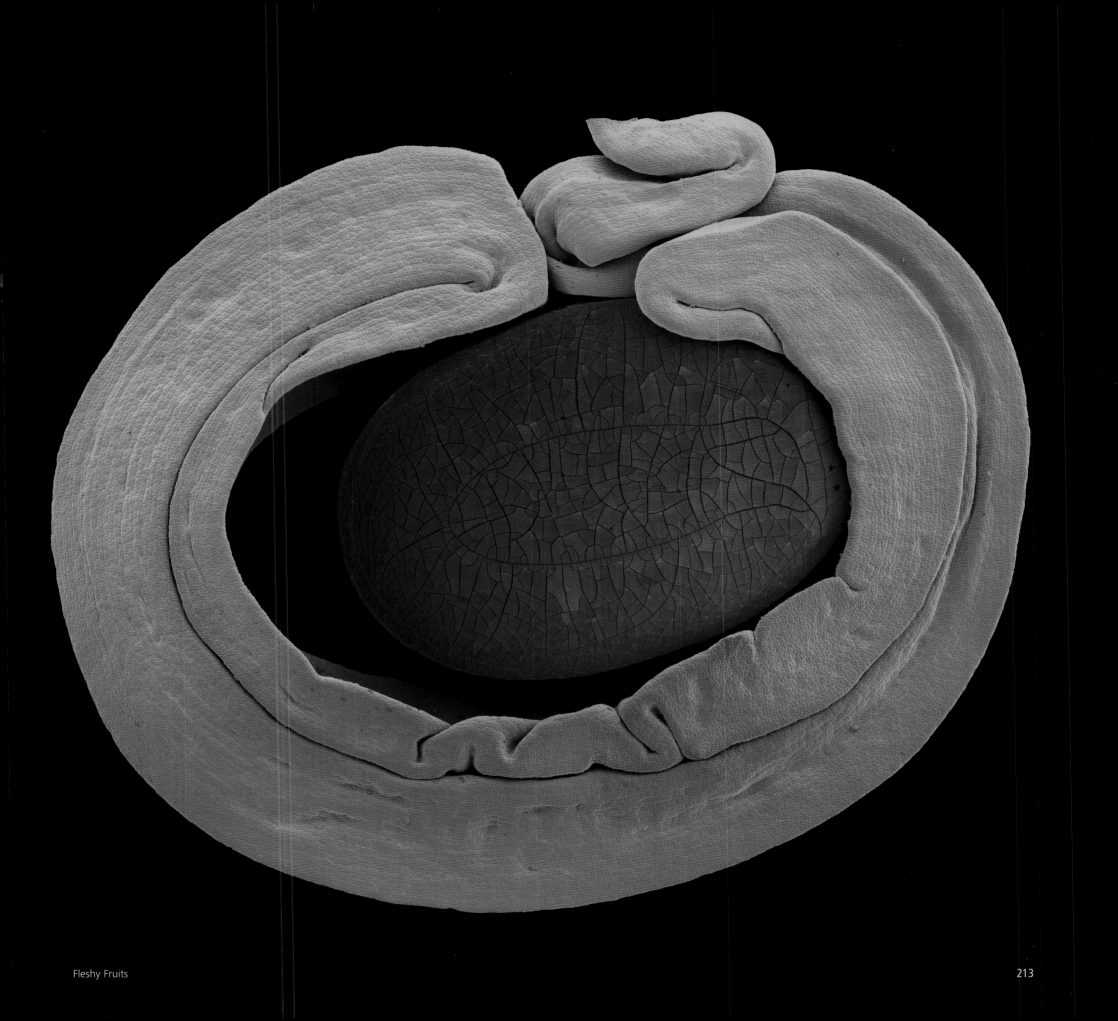

...rds harvest the seeds proves that the red arils of the ackee are at least as popular as the pale ones of its cousins the ackee and the guarana. In fact, red arils are far more common than pale ones in ornithochorous fruits. One of them even changed the course of history.

Arillate seeds and the fate of New York

The most precious and desirable red aril in all of human history is borne inside a rather humble fruit. When the thick-walled coccum of the nutmeg (*Myristica fragrans*, Myristicaceae) – initially green and later pale yellow to light brown – is split down the middle, a single large seed with a spectacular, lacy, crimson-red aril is revealed. Both the seed and the aril, known as nutmeg and mace, have been one of the most precious commodities of the spice trade for many hundreds of years. Originally, *Myristica fragrans* was indigenous to the Banda Islands, a tiny group of islands that are part of the famous Spice Islands or Moluccas in the Indonesian archipelago. The precise origin of the nutmeg had been a well-kept secret for centuries, until, in 1512, the Portuguese became the first Europeans to set foot on the Banda Islands. In Europe at the time, nutmeg was considered a panacea for all kinds of ailments, including plague and impotence. Even though the Portuguese purchased nutmeg and other spices from the native people at inflated prices, they still made enormous profits when they sold them in Europe, where a nutmeg was worth its weight in gold. Naturally, the lucrative spice trade also came to the attention of other European sea powers, especially the British and the Dutch. By the seventeenth century the Dutch controlled all but one of the Banda Islands. Somehow the British managed to establish a presence on Run, the westernmost Island of the Bandas, much to the displeasure of the Dutch. After several clashes, the British and the Dutch settled their disputes with the Treaty of Breda on 31 July, 1667. Historically, the most interesting part of the deal was that the British returned Run Island to the Dutch in exchange for Manhattan Island, then just a small Dutch trading post in the New World. With the exchange of territories the Dutch gained a monopoly over the nutmeg trade, which they defended ruthlessly. Every nutmeg leaving the Bandas had to be sterilized with lime so that no one could set up a rival plantation elsewhere. Any attempt to smuggle viable nutmeg seeds was punishable by death. The Dutch monopoly ended in the early nineteenth century when the British temporarily regained control and used the opportunity to establish nutmeg plantations in some of their colonies, including the island of Grenada in the eastern Caribbean. Today, Grenada is the second-largest producer of nutmeg after Indonesia.

Nutmeg and mace have a similar, highly aromatic taste but the aroma of the aril is considered more refined and delicate. Nutmeg was once believed to possess magical powers and was used as a medicine, aphrodisiac and hallucinogen. In fact, the seed contains

below: *Euonymus europaeus* (Celastraceae) – spindle tree; native from Europe to western Asia – fruits (loculicidal capsules); the bright red capsules open to expose three or four seeds wrapped in an orange aril. The bird-dispersed fruits are poisonous to humans; diameter c. 1-1.5cm

bottom: *Myristica fragrans* (Myristicaceae) – nutmeg; native to the Moluccas – fruit (coccum) with arillate seed; borne within a single dehiscent carpel, the arillate seed is laced with a deep crimson red, fleshy aril. The seeds are dispersed by birds, such as imperial pigeons (*Ducula* spp.) and hornbills (family Bucerotidae); seed c. 3cm long

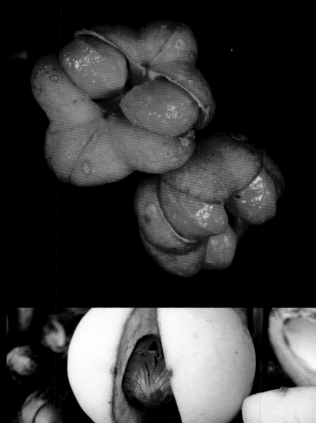

Alectryon excelsus (Sapindaceae) – titoki tree; native to New Zealand – fruits (camaras); although the gynoecium of titoki flowers consists of three or four joined carpels, all but one are aborted. Only one carpel remains fertile and develops into a single-seeded, 8-12mm long fruit (camara) that dehisces with an irregular split from which a black seed wrapped in a scarlet, fleshy aril emerges. The speed with which birds discover and collect the seeds proves the success of the titoki's ornithochorous strategy. The fruit also made an impression on Joseph Gaertner who was reminded of a cockscomb when he described the genus in 1788. In Greek mythology, Alectryon was a boy, instructed by Ares, the god of war, to stand guard by his door while he indulged in illegitimate love with Aphrodite, the goddess of love, lust and beauty. However, Alectryon fell asleep and Ares was discovered by Helios, the god of the sun. To punish the boy, Ares turned him into a cockerel, which since that day has never forgotten to announce the rising sun

myristicin, a phyenylpropanoid with hallucinogenic effects. Between half a seed and two seeds are needed for a trip. Slaves on board spice-laden vessels are said to have helped themselves to the drug to soothe their suffering and evoke a pleasant, euphoric feeling. However, overindulgence in nutmeg has some very unpleasant side effects including palpitations, blurred vision and extreme nausea. Severe poisoning can even lead to coma and death. Today, nutmeg and mace are mainly used as spices by the food industry and in domestic cooking. Harmless in moderate amounts, ground nutmeg adds a delicate flavour to such mundane dishes as spinach, mashed potatoes, soups, and German Bratwurst.

As the contrasting colour scheme of the nutmeg fruit leads us to predict, its natural dispersers are birds. In Indonesia, pigeons from the genus *Ducula* and hornbills (family Bucerotidae), both able to gulp down large seeds, are probably the most important natural dispersers. In South America, guans (*Penelope* spp.), trogons (*Trogon* spp.) and toucans (*Ramphastos* spp.) disperse the very similar seeds of *Virola*, a myristicaceous genus closely related to *Myristica fragrans*.

Dispersal by mammals

Birds are by far the most important vertebrate seed dispersers worldwide. The wonderful diversity of colourful ornithochorous fruits reflects the pivotal role they play in the life of many plants. Fruit-eating mammals have also entered into distinct co-adaptive relationships with fruits, particularly in the tropics. Visitors to tropical countries, from the early explorers to today's holidaymakers, have always been fascinated by the diversity of tropical fruits, which look and smell so very different from anything growing in the temperate northern hemisphere. People are impressed by their large size, curious appearance and fruity aroma (which can be disagreeable to European noses), but most are not aware that the exotic appearance of many tropical fruits is part of a plot to bribe equally exotic mammals into devouring them. Being mammals themselves, humans, too, fall for the seductive ploys of many of these fruits, which include avocados, bananas, custard apples, dates, figs, guavas, jakfruits, lychees, mangoes, mangosteen, melons, papayas, passionfruits, pineapples, rambutans and soursops.

The general syndrome of diaspores adapted to mammal-dispersal (*mammaliochory*) is similar to the bird-dispersal syndrome. However, the different life-styles and sensory physiology of mammals have influenced the evolution of co-adapted fleshy fruits in different ways. Leaving aside larger flightless birds such as ostriches, emus, cassowaries and other ratites, on average mammals have a larger body mass than birds. Most mammals are colour-blind night-feeders with a strong sense of smell, and lead a terrestrial rather than an arboreal life.

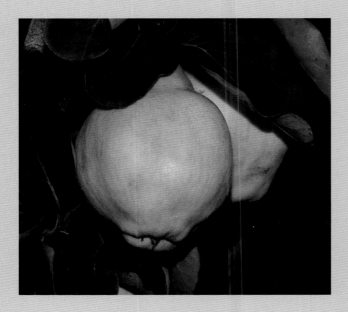

Their claws and teeth allow them to manipulate and masticate their food. The co-adaptive response of mammaliochorous fruits includes large size, the possession of a thick skin or tough rind impregnated with repulsive chemicals (e.g. essential oils in citrus peel), and strong physical or chemical protection of the seeds to discourage their destruction by mastication. Colour, the most powerful signal of bird-dispersed fruits, is lost upon colour-blind creatures and replaced by a strong olfactorial signal tuned to charm their sensitive noses. Therefore, mammal-dispersed fruits tend to have drab colours ranging from various shades of green, brown and pale yellow to dull orange, and emit a heavy, sweet, often musty, sour or rancid odour. Those fruits aiming at collection by ground-dwelling dispersers tend to be dropped as soon as they are ripe to provide easy access. Temperate examples of the latter grow in our gardens, among them our all-time favourite, the apple (*Malus pumila*) and its cousin the sweet-smelling quince (*Cydonia oblona*). Both originated in western Asia. Despite their long history of human domestication they still display the dull green or yellow coloration and smell of a typical mammal fruit (except for some red apple cultivars). Like their wild ancestors, their fruit ripens to coincide with the time when consumers of larger fruits, such as bears, are eating as much as they can to prepare for their long winter hibernation.

Despite the fact that many mammaliochorous fruits can easily be identified as such, the full suite of characters defining the ideal mammal-dispersal syndrome is not always expressed. As usual, genetic inheritance sets limits, and diffuse selection pressures push for trade-offs to ensure a balance between attracting legitimate dispersers and repelling unwanted seed predators and pulp thieves. The fact that many fruits are eaten by birds, bats, monkeys and other mammals proves the impossibility of drawing clear borderlines between the different dispersal syndromes. This problem is even more evident in the impoverished faunas of the temperate regions, where many predominantly carnivorous animals include a substantial amount of fruit in their diet. In fact, bears, racoons, ringtails, weasels, ferrets, martens, otters, badgers, dogs, wolves and foxes are among the most important mammalian dispersers in the temperate northern hemisphere.

In spite of the difficulty in quantifying and thereby scientifically proving the adaptational nature of many traits linked to distinct animal dispersal syndromes, ecologists have ventured to propose even finer distinctions within the mammal-dispersal syndrome.

The bat dispersal syndrome

Along with birds and monkeys, fruit bats are the most important seed-dispersing animals in tropical rainforests. Like birds, their ability to fly renders them excellent candidates for highly effective seed dispersal. When feeding, fruit bats do not necessarily devour the entire fruit. Any harvested fruit is usually taken to a roost or another safe place nearby before being eaten. Most of the time, they suck out the juice from the pulp and discard the remnants, including the seeds. Consequently, the average dispersal distance is only a few hundred metres, despite the fact that some species of flying foxes are known to visit feeding grounds up to 40km away from their roosting sites. Only very small seeds such as those of figs, which are swallowed with the pulp, have any chance of being transported for many kilometres before being deposited with the faeces.

Within the zoological order Chiroptera (bats), frugivory has evolved independently among Old and New World bats. The Old World fruit bats all belong to the Pteropodidae, the only family in the superorder Macrochiroptera (= large bats), so named because it boasts the largest bats in the world. Although the smallest members of the family measure only 6–7cm from head to tail, flying foxes (*Pteropus* spp.) can be as long as 40cm with a wingspan of 1.7m. The Pteropodidae are widely distributed throughout the tropical and subtropical regions of Africa, Asia and Australia, and include more than 160 species. Their frugivorous counterparts in the New World are generally smaller and belong to the family Phyllostomatidae (American leaf-nosed bats) in the order Microchiroptera (= small bats). Contrary to Old World fruit bats, which have relatively simple ears, New World fruit bats use sophisticated echolocation for their navigation. With just one exception, the Pteropodidae lack the ability for echolocation and rely on vision to avoid obstacles, and their sense of smell to locate fruit. The two groups also differ slightly in their dietary preferences. Whereas Old World fruit bats live entirely on nectar and fruits, their New World cousins are less strongly co-adapted to fruits and get a large amount of their protein from insects. Despite these differences, the postulated bat dispersal syndrome applies to both groups of frugivorous bats.

Fruits adapted to bat dispersal (*chiropterochory*) have many traits of the general mammal-dispersal syndrome. A stale, sour, even rancid odour reminiscent of fermenting fruit is specifically bat-related; the hint of butyric acid in the fruits of the Mediterranean carob tree (*Ceratonia siliqua*, Leguminosae) is a good example. The fondness of bats for this kind of odour is probably inspired by the animals' own smell. This is not meant to imply that all bat-dispersed fruits stink. There are plenty of fragrant examples, including the Malabar plum or rose apple (*Syzygium jambos*, Myrtaceae). This popular fruit tree from south-east Asia is planted as an ornamental throughout the tropics, where it sometimes becomes invasive. Its small, pale yellow, pear-shaped fruits have a pleasant taste reminiscent of water melon, and a fine aroma of rose water, hence its Sanskrit name "jambu" (rose-apple tree). Other fruits popular with bats with a pleasant though faint perfume are figs (*Ficus carica*, Moraceae), dates (*Phoenix dactylifera*, Arecaceae) and cashew apples (*Anacardium occidentale*, Anacardiaceae).

Pteropus ualanus (Pteropodidae) – flying fox, feeding on the fruit of a screwpine (*Pandanus* sp., Pandanaceae); photographed on Kosrae Island, Micronesia – members of the genus *Pteropus*, commonly known as flying foxes or fruit bats, are the largest bats in the world. After birds, fruit bats are the most important consumers and dispersers of fruits in the tropics

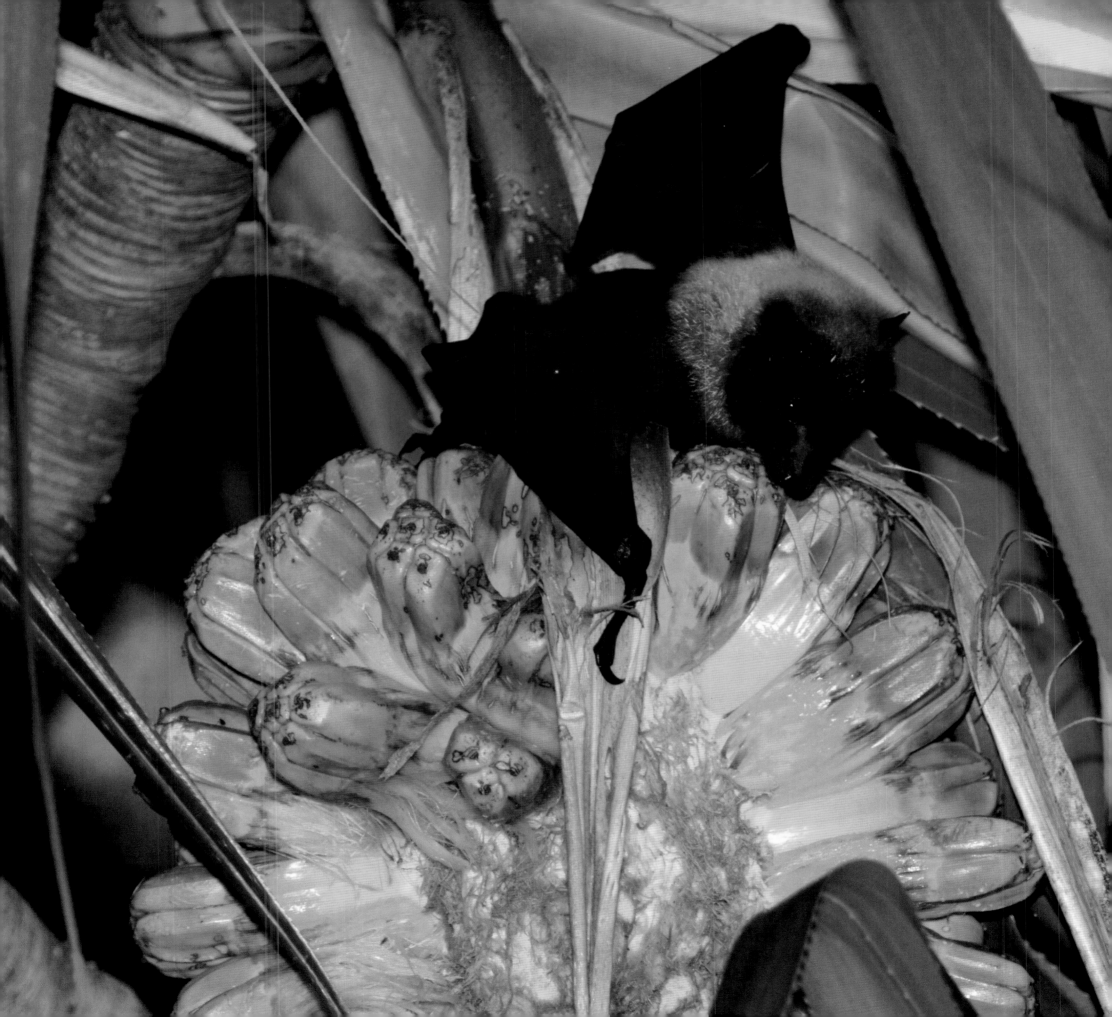

To increase their chances of being spotted from the air by the colour-blind nocturnal creatures, bat fruits tend to be permanently attached to the mother plant and displayed in an exposed position outside the dense foliage of the tree canopy. Figs borne directly on the trunk and major branches rather than on young leafy shoots, and mangoes dangling from long stalks are considered specific bat-related adaptations that help increase visibility. Although variable in size and texture, bat-dispersed fruits are often large, weakly protected physically, and contain big seeds. Whereas ornithochorous fruits tend to be oily, chiropterochorous fruits win the affection of bats by being sweet, soft and watery.

Apart from the sometimes unsavoury smell, the syndrome ascribed to bat fruits is also mouthwateringly attractive to humans. In fact, we owe the existence of many of our exotic favourites to the co-adaptive influence exerted by fruit bats over millions of years. Among these chiropterochorous delicacies are bananas (*Musa* spp., Musaceae), rose apples (*Syzygium jambos*, Myrtaceae), custard apples (*Annona reticulata*, *A. squamosa*, Annonaceae), jakfruit (*Artocarpus heterophyllus*, Moraceae), guavas (*Psidium gajava*, Myrtaceae), and passionfruits such as the purple granadilla (*Passiflora edulis*, Passifloraceae), sweet granadilla (*Passiflora ligularis*) and the maracuyá (*Passiflora quadrangularis*). Being such tasty, nutritious and easily digestible morsels, it is not surprising that bats have to share their fondness for many of these fruits with other animals, including birds and primates, as well as humans.

Monkey fruits – the primate-dispersal syndrome

Next to fruit bats, primates are the most important seed-dispersing mammals in tropical rainforests, even though their foraging behaviour is mostly destructive. Monkeys and apes devour everything they can identify as edible, including leaves, flowers, ripe and unripe fruits, and seeds, as well as insects, eggs, and even meat. Besides sharing a similar dietary preference with humans, they also possess human-like colour vision, a less strongly developed sense of smell, and hands with opposable thumbs that afford them great dexterity. Despite the fact that primates eat all kinds of soft- and hard-skinned fruits that are primarily ornithochorous or chiropterochorous, a separate primate-dispersal syndrome has been identified. This includes the typical dull coloration and potentially unpleasant, heavy smell of mammaliochorous fruits, in combination with a hard outer husk that requires strength and skilful manipulation before the edible parts can be accessed. Examples of such armoured fruits eaten by primates include the monkey apple (*Strychnos spinosa*, Loganiaceae), mangosteen (*Garcinia mangostana*, Clusiaceae), cacao pods (*Theobroma cacao*, Malvaceae), baobab (*Adansonia digitata*, Malvaceae), and the famous (or rather infamous) durian (*Durio zibethinus*, Malvaceae).

Saimiri oerstedii citrinellus (Cebidae) – Costa Rican squirrel monkey; endemic to Costa Rica where the species is critically endangered. Squirrel monkeys and other primates eat primarily fruits and insects but generally devour anything they can identify as edible, including flowers, leaves, buds and seeds, as well as eggs and even meat. Because of their predominantly "fruity" diet, apes and monkeys, alongside birds and fruit bats, are the most important seed-dispersers in tropical rainforests. Although a specific primate-dispersal syndrome has been identified, their intelligence and dexterity allow them to distinguish and remove indigestible parts of all kinds of fruits, even if they are primarily bird or bat dispersed

Monkey apple

The monkey apple (also called kaffir orange or Natal orange) is native to tropical and subtropical Africa, where its leaves, roots and fruits are used medicinally. Numerous, physically weakly protected but poisonous, large, flat seeds embedded in an edible soft, juicy pulp occur inside the smooth, green to yellow, hard-shelled fruit (an amphisarcum). The fruits are the size of a small orange, and are eaten by monkeys and baboons as well as bush pigs and eland antelopes. Local people in Africa turn the emptied woody shells of the monkey apple into artefacts including decorative balls and musical instruments such as the marimba.

The Queen of Fruits

With its soft but very thick, dark purple, bitter-tasting outer husk with 5-10 flat seeds embedded in a juicy, snow-white to pinkish pulp inside, the mangosteen presents the typical features of the primate-dispersal syndrome. What appears to be an aril or a sarcotesta around the seeds is, in fact, the inner fruit wall (endocarp), which has wrapped itself around each seed and turned pulpy. Even though developmentally not part of the seeds, the pulp is tightly attached to them and difficult to remove, a strategy that aims at manipulating potential pulp thieves into swallowing the seeds. Although not related, the mango (*Mangifera indica*, Anacardiaceae) applies the same cunning ruse by mixing the long hairs that cover the surface of its single-seeded stones into the pulp.

As predicted by the mangosteen's appearance, monkeys love its fruits, as do humans. Praised for centuries as the Queen of Fruits, the mangosteen is one of the most delicious tropical fruits. It is enjoyed by people in its native south-east Asia and other tropical countries where it is cultivated. And now Europeans have learned to value it too, especially as it lacks the typical musty smell of many other tropical fruits. Unfortunately, the shelf-life of the tennis-ball-sized fruits is brief, only two or three days when fully ripe. If they are to reach Western supermarkets they have to be harvested while still unripe, which is good for business but bad for their taste. Whoever is fortunate enough to enjoy naturally ripened mangosteens straight from the tree will be delighted by the deliciously fruity smell and exquisite taste – described variously as a blend of pineapple and peaches or reminiscent of strawberries mixed with oranges. Queen Victoria heard about the legendary fruit and allegedly once offered a large reward to anyone who would bring her a mangosteen to taste.

Cacao – food of the gods

Monkeys are the main dispersers of the indehiscent pods (amphisarca) of the chocolate tree (*Theobroma cacao*, Malvaceae) in the Amazonian rainforest. The rich yellow- to orange-

coloured fruits are protected by the thick, tough rind that is typical of primate–dispersed fruits. Like the mangosteen, the inner edible part consists of a sweet, white, juicy pulp that surrounds the large seeds. However, unlike in *Garcinia*, the nutritious reward represents a true sarcotesta. Although monkeys avoid swallowing the soft, bitter–tasting seeds that rely on chemical rather than physical protection, they still disperse them. They remove the cohesive lump of up to fifty fleshy seeds from the fruit, and then take their bounty to a safe place in the canopy where they nibble off the luscious sarcotesta. The fact that cacao trees bear their fruits on the trunk and larger branches can be interpreted as an indicator of chiropterochory. However, although bats attack the fruits they are unable to break through the husk. A possible explanation for the caulicarpous position of the cacao pods could be their large size and weight.

Theobroma cacao has its origin in the Amazon rainforest, but for millennia native peoples have cultivated and distributed the valuable tree throughout Central and South America. Its fruits are the source of chocolate, which was already an important commodity in pre-Columbian times. The emperor of the Aztecs, Montezuma II, is said to have taken no other drink than hot, frothy chocolate flavoured with vanilla and spices. The Spaniards shared the Aztecs' fondness for chocolate and introduced it to Europe, where it quickly became a much-loved treat. Chocolate has lost nothing of its appeal today. Our admiration for this great delicacy is probably best reflected in the plant's Latin name chosen by the famous Carl von Linné himself. He called it *Theobroma*, which means "food of the gods."

Cocoa, the main ingredient of chocolate, is obtained from the seeds of the cacao tree. The so-called cacao beans are first fermented, then crushed and ground into a powder from which the excess fat, the cocoa butter, is extracted. The resulting dry cocoa powder forms the basis of chocolate drinks. It is also the main ingredient of solid chocolate, which is prepared by mixing cocoa powder with additional cocoa butter, sugar and flavouring. Cocoa butter, an expensive vegetable fat, is also an ingredient in sweets, soap, cosmetics and ointments. Because of its low melting point it is also used as a base for suppositories.

In addition to healthy antioxidants, several psychoactive chemicals have been identified in cocoa, which may well account for the legendary exhilarating powers of chocolate. However, recent research suggests that the concentrations of these chemicals are too low to have a significant effect. The most likely reason that chocolate makes us happy is simply the unique combination of taste, texture and aroma, which stimulates the release of endorphins in the brain, in the same way that other sweet foods do. Endorphins are the body's own pleasure hormones: some alleviate pain while others evoke feelings of happiness, well-being and ecstasy.

The baobab

With their huge swollen trunks that store water to get them through dry periods, baobabs (*Adansonia* spp., Malvaceae) are the most iconic trees in the arid landscapes of Africa, Madagascar and Australia. Only eight species of baobabs are known. They all produce large, velvety-brown, indehiscent woody pods (amphisarca) filled with numerous seeds embedded in plenty of white, mealy pulp. The tasty pulp is nutritious and rich in vitamin C. It is either eaten as a sweet treat or used to prepare a refreshing drink evocative of the sour-fruity flavour of dried apples.

The most famous species of all is the African baobab (*Adansonia digitata*). Mature trees boast enormous trunks that to early European travellers suggested great age and historic significance. However, they consist mostly of soft, water-storing tissue and are much younger than an English oak of comparable size. Its grotesque outline has not only gained the African baobab great cultural importance among the native people of that continent but also the name "upside-down tree." In their natural environment, many smaller animals indulge in the soft inner part of the fruits once the hard shell has been broken. Intact ripe fruits are eaten predominantly by the baobab's legitimate dispersers, which include monkeys, baboons and allegedly elephants, eland and impalas.

Durian – the King of Fruits

The south-east Asian genus *Durio* comprises 28 species, eight of which are edible. The economically most important species of durian (Malay for "thorny fruit"), *Durio zibethinus*, has been cultivated in south-east Asia for centuries. Its highly prized fruits can be the size of a football and weigh up to three kilograms. On the outside they are protected by a ferociously spiny, dull green to yellowish husk. When ripe, the fruits drop off the tree and split slightly from the apex down, along distinct preformed lines, revealing themselves morphologically as loculicidal capsules. At this stage the fruits also emit their infamously foul smell, resembling a blend of faeces, unwashed socks and rotten garlic. Understandably disliked by most uninitiated Europeans and banned from the Underground in Singapore, the durian is savoured by people all over Asia, who call it the King of Fruits. This apparently dubious admiration does not reflect a twisted sense of smell but fondness for the extraordinarily delicious taste of what lies inside the malodorous capsule. The edible part of the durian consists of several large, chestnut-brown seeds wrapped in a white or cream to golden-yellow funicular aril. As long as the fruits are immature, the arils are hard and unpalatable; but by the time they are shed from the trees, the hard funicular tissue has turned into a custard-like cream with the consistency and flavour of a tantalising mix of nuts, spices, bananas, vanilla

Durio zibethinus (Malvaceae) – durian, cultivated variety; native to south-east Asia – fruit (loculicidal capsule); below: entire fruit; opposite: fruit opened to reveal the large seeds wrapped in a delicious creamy aril. Typical of mammal-dispersed fruits, the durian drops off the tree when ripe and begins to emit a heavy, musty odour. Carpologically, the durian is between an indehiscent pod (amphisarcum) and a loculicidal capsule. It opens only after shedding with just a few very short, narrow splits at the apex, which offer animals a toehold to break the fiercely spiny capsule further apart into its five valves. In the wild, durians attract a range of large mammals, including orang-utans, bears, panthers, tapirs, rhinos and elephants. Despite its smell, which most western people find revolting, in its native south-east Asia, the durian is considered the "King of Fruits." In a fully ripe durian, the initially rubbery, hard arils have turned into a custard-like cream with an indescribably exquisite flavour, reminiscent of a mixture of custard, nuts, spices, bananas and onions; c. 25cm long

and – bizarrely – onions. The famous nineteenth-century naturalist Alfred Russel Wallace once wrote of the durian, "*Its consistence and flavour are indescribable. A rich butter-like custard highly flavoured with almonds gives the best general idea of it, but intermingled with it come wafts of flavour that call to mind cream-cheese, onion sauce, brown sherry and other incongruities.*"

The durian is a prime example of a fruit adapted to mammal-dispersal. Its weight and armour prevent smaller animals that are unable to swallow the seeds from becoming pulp thieves or seed predators. In its natural habitat only the largest, most charismatic animals of south-east Asia have the skill and strength to pull the capsules apart into their seed-bearing valves. If an orang-utan catches a whiff of the far-reaching odour, it will travel a long way through the rainforest to be the first to reach the treasured delicacy. But the orang-utan has to be quick. The tasty arils are equally irresistible to bears, panthers and tapirs, as well as Asian rhinos and elephants.

Although they are edible after boiling or roasting, the large seeds of the durian rely on poison rather than physical protection. Therefore, animals either discard the seeds (e.g. orang-utans) or swallow and defecate them intact.

Perhaps coincidentally, giants such as rhinos and elephants share a taste for certain fruits with primates and other animals. However, there are fruits with traits that strongly suggest co-adaptation specifically to large-bodied mammals.

Big fruits need big mouths – the megafaunal dispersal syndrome

Overall, fruits adapted to dispersal by large mammals conform to the mammal-dispersal syndrome. However, they do display certain specializations that reflect their preference for large-bodied dispersers. In 1982, Daniel Janzen and Paul Martin summarized these particular traits as the megafaunal dispersal syndrome. Megafauna was defined to include all animals with a body weight of more than 100 pounds (45kg). The most obvious indicator of the megafaunal dispersal syndrome is large fruit size. Fleshy fruits have evolved to be eaten without seed loss by creatures that are able to fit them into their mouths whole rather than in pieces. Therefore, indehiscent, fleshy fruits that seem too large for dispersal by small mammals are likely candidates for the megafaunal dispersal syndrome. Their seeds are sometimes large, as in avocados (*Persea americana*, Lauraceae) and mangoes (*Mangifera indica*, Anacardiaceae), or small, as in papayas (*Carica papaya*, Caricaceae). In any case they are either physically protected against grinding molars by a thick, hard endocarp or seed coat (e.g. custard apples and their relatives, *Annona* spp, Annonaceae), or they are chemically protected by sharp- or bitter-tasting toxins that discourage mastication (avocado, papaya). The former is usually the case in hard-shelled fruits that require lengthy chewing (e.g. tamarinds,

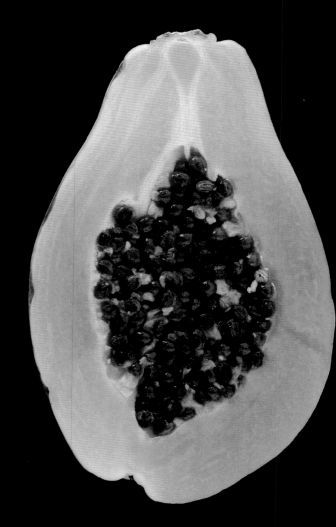

Carica papaya (Caricaceae) – papaya, cultivated variety; native to tropical America – fruit (pepo); below: longitudinal section of seed; opposite: seed (sarcotesta removed); large size, a leathery skin and a central cavity with many small seeds whose sharp taste discourages mastication, are indicators of the megafaunal dispersal syndrome. The seed coat is differentiated into a lubricating gelatinous sarcotesta, making the seeds easier to swallow, and a hard inner layer with sharp ridges and spines that further discourage mastication; sectioned fruit 12cm long, seed 6mm long

Tamarindus indica, Leguminosae), whereas the latter is typical of soft fruits that are easily squashed by pressing the tongue against the palate (e.g. papayas). Often, the germination of seeds of plants adapted to attract megafaunal dispersers is greatly enhanced after they have passed through the gut of an animal. When catering for large, terrestrial dispersers, the best way for trees to present their fruits is to drop them when they are ripe, or even shortly before. In habitats where ground-dwelling predators such as rodents pose a substantial threat, small megafauna-dispersed trees often hang on to their fruits for several months.

Africa's large mammals and their fruits

The biggest terrestrial animals alive today are the elephant, rhinoceros, and hippopotamus. It is therefore no coincidence that the megafaunal dispersal syndrome is best expressed in Africa and Asia where these animals still occur. Africa is especially rich in large herbivores who also include fruits in their diet. In fact, ruminants such as giraffes and antelopes, and non-ruminants such as elephants and rhinos are among the most important seed dispersers in the African savannah.

The fruits of many legumes, most notably acacias (*Acacia* spp., a huge genus soon to be split into five separate genera), are specifically adapted to attract these animals. Usually at times when little grass is available, the trees produce large, indehiscent pods (camaras) weighing more than 50g, with a brown, leathery husk and a distinct smell that attracts even cattle. Since herbivores are colour-blind, the brown fruits are visually inconspicuous but rich in digestible carbohydrates and protein; they contain extremely hard, smooth seeds that can withstand the grinding of strong molars. The fruits can remain on the tree but are often dropped to the ground as soon as they are ripe to provide easy access for their large terrestrial dispersers. By eating and digesting the indehiscent fruits, the animals not only free the seeds from their envelope, they also kill any seed-eating insects (e.g. weevils, bruchids) that may have attacked the fruits prior to dispersal. Some legumes dispersed by elephants are known to be dependent on the biological pest control service provided by the beast's gut. Without it, most seeds would fall prey to insect infestation. Seeds that have passed through the gut of an elephant also show much improved germination, a phenomenon that can be observed in many other endozoochorous fruits whether dispersed by birds or mammals. Digestive acids and enzymes remove perishable pulp, which would otherwise encourage fungal and bacterial infections and weaken the hard seed coats, making it easier for the embryo to protrude.

In their native habitat, the African forest elephant (*Loxodonta cyclotis*) and the African bush elephant (*Loxodonta africana*) are both major terrestrial dispersers. The fruits of some

plants are primarily or even exclusively dependent on these increasingly threatened species for their dispersal.

Sausages that grow on trees

In the dry savannahs and woodlands of tropical Africa there is a tree with very curious-looking diaspores. Known as the sausage tree, *Kigelia africana* (Bignoniaceae) dangles its large fruits from the canopy like giant sausages suspended from long, stout cords. Its strange fruits can reach 1m in length with a diameter of 18cm, and weigh 10kg. Their enormous size, greyish-brown colour and fibrous, cellulose-rich pulp indicate dispersal by very large herbivores with an ability to digest cellulose. The fruits are allegedly eaten by elephants and hippopotamuses, as well as bushpigs, porcupines, monkeys and baboons. Among this illustrious coterie of potential dispersers, African bush elephants are probably the most effective owing to their large home ranges and long gut-retention time. To date there are no detailed scientific investigations documenting the degree of dependency of *Kigelia africana* on elephant dispersal.

However, there is one African tree that depends entirely on elephants for the dispersal of its fruits. It is called *Balanites wilsoniana*, a member of the bean caper or caltrop family (Zygophyllaceae).

Fruits that only elephants like

Balanites wilsoniana is a tall deciduous tree from the rainforests of Africa, ranging from Côte-d'Ivoire to Kenya. With a height of up to 40m and an extensive crown, the species forms part of the upper rainforest canopy. During the fruiting season the trees produce large, greenish-brown drupes approximately 9 x 6cm, which drop to the ground when ripe emitting an unpleasant, yeasty smell. After shedding, the drupes stay fresh for about a month, lying under the parent tree awaiting dispersal. With respect to the target disperser, the fruit does not have many options. The single-seeded kernel inside each drupe is 8.8 x 4.7cm, too large for most frugivores to be potential endozoochorous dispersers. In addition to the prohibitive size of their stones, the fruits also contain toxic chemicals that keep potential pulp thieves and seed predators at bay.

Several scientific studies using indirect observation as well as camera traps have proved that the only animals that eat and disperse the fruits of *Balanites wilsoniana* are African forest elephants (*Loxodonta cyclotis*). Fruiting coincides with the summer dry season, when fresh leaves and grass are less abundant. It is then that the elephants actively seek out one clump of trees after another to savour the nutritious fruits, their pulp rich in fat and protein.

opposite: *Kigelia africana* (Bignoniaceae) – sausage tree; native to tropical Africa – fruit; with its thin, hard skin and solid, fibrous pulp the strange fruit is probably best classified as an amphisarcum. It can reach 1m in length and weigh up to 10kg. Its enormous size and cellulose-rich pulp indicate adaptation to dispersal by very large herbivores such as elephants and hippopotamuses; fruit 60cm long

African forest elephants (*Loxodonta cyclotis*) feeding on the fruits of *Balanites wilsoniana* (Zygophyllaceae); scientific studies have proved that elephants are the only effective dispersers of the fruits of this rare African tree. They are the only animals able to swallow the very large stones of the greenish-brown drupes, ensuring both long-distance dispersal and pulp removal. The latter plays a significant role in seedling survival, since plants germinating from undispersed seeds are likely to fall prey to bacteria and fungi that thrive on the rotting pulp

Although a small percentage (3 per cent) of undispersed seeds also germinate, the seedlings have a very low survival rate (16 per cent); but after passing through an elephant's gut, seed germination is not only accelerated but also vastly improved (55 per cent). Moreover, seedlings growing from dispersed seeds have a much higher chance of survival. Scatter-hoarding rodents also help to disperse a few *Balanites wilsoniana* seeds but their contribution is insignificant. Only forest elephants provide this rare tree with a highly effective dispersal service that guarantees the regeneration of existing populations and possibly also the establishment of new ones. The limited recruitment from undispersed seeds alone could not maintain populations at their present densities.

When the elephants are gone

The great reliance of *Balanites wilsoniana* on forest elephants for the dispersal of its seeds strongly suggests that the long-term survival of the species depends on the continued presence of these charismatic animals. Elephants and their extinct relatives have been part of Africa's fauna for more than fifty million years. Over the past century, African elephant populations have been in rapid, continuous decline thanks to the destruction of their habitats, hunting, and other hardships imposed on these majestic animals by an ever-growing population of *Homo sapiens*. The disappearance of elephants from many habitats has left populations of *Balanites wilsoniana* without the vital services of their co-adapted disperser. There are several other plant species whose decline has been linked to the demise of elephants. The Liberian cherry (*Sacoglottis gabonensis*, Humiriaceae) and *Irvingia gabonensis* (Irvingiaceae), both West African trees, depend on elephants for the same reasons as *Balanites wilsoniana*. They produce drupes with stones so big that they can hardly pass through a pharynx smaller than that of an elephant. Interestingly, as in *Balanites wilsoniana*, their ripe fruits give off a yeasty aroma as they lie on the ground, suggesting that this specific olfactory signal, as well as fibrous pulp and large seeds (or stones), are indicative of elephant dispersal in general.

Reflecting the million-year-long presence of elephants in the region, there are many more examples of elephant fruits with oversized seeds in tropical Africa. Among them are members of the Sapotaceae (*Tieghemella heckelii*, *Baillonella toxisperma*), Arecaceae (e.g. *Borassus aethiopum*, *Phoenix reclinata*), *Klainedoxa gabonensis* (Irvingiaceae), *Panda oleosa* (Pandaceae) and the African mammey apple (*Mammea africana*, Clusiaceae; a relative of the mangosteen). Field observations have shown that populations of African trees with elephant-dispersed fruits begin to decline within decades if they are deprived of their disperser. In view of the dwindling elephant populations, understanding the relationship between these plants and their mutualists is vital for their conservation.

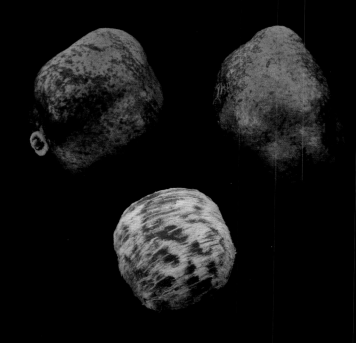

Borassus aethiopum (Arecaceae) – elephant palm; native to Africa's dry savannahs – fruits (drupes) and stone; with a height of up to 30 metres this species is Africa's tallest palm. It produces large orange drupes which, when ripe, emit a strong smell that has been likened to turpentine. The peculiar odour and the fibrous-fleshy pulp that covers a massive stone indicate co-adaptation to elephants as dispersers. African bush elephants (*Loxodonta africana*) are indeed fond of the fruits and play an important dispersal role. Baboons (*Papio anubis*) also eat the fruits but owing to their small gape size they are more likely to act as pulp-thieves rather than providing an effective dispersal service. The fruits are also eaten by local people. The dry fruits shown here have lost their orange colour; the fruits are 12-16cm long when fresh; the stone is c. 6cm long

Elephants are not the only members of Africa's megafauna that have great importance as seed dispersers during the course of evolution. At 40-65kg the aardvark may be a megafaunal lightweight, but this peculiar-looking mammal is literally the only chance of survival for the aardvark cucumber.

The aardvark and its cucumber

The aardvark cucumber or aardvark pumpkin (*Cucumis humifructus*), at home in the dry savannah regions of southern Africa, does something very unusual. It is the only member of the gourd family (Cucurbitaceae) that buries its fruits underground. As soon as the flowers have been pollinated, their stalks rapidly elongate, pushing the ovary down into the soil, where the fruit matures. This alone would not be exceptional but *Cucumis humifructus* appears to have developed a rather exclusive relationship with the enigmatic aardvark (*Orycteropus afer*, meaning "digging foot"). The aardvark cucumber is an annual plant that buries its knobbly, large (5cm), pale biscuit-coloured fruit to a depth of between 10 and 30cm. Carefully timed to coincide with the dry season, the aerial parts of the plant die and the fruits ripen. Aardvark cucumbers are equipped with a tough, water-resistant skin that enables them to stay intact in their underground hideaway for many months without rotting. They emit a scent that percolates through the soil. Only the aardvark (Afrikaans for "earth pig") has the nose to detect the scent and claws powerful enough to dig up the fruits from the dry soil. Outside the fruiting season of *Cucumis humifructus*, the aardvark feeds exclusively on ants and termites. However, in the dry season, when a visit to one of the rare remaining waterholes becomes a perilous venture, the juicy fruits of the cucurbit provide the aardvark with a precious source of water. In return, the aardvark's habit of burying its dung ensures that the seeds ingested intact with the juicy pulp are planted with a good dose of fertilizer. Whether the result of strict co-adaptation or an evolutionary cul-de-sac, by relying on a single animal species for the dispersal of its seeds, the aardvark cucumber has developed one of the least wasteful strategies of dispersal. However, this does not come without risk. If the aardvark becomes extinct, *Cucumis humifructus* will almost certainly meet the same fate.

The aardvark cucumber and the few strictly elephant-dispersed fruits are clearly rare examples of extremely specialized fruit-frugivore relationships involving megafauna. A few more remarkable cases can be found outside Africa.

Mallotus nudiflorus and the Indian rhinoceros

Mallotus nudiflorus (formerly *Trewia nudiflora*, Euphorbiaceae) is a tall deciduous tree commonly found in the riverine forests of India, Nepal and southern China. Its large, hard,

Orycteropus afer (Orycteropodidae) – aardvark; native to Africa – despite its name, which translates from Afrikaans as "earth pig," this distinctive, nocturnal mammal is not related to pigs. Its diet, which consists almost exclusively of ants and termites, also includes one particular fruit, the aardvark cucumber. An annual herb from the pumpkin family (Cucurbitaceae), the aardvark cucumber (*Cucumis humifructus*) buries its fleshy fruits underground where they can stay for months without rotting. During the dry season, when water is scarce, the aardvark cucumber emits a scent that specifically attracts the aardvark, for which the succulent fruits provide a precious source of water. Since the aardvark is the only animal that can locate and unearth its fruits, *Cucumis humifructus* is entirely dependent on this animal for the dispersal of its seeds

dull-coloured fruits are unattractive to most frugivores in the area, including monkeys, bats and birds. To solve the riddle of the natural disperser of the fruits of *Mallotus nudiflorus,* two zoologists, Eric Dinerstein and Chris Wemmer, undertook a detailed scientific study. Their results, published in 1988, demonstrated that the only animal with a taste for the bitter drupes is the Indian rhinoceros (*Rhinoceros unicornis*).

Mallotus nudiflorus drops its fruits during the rainy monsoon season (June–October) when they become the preferred food of the Indian rhinoceros. A shade-intolerant tree, *Mallotus nudiflorus* greatly benefits from the rhinoceros's custom of defecating in habitual latrine sites. For the latter the animal mostly chooses open grassland, where plenty of light and a generous helping of nutritious dung guarantee rapid germination and seedling growth straight from the dung piles. Uningested fruits usually rot underneath the parent tree; even if their seeds germinate, their prospects of survival in the shadow of the forest canopy are slim. Therefore, the distribution, regeneration and maintenance of *Mallotus nudiflorus* populations are almost entirely attributable to the activity of a single animal species. The vital, exclusive service provided by the Indian rhinoceros proves once more the significance of megafaunal dispersal.

The nitre bush and emus

The nitre or Dillon bush (*Nitraria billardierei*, Nitrariaceae) is a salt-tolerant shrub from Australia. It grows on saline soils and so the pulp of the small, red or yellow drupes is rich in salt. The edible fruits, which taste like salty grapes, were bush tucker of the Aborigines. Emus (*Dromaius novaehollandiae*) are reportedly the primary dispersers of the nitre bush. Mammals also eat the fruits but passage through the gut of the flightless emus has by far the most beneficial influence on the germination of the seeds.

Galápagos tomatoes and giant tortoises

The Galápagos Islands are home to two endemic species of tomato, *Solanum cheesmaniae* and *Solanum galapagense.* A study in the 1960s, when both taxa were treated as a single species, found that the seeds of these Galápagos tomatoes are subjected to indefinite physical *dormancy* owing to an extremely thick seed coat that prevents the seeds from germinating. Although various animals eat the berries, the seeds only germinate after passage through the gut of a giant tortoise (*Geochelone elephantopus*), a journey that can take between one and three weeks. Only exposing the seeds to a strong solution of sodium hypochlorite, which erodes the seed coat, could simulate a comparable improvement in germination. The study therefore concluded that the giant tortoise might be an important partner of the Galápagos tomatoes, both for breaking the dormancy of their seeds and for seed dispersal.

Mallotus nudiflorus (Euphorbiaceae) – native to east and south-east Asia – fruits (drupes); when the 4.5cm large, hard, bitter-tasting drupes of this species ripen in the monsoon season, they become the preferred food of the Indian rhinoceros (*Rhinoceros unicornis*). The fact that fruits that are not eaten by rhinos usually rot underneath the trees would suggest that these charismatic animals are the only effective dispersers of *Mallotus nudiflorus*; the photograph shows dried immature fruits of a Cambodian specimen in the Kew Herbarium

to Australia – plant with fruits (drupes); the main dispersers of the fleshy fruits are emus (*Dromaius novaehollandiae*). The edible fruits are bush tucker of the aborigines. Since the nitre bush mostly grows on saline soils, its fruits, which turn dark red when ripe, have a pleasant, sweet but slightly salty taste

bottom: *Geochelone elephantopus* (Testudinidae) – Galápagos tortoise; endemic to the Galápagos Islands; owing to its vegetarian diet, the giant Galápagos tortoise, the largest living tortoise on Earth, is an important seed disperser on the islands of the archipelago

Even on a global scale, there are less than a handful of further examples where a plant species is known to depend on a single animal species, whether large or small, for the dispersal of its seeds. The Brazil nut (*Bertholletia excelsa*, Lecythidaceae) and the agouti (*Dasyprocta agouti*) from South America were mentioned earlier. Coincidentally, another remarkable case of obligate mutualism was recently reported from South America. The odd couple consists of a peculiar mistletoe and an even more peculiar mammal. The majority of mistletoes are aerial hemiparasites on the branches of trees. Taxonomically they belong to the order Santalales where they are distributed across three families, Santalaceae (including Viscaceae), Loranthaceae and Misodendraceae. If mistletoes rely on animal dispersal – and most of them do – they are exclusively ornithochorous. At least, that was the general view until one very unusual exception was described in the December 2000 issue of the highly regarded scientific journal *Nature*. In the temperate forests of southern Argentina's Lake District, scientists discovered that the green fruits of the loranthaceous mistletoe *Tristerix corymbosus* are distributed solely by an endemic nocturnal marsupial called *Dromiciops australis*. Hardly bigger than a mouse, *Dromiciops australis* not only helps to disperse the mistletoe's seeds, passage through the tiny frugivore's gut has also proved critical for seed germination. Whereas seeds cleaned by hand failed to sprout, over 90 per cent of seeds collected from the marsupial's faeces germinated. Interestingly, *Dromiciops australis* is the only living representative of the family Microbiotheridae, a marsupial lineage whose origin is presumed to date back to the time of the southern supercontinent called Gondwana that formed some 500 million years ago and started to break up again some 165 million years ago. *Tristerix* is considered to be one of the most primitive extant genera in the Loranthaceae, a family that dates back to the middle of the Cretaceous period (142–65 million years ago). The combination of two such antediluvian partners is unlikely to be coincidental. The close mutualism between *Dromiciops australis* and *Tristerix corymbosus* rather represents the primitive remains of a once prosperous co-adaptational relationship between marsupials and loranthaceous mistletoes. Considering that the bird lineages involved in the dispersal of loranthaceous seeds originated only some 20–25 million years ago, the ancestors of *Dromiciops australis* could have been dispersing mistletoe seeds for many millions of years before birds took over.

Till death do us part

As described earlier, co-evolution between fleshy fruits and frugivores has always been diffuse and consequently prevented development of strict one-to-one fruit-animal mutualisms.

Therefore, all cases in which a plant species has come to depend on a single animal species for fruit dispersal are almost certainly remnants of formerly more diverse partnerships involving not just one but several, now extinct, species of animal dispersers. Naturally, this brink scenario raises the question as to what would happen if the last remaining disperser species of a plant also became extinct. In extreme cases such as *Cucumis humifructus*, whose underground fruits make it impossible for undispersed seeds to establish a new plant, extinction of the aardvark would most probably also mean the end of the aardvark cucumber. However, would other animal-dependent plant species with less complicated reproductive biologies have a better chance of survival if they lost all their natural dispersers?

The dodo and the tambalocoque – a textbook fairy tale

On the island of Mauritius in the western Indian Ocean grows a rare endemic tree called tambalocoque. With reputedly only thirteen mature trees left in 1973, the tambalocoque, a member of the sapote family (Sapotaceae), was considered a doomed species. In 1977, local folklore and the woeful state of the rare tree inspired Stanley Temple to hypothesize that the extinction of the dodo (*Raphus cucculatus*) in the seventeenth century was to blame for the demise of the tambalocoque, botanically *Sideroxylon grandiflorum* (formerly *Calvaria major*). The dodo was one of many odd flightless birds known only from Mauritius and adjacent islands. Its long isolation on these predatorless islands, led the unsuspicious bird to lose not only the ability to fly but also to defend itself. Its name derived from the Dutch word *dodoor* (sluggard), indicating that at a weight of up to 23kg the stocky bird was hardly able to flee, making it an easy prey for hungry (and bored) sailors. Slow, and unprepared for predators, the dodo had to watch its eggs being eaten by non-native animals such as cats, rats and pigs. Before its final extinction, which has been dated to the year 1690, the dodo allegedly ate the fruits of the tambalocoque tree as part of its diet. The green drupes, 5cm long, consist of an extremely hard stone covered by a thin (5mm) layer of a tenacious, fleshy pulp. Mauritian folklore claimed that the dodo was the sole disperser of the tambalocoque fruits, whose seeds were able to germinate only after the stones had passed through the gut of a dodo. In an attempt to turn folklore into science, Temple carried out a simple experiment during which he force-fed thirteen tambalocoque stones to turkeys. Of ten seeds that remained intact, three germinated. From a scientific viewpoint, small sample size and absence of a control (undigested stones) should have rendered his results meaningless. Nevertheless, Temple proposed a co-evolved obligate mutualism between the dodo and the tambalocoque tree and argued that the tree's endocarps had to undergo abrasive treatment in the gizzard of a dodo before the seeds could germinate. Probably because it involved the

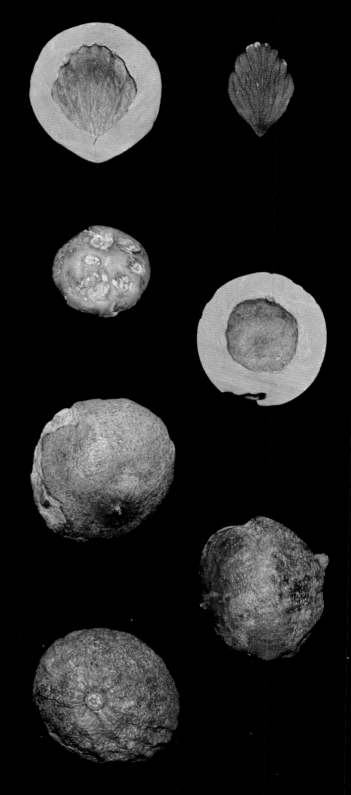

opposite: *Sideroxylon grandiflorum* (Sapotaceae) – tambalocoque tree; endemic to the island of Mauritius in the Indian Ocean – three dried drupes (bottom), stone (cut into three pieces) and seed (top right); the extremely thick-walled, woody stones were once believed to have depended on abrasion in the famed dodo's gizzard before the seeds could germinate. However, the dodo's gizzard was so strong that it would probably have destroyed the stone including the seed. It seems more likely that other members of Mauritius's extraordinary fauna, most of which have become extinct since the island was discovered, once helped disperse the tambalocoque's seeds

Raphus cucculatus (Columbidae) – dodo; once endemic to Mauritius, now extinct – related to doves and pigeons, the extinct flightless bird stood almost a metre tall. Ever since Lewis Carroll featured the mythical bird in his book *Alice's Adventures in Wonderland*, the dodo has risen to fame and become an iconic symbol of extinction. Little is known about the dodo's diet but it was probably a granivore rather than a seed-dispersing frugivore

famous dodo, Temple managed to publish his intriguing but flawed hypothesis in one of the world's most prestigious scientific journals, ensuring that it became a standard example of an obligate fruit-frugivore mutualism quoted in numerous textbooks. Before and after Temple's bold claims, others have reported the germination of seeds from unabraded tambalocoque stones, as well as the occurrence of a number of younger trees in the wild. The latter indicates that at least some dispersal must still take place, perhaps by introduced animals. Without removal of the fleshy mesocarp, bacterial and fungal infections attracted by the rotting pulp reliably destroy the tambalocoque seeds, a vulnerability not restricted to *Sideroxylon grandiflorum*. The importance of thorough pulp removal for successful seed germination has been demonstrated for many fleshy-fruited species. Assuming that the dodo did eat the fruits of the tambalocoque, its extremely strong gizzard could have destroyed most of the stones it swallowed. Little is known about the dodo's diet, and it might well have been a destructive granivore rather than a helpful frugivore. Besides, there are other extinct animals such as large-billed parrots (*Lophopsittacus mauritianus*) and giant tortoises (*Geochelone* sp.) that could have been important dispersers of the tambalocoque. Although nowadays generally rejected as unsubstantiated, Temple's theory can never be entirely disproved owing to the permanent absence of the dodo and many other extraordinary Mauritian endemics that shared its deplorable fate.

Temple might have chosen a bad example with the dodo and the tambalocoque tree. However, his core contention, namely that the tree is in decline because it has lost its natural dispersers, seems compelling in the face of the mass extinction that has devastated the native Mauritian fauna. Ever since, the frightening idea that the decline of a plant species could be linked to the total extinction of an animal has generated much debate. Today, more than ever, we are aware of the human-induced extinction crisis brought about by an exponentially growing world population. The speed of human expansion is rivalled only by the pace at which natural habitats vanish, the atmosphere is polluted, and resources are overexploited, to name but a few of the side effects accompanying the Earth's growing overpopulation by *Homo sapiens sapiens*. We should therefore not be surprised if the general scenario depicted by Temple applies to an increasing number of plants and their animal mutualists. Large mammals are particularly vulnerable to human disturbance on account of their long generation time and high demand for food, which requires a large home territory and entails a low population density. As a consequence, in any given habitat both species diversity and abundance of animals drop with increasing body size. Therefore, plants co-adapted to megafauna have a limited chance to substitute the loss of an important disperser species, as demonstrated by elephant-dispersed trees in Africa.

Anachronistic fruits

Evolution is a slow process. Depending on their degree of complexity, the development of co-adaptational traits in response to a specific guild of dispersers has taken between hundreds of thousands and many millions of years. If key animal dispersers become extinct, they may be followed shortly afterwards by their green dependents. If a plant manages to survive somehow without its animal mutualists, it will still keep expressing the co-adaptive traits that once made their ancient pact such a success. In fact, owing to the sluggishness of evolutionary response, this anachronistic behaviour is bound to persist for a long time.

In 1982 the ecologists Daniel Janzen and Paul Martin proposed a theory that could explain why certain endozoochorous fruits in the New World have no apparent natural dispersers. Sounding more like a botanical version of Sir Arthur Conan Doyle's *The Lost World*, Janzen and Martin hypothesized that many species of American trees produce fruits that are adapted to dispersal by the long-lost beasts of the Ice Age. Until 13,000 years ago, towards the end of the Pleistocene epoch (1.8 million to 11,550 years ago) when the last Ice Age drew to a close, North America boasted a megafauna far richer than Africa's today. The ancient menagerie of potential megafaunal seed dispersers featured several species of native wild horses, camels and tapirs alongside fantastic creatures such as gomphotheres (four-tusked elephant-like creatures), mastodons and woolly mammoths weighing up to ten tons, giant ground sloths, the largest the size of a modern elephant, glyptodonts, giant armadillo-relatives the size of a small car, giant short-faced bears nearly twice the size of a grizzly, giant bisons, giant peccaries, giant beavers and giant tortoises. By the end of the last Ice Age, these great beasts had inhabited the whole of the Western Hemisphere for much of the Caenozoic era, which spans the last 65 million years until the present. The Caenozoic began with the demise of the dinosaurs at the Cretaceous-Tertiary boundary. The so-called K-T (Cretaceous-Tertiary) event, which most probably involved a large comet colliding with the Earth, caused the extinction of 70 per cent of all species on the planet, including the dinosaurs and most other large land animals. This catastrophe opened up the opportunity for mammals to radiate from small, rather nondescript forms into a hugely diverse group of animals, conquering land, air and sea, making the Caenozoic era the Age of Mammals.

Despite their phenomenal success that lasted for many million years, three quarters of all the species and individuals of the Pleistocene megafauna in the New World died out around 13,000 years ago within a period of just a thousand years, not even the blink of an eye geologically. All the great megaherbivores exceeding 1,000kg in weight and their exceptionally large predators such as the giant short-faced bear (*Arctodus simus*), the American lion (*Panthera leo atrox*) and at least two species of American cheetahs (*Miracinonyx*

Fruits of North American Leguminosae whose co-adapted dispersers are among the extinct Pleistocene megafauna: bottom: *Gleditsia triacanthos* (Leguminosae) – honey locust; native to eastern North America – fruits (camaras); very similar to megafauna-dispersed legume fruits in Africa, the large indehiscent pods of the honey locust have a leathery skin followed by a thin layer of edible pulp around numerous hard seeds. Adapted to dispersal by large Ice Age herbivores, today the fruits are eaten only by introduced livestock such as cattle and horses. Another anachronistic adaptation, the fierce, 10-20cm long thorns on trunk and branches of the 20-30m tall tree once served as a defence against tall browsers

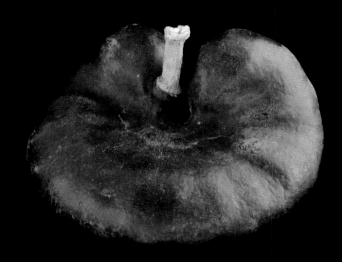

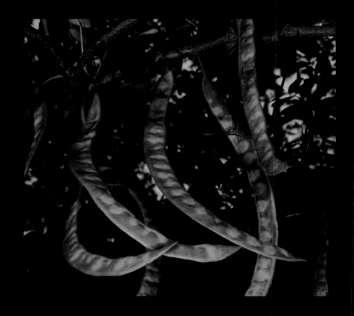

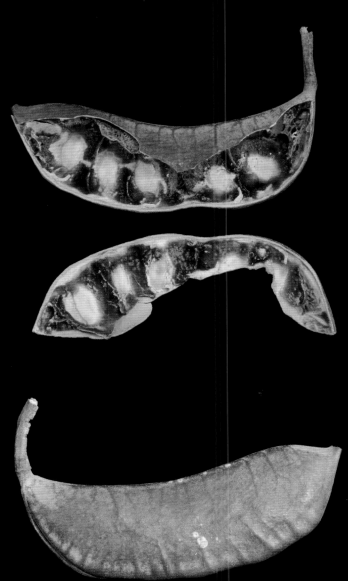

opposite top: *Enterolobium cyclocarpum* (Leguminosae) –
guanacaste; native to tropical America, the national tree of
Costa Rica – fruit (camara); attractive to modern introduced
horses, the guanacaste's fruits were once probably dispersed
by now extinct, horse-like herbivores; diameter 8cm

Gymnocladus dioica (Leguminosae) – Kentucky coffee; native
to North America – fruit (camara); typical of megafauna-
dispersed fruits, the woody pods are filled with a dark, sweet
pulp. They are poisonous to livestock, so today only scatter-
hoarding rodents occasionally eat the pods and facilitate
limited dispersal; fruit 15cm long

inexpectatus, M. trumani), several species of sabre-toothed cats, including the sabre-toothed tiger (*Smilodon*), and dire wolves (*Canis dirus*), all disappeared. The sudden extinction of the American megafauna must have left many co-adapted plants without their dispersers. Today, we are faced with many New World plants whose puzzling fruit and seed traits can be explained only in the light of the extinct Pleistocene megafauna. Somehow they managed to survive without their dispersal partners with whom they shared the same habitat for millions of years. In the short time that has passed since the extinction of the big beasts, these plants have not yet developed an extensive response to the absence of their dispersers.

Janzen and Martin's comparison of New World fruits that they suspected were anachronistic with fruits that "look, feel and taste like those eaten by large seed-dispersing mammals in Africa" led them to define the general "megafaunal dispersal syndrome." According to their hypothesis, anachronistic fruits qualify for the megafaunal dispersal syndrome, while at the same time certain ecological indicators suggest that their dispersal partners are missing. The most obvious signs are fruits rotting underneath the parent tree or inefficient dispersal by extant animals such as small rodents for which the fruits seem overbuilt. The feeding of introduced livestock on fruits that would otherwise remain uneaten is also considered a sign of anachronism, horses and cattle apparently filling the gap left by mastodons and others. Plants with anachronistic fruits that are left to be dispersed solely by gravity (including flowing water) have a patchy or restricted distribution that is often limited to floodplains.

The most likely candidates for megafaunal dispersal are members of the legume family (Leguminosae), which produce large, tough-rinded, indehiscent, pulp-filled pods (camaras) with hard seeds, similar to their relatives in Africa that have adapted for elephants and antelopes (e.g. *Acacia* spp.). Among these American anachronists are honey locust (*Gleditsia triacanthos*), mesquite (*Prosopis* spp.), Kentucky coffee (*Gymnocladus dioica*), guanacaste (*Enterolobium cyclocarpum*), horse cassia (*Cassia grandis*) and stinking toe tree (*Hymenaea courbaril*). Today's native fauna largely ignores their indehiscent sugary pods but many of them, especially those of mesquite and honey locust, attract the tastebuds of introduced horses and cattle. Only the fruits of the Kentucky coffee (up to 25cm long and 5cm wide), though rich in a dark greenish, sweet pulp, are poisonous to livestock. Occasionally eaten by scatter-hoarding rodents who facilitate limited dispersal, the seeds of the Kentucky coffee were once used by European settlers in Kentucky to prepare a coffee substitute, hence its name. Very similar to those of the Kentucky coffee are the chunky pods of *Hymenaea courbaril*, a tall tropical hardwood tree from Central and South America and the Antilles. Edible but endowed with a particularly pungent mammal smell, the fruits are responsible for the stinking toe tree's unflattering name. Inside the 15cm long and 8cm wide, slightly flattened, reddish brown camaras are three to four large black seeds embedded in a dry, fibrous brown powder. The strange pulp has a sweet, date-like flavour and is sometimes used to prepare drinks. Typical of anachronistic fruits, the ripe pods drop on the forest floor where the vast majority of them remain unopened. Feral pigs, agoutis and pacas (medium-sized rodents) are the only animals that manage to gnaw open a few pods to access the pulp. As the seeds are very large (3.5 x 2.5cm) and hard as stone they are usually discarded *in situ*. If accidentally swallowed, the seeds would pass through the animals unharmed and could be dispersed some distance from the parent plant. Still, most fruits remain untouched and eventually fall prey to fungi and bacteria. However, in north-eastern Brazil, in the arid areas of caatinga vegetation that are exposed to long periods of drought interspersed with irregular storms and associated flash floods, fruits of *Hymenaea courbaril* sometimes pile up underneath the trees for a number of years. When the rains come the buoyant hard-shelled pods get carried along in the fast-flowing, but temporary rivers. As the floodwaters subside, the fruits are deposited on the river banks where it is likely that at least some seeds will germinate, thus effecting dispersal.[7]

Janzen and Martin's list of presumed New World anachronisms is long and includes not only legumes but many plants from a wide variety of families that appear to have evolved the size, colour, odour and texture of their fruits through co-adaptational interaction with ancient big mammals. Among them are popular subtropical and tropical fruits such as persimmon (*Diospyros* spp., Ebenaceae), bullock's heart (*Annona reticulata*, Annonaceae), hog plum (*Spondias mombin*, Anacardiaceae), genipap (*Genipa americana*, Rubiaceae) and sapodilla (*Manilkara zapota*, Sapotaceae). Less palatable anachronistic suspects include the tropical calabash tree (*Crescentia cujete*, Bignoniaceae) and its relative the jicaro (*Crescentia alata*), together with two oddities from temperate North America, the pawpaw (*Asimina triloba*, Annonaceae) and the osage orange (*Maclura pomifera*, Moraceae).

Size no longer matters

With its grotesquely oversized and formidably armoured fruits, the calabash tree would seem out of place in its native Central America, were it not meant to feed only the largest mouths of the Pleistocene. As if anticipating the heavy weight of the fruit, the pale, cabbage-scented, bat-pollinated flowers are borne directly on the trunk and major branches. Once fertilized, the ovary grows into a huge, woody globe – first green and then yellowish – up to 30cm in diameter. Inside, the fruit (an amphisarcum) contains a large number of small seeds embedded in a white, floury pulp. The fruits of the calabash tree have several uses.

below: *Hymenaea martiana* (Leguminosae) – jatoba, native to South America – fruits (camaras); a close relative of the stinking toe-tree with very similar fruits, the pods contain an edible pulp but emit a pungent smell typical of mammal-dispersed fruits; fruit c. 15cm long

bottom: *Hymenaea courbaril* (Leguminosae) – stinking toe-tree; native to tropical America – fruits (camaras); in Brazil's caatinga, flash floods occasionally help disperse dry pods that have piled up underneath their parent trees. The chunky pods have long lost their co-adapted animal dispersers; fruit c. 15cm long

below: *Asimina triloba* (Annonaceae) – pawpaw; native to eastern North America – fruit (baccetum); several free carpels produce a cluster of large berrylets with a delicious, custard-like pulp reminiscent of tropical fruits. The paw-paw is the largest edible native fruit of North America; fruitlet 7-15cm long

bottom: *Crescentia cujete* (Bignoniaceae) – calabash tree, native to tropical America – fruit (amphisarcum); too large to fit any contemporary mouth, the woody, ball-shaped fruits must have been dispersed by members of the Pleistocene megafauna; diameter c. 20cm

Ground and mixed with water, fresh calabash seeds yield a sweet refreshing drink. The pulp is applied as a medicine to treat asthma, diarrhoea, stomach-ache, bronchitis and colds. Once emptied, the very hard shells are turned into containers for water, salt and tortillas, as well as musical instruments and other handicrafts. According to Christopher Columbus, native Americans used the hollowed-out fruits to camouflage their heads while swimming in the hunt for waterfowl, allowing them to pull individual birds under water without upsetting the rest of the flock.

The jicaro is a close relative of the calabash tree, with similar but much smaller fruits 6-15cm in diameter. The small shrubby tree is common in dry, grassy habitats of the Pacific side of Central America. Like the calabash, the woody fruits of the jicaro each contain several hundred seeds embedded in a slippery, fibrous pulp. Initially astringent and pale in colour, the pulp turns into a slimy black mass with a penetrating foetid aroma typical of many mammal-dispersed fruits. Despite its repulsive smell, the pulp tastes quite sweet and is perfectly palatable to humans. Modern introduced horses (*Equus caballus*) avidly consume the ripe pulp, which they swallow only after chewing it slightly, leaving most seeds intact. In the absence of horses the fruits simply rot on the ground in the rainy season, giving the seeds no chance to germinate. The fact that range horses are the sole dispersers of the jicaro fruits led Janzen to assume that these animals filled the gap left by extinct Pleistocene horses.

The largest fruit of America

The pawpaw is North America's largest native edible fruit. A member of the Annonaceae family, *Asimina triloba* possesses an apocarpous gynoecium consisting of seven to ten free carpels. Hence, each flower produces not just one berry but a whole cluster of up to (but usually less than) ten soft-skinned berrylets (baccetum). The individual fruitlets are 7-15cm long and weigh between 150 and 450g. Since they look a bit like plump green bananas, pawpaws are sometimes called poor man's bananas. Embedded in the fragrant, yellow custard-like pulp are 10-14 large, dark brown or black seeds, 15-25mm in diameter, arranged in two rows. With the rich, complex flavour of a tropical fruit reminiscent of a blend of banana, pineapple, mango and custard, the pawpaw is unique among temperate fruits. Although a delicacy, ripe pawpaws have a shelf life of just two or three days, which rules them out from being put on sale in the supermarket chains.

The seeds of *Asimina triloba* are too large to fit the gape of any native bird but the faeces of racoons (*Procyon lotor*), red foxes (*Vulpes vulpes*) and opossums (*Didelphis virginiana*) were allegedly found to contain intact viable seeds. Nevertheless, the pawpaw's large size, the fact that it drops off the tree as soon as it is ripe, its rarity in the wild, and the occurrence of wild pawpaw trees in floodplains (indicating haphazard dispersal through flowing water only), can together be interpreted as indicators of anachronism. Although an individual tree is relatively short-lived (25-50 years), the pawpaw continuously produces new stems from root suckers. Over time, a single pawpaw tree can give rise to a huge clonal patch of trees that can persist for hundreds if not thousands of years. The pawpaw thus ensures not only extreme longevity, but also the capacity to produce sufficient fruit in one site to cater for the large appetites of big Ice Age mammals.

Osage orange

The fruits of the osage orange (*Maclura pomifera*) are another North American curiosity. Bright green, knobbly balls the size of an orange or grapefruit, they look a bit like miniature breadfruits, which, in fact, they are. Like *Artocarpus altilis*, *Maclura pomifera* is a member of the mulberry family (Moraceae). Female trees produce spherical inflorescences that grow into compound fruits that are morphologically very similar to breadfruits. However, unlike a breadfruit, the four decussate tepals forming the fleshy perigone of the individual female flowers are still separate, as in a mulberry. In the peculiar brain-like surface pattern of an osage orange, a female flower is therefore not represented by a single round bump but by four crosswise ridges.

By the time the Europeans arrived in America, *Maclura pomifera* was restricted to a few river valleys in eastern Texas, Oklahoma and Arkansas, an area that approximately corresponds to the home territory of the Osage people. To native Americans, the trees were valuable for their wood, which is unsurpassed anywhere in the world for making bows. European explorers reported that native people would travel many hundred miles in the quest for the long-lived trees. Osage orange heartwood is also of superior hardness, extremely resistant to decay, and immune to termites. In the nineteenth century, it yielded valuable timber for railway sleepers, fence posts, and the hubs and rims of wagon wheels. Nowadays, osage orange trees are widely cultivated ornamentals, both in the United States and elsewhere.

The trees drop all their large fruits within a couple of days in autumn, a typical sign of mammal dispersal. Whilst lying on the ground waiting to be eaten, the bright green balls emit a pleasant fragrance similar to air freshener. However, the pleasant smell is somewhat deceiving. The fibrous flesh has the consistency of a raw potato, is mildly poisonous, and not very palatable. Unsurprisingly, neither man nor beast is tempted to accept the osage orange's gift in exchange for the dispersal of its seeds. Although the Carolina parakeet (*Conuropsis carolinensis*), the only parrot species native to the eastern United States, is said to eat the fruits, it was probably more interested in the seeds than the pulp. Unfortunately, whether or

overleaf left: *Maclura pomifera* (Moraceae) – osage orange; native to North America – left: fruit (sorosus); with no effective dispersers among today's living native animals, the osage orange must have evolved to suit the palate of some members of North America's extinct Ice Age megafauna, probably mastodons; diameter 8cm

overleaf right: Microscopic detail of fruit surface: the fleshy perigone of each female flower consists of four tepals arranged crosswise. The long, thread-like stigma emerges between them. Whereas in its relative the breadfruit the four tepals are fused to form a single round bump, in the osage orange the tepals remain separate, creating a brain-like surface pattern

Maclura pomifera (Moraceae) – osage orange, native to North America – young fruits (sorosus), the long stigmas marking the individual female flowers still attached; diameter 5cm

Tamarindus indica (Leguminosae) – tamarind; only known in cultivation, probably originating in tropical Africa – fruits (camaras); with their woody outer husk, nutritious pulp and extremely hard seeds, tamarinds are typical megafauna fruits. In Africa, the continent with the most diverse extant mega-fauna, they are dispersed by large ruminants whereas in Asia, monkeys have become important dispersers.

not any seeds passed through its beak and gut intact will never be proved, since the last representative of the species died in Cincinnati Zoo in 1918. Other anecdotal evidence claims that horses, fox squirrels, opossums and possibly also racoons and foxes occasionally eat the pulp. Even if these claims are true, most osage oranges are left to rot on the ground and seed dispersal seems to be rare and serendipitous. The circumstantial evidence renders the possibility highly unlikely that any of these animals are the co-adapted primary dispersers of *Maclura pomifera*. A more plausible explanation is that the fruits evolved to repel all animals except some extinct members of the Pleistocene megafauna, perhaps giant ground sloths, mammoths, mastodons or native horses.

How can it be true?

Ever since Dan Janzen and Paul Martin published their fantastic-sounding megafauna dispersal theory involving extinct Pleistocene megafauna they have encountered both support and criticism. It is true that when scientific rigour is applied, their hypothesis seems based on precarious ecological extrapolations and assumptions supported by little direct evidence. The biggest question is how the presumed anachronistic species managed to survive without their dispersers for more than ten thousand years. A long generation time, limited and fortuitous dispersal by pulp thieves, seed predators and surface water movements, as well as usefulness to humans, could be the reasons that allowed extreme anachronists their reduced but continued existence. Milder cases of anachronism are still dispersed by living mammals and birds, a circumstance that has been used as an argument against Janzen and Martin's theory. However, it would be natural that many of the gaps left by the ancient megafaunal dispersers would be filled by smaller frugivores, for which the fruits seem overbuilt. In the absence of their much larger competitors, monkeys, bats and others may opportunistically feed on such fruits. Often, though, they eat only the pulp or the arils of large fruits without swallowing the seeds. Being both intelligent and dexterous, primates are particularly known for exploiting all kinds of fruits, irrespective of whether they are primarily ornithochorous or mammaliochorous. For example, tamarinds (*Tamarindus indica*, Leguminosae) with their hard outer shell containing a nutritious pulp and extremely hard seeds are typical megafauna fruits. In Africa ruminants mainly disperse tamarinds, whereas in less megafauna-rich south-east Asia, monkeys have become important dispersers. In Madagascar, five big-seeded tree species seem to be critically dependent on dispersal by the red-collared lemur (*Eulemur fulvus collaris*). However, what appears to be a case of tight co-evolution is in fact the consequence of the extinction of larger frugivorous birds and lemurs that were probably the primary dispersers of these fruits.

The fact that trees with anachronistic fruits often produce crops that vastly exceed the appetite of small mammal-dispersers is yet another indication that they evolved to attract much bigger creatures. Another argument questions the very existence of a megafaunal dispersal syndrome. Since large herbivores mostly live on general plant matter rather than concentrating on fruits, some people find it hard to accept that a specific megafaunal dispersal syndrome could have evolved. The scientifically proven decline of elephant-dispersed African trees that followed the disappearance of their disperser should put sceptic minds at rest. One more question should be asked, though. What caused the American megafauna to become extinct within just a thousand years?

Where have all the mammoths gone?

Various theories have been forged as to why the Age of Great Mammals ended in North America about 13,000 years ago. Some suggest that diseases are to blame. Others believe that climatic changes are the primary reason. One possibility is that the furry animals, adapted to thousands of years of cold, could not handle the warmer temperatures of the Holocene, the time period that followed the Pleistocene from some 11,500 years ago to today. More mysteriously, the warming at the end of the Pleistocene that brought the last Ice Age to an end was interrupted by a sudden reversal of temperatures that briefly threw the earth back into a bitterly cold climate. This short, 1300 year period of cooling, scientifically referred to as the Younger Dryas, or more graphically the Big Freeze, began some 12,700 years ago and took many animals by surprise. A generally accepted theory is that meltwater running off the northern ice cap created a vast freshwater lake in the centre of North America. When a natural dam eventually broke, Lake Agassiz emptied into the North Atlantic, where it interrupted the Gulf Stream, the ocean current that carries warm water from the tropical south to northern latitudes. As Lake Agassiz drained away, the Gulf Stream slowly went back to normal and the Younger Dryas ended.

More recently, scientists have found evidence to suggest that it was the impact of a comet or asteroid that caused the abrupt onset of the Younger Dryas cooling. Dubbed the Younger Dryas impact event, the extraterrestrial assault supposedly occurred 12,900 years ago in the area of the Great Lakes of North America. The shockwave and thermal pulse would have caused a flood of meltwater that had the same effect as the emptying of Lake Agassiz. In the comet's aftermath, the swift change of climate and other impact-related effects are assumed to have wiped out the magnificent Pleistocene megafauna.

Much earlier, in the 1960s, Paul Martin had proposed another – highly controversial – theory that explained why all the big mammals of North America vanished so suddenly.

opposite: *Macrozamia moorei* (Zamiaceae) – native to eastern Australia (Queensland) – seed cone; the fleshy seeds of Australia's cycads are also assumed to have been part of the mihirungs' diet. When ripe, the cones disintegrate and scatter their brightly coloured seeds around the plant where they can be easily picked up by ground-dwelling animals. Dromornithid birds, Australia's dominant frugivores during the Tertiary (65-2 million years ago), would have been able to swallow entire megasporophylls including the seeds

Pleiogynium timoriense (Anacardiaceae) – Burdekin plum; native to central Malesia and the Pacific area, including Australia – stones; like North America, Australia lost 94 per cent of its large animals sometime during the Pleistocene. The drupes of the Burdekin plum are one of several Australian fruits assumed to be anachronistic. As they were borne at a low height and accumulated on the ground, the co-adapted dispersers of the Burdekin plum may have been among the mihirungs, a uniquely Australian group of very large, flightless, frugivorous birds that died out some 30,000-50,000 years ago; diameter 2.3cm

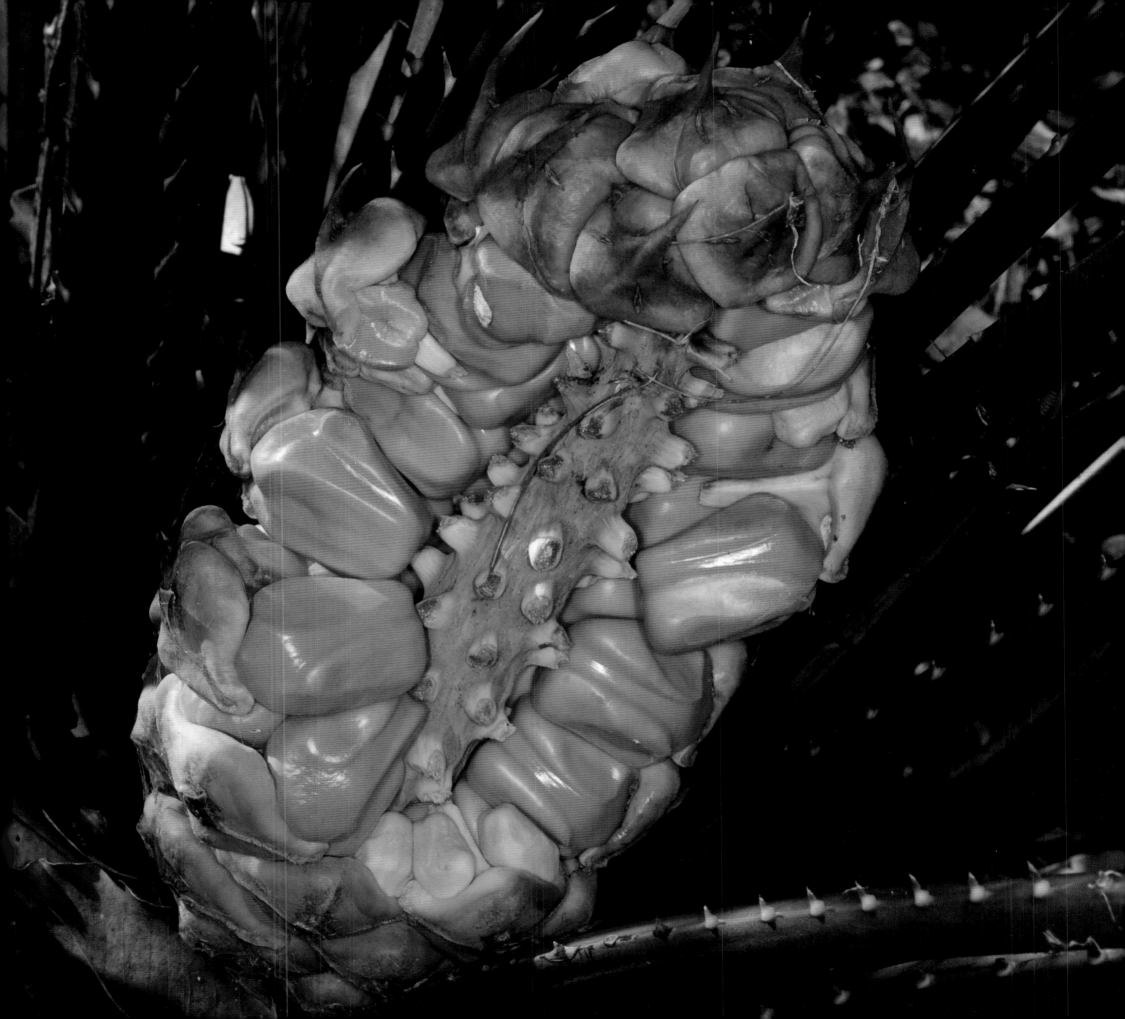

When considering the Age of Great Mammals on a global scale, it turns out that the New World is not the only place to have experienced the loss of its megafauna. Australia was once home to giant marsupials, including 2-3m tall kangaroos (*Procoptodon goliath*), tapir-like creatures (*Palorchestes azael*) and the largest marsupial known to exist, a two-ton, hippopotamus-sized relative of the wombat, *Diprotodon optatum*. There was also a giant, two-ton carnivorous lizard (*Megalania prisca*). Other "mythical creatures" from downunder include the mihirungs, flightless birds of the now extinct family Dromornithidae whose closest living relatives are goose-like birds. During the Tertiary, the uniquely Australian mihirungs were the dominant frugivores and browsers on the continent. Among these "gigantic geese" were the largest birds that ever lived, such as the 3m tall, 500kg Stirton's Thunder Bird (*Dromornis stirtoni*) and the smaller but still massive *Bullockornis planei*, nicknamed Demon Duck of Doom for its suspected carnivorous inclinations. Although exact dates for their extinctions are missing, Australia lost 94 per cent of its large animals sometime during the Pleistocene, around 30,000-50,000 years ago. Today, many plants in Australia's Northern Territory show evidence of anachronism. Their apparently orthno-chorous fruits bear a single, very large and extremely hard seed or stone covered by a thin fleshy pulp (e.g. *Pleiogynium timoriense*, *Owenia reticulata*). The fact that these fruits are often borne at a low height on the trees and that they accumulate on the ground where they are left to rot or succumb to insect infestation suggests that their co-adapted dispersers were among the extinct dromornithid birds, of which the last ones died out between 50,000 and 35,000 years ago. In temperate Eurasia, Pleistocene megafauna, which included the straight-tusked elephant (*Elephas antiquus*), woolly mammoth (*Mammuthus primigenius*) and woolly rhinoceros (*Coelodonta antiquitatis*), disappeared gradually between 50,000 and 12,000 years ago, with one remarkable exception. As melting glaciers caused sea levels to rise, a population of mammoths became isolated on the piece of Siberia that is now Wrangel Island. Typical of island populations faced with limited resources, the stranded mammoths gradually evolved into a dwarf form. Out of harm's way, the Wrangel Island mammoths managed to outlive their mainland cousins by more than 7,000 years, until humans arrived. The last surviving dwarf mammoths died only 3,700 years ago, 850 years after the Egyptians built the great pyramid of Giza.

More recent extinction events hit Madagascar. Within the last 2,000 years the island has lost its largest lemurs, two native species of hippopotamus, and all eight known species of the famous elephant birds (*Aepyornis* spp., *Mullerornis* spp.). Likewise, New Zealand was once home to ostrich-like ratites, 11 species of moas including 2-4 species of giant moa (*Dinornis*), which stood 3.6m tall and weighed up to 300kg. All died out as recently as 1200-

Dromornis stirtoni (Dromornithidae) – Stirton's thunderbird (artist's impression by Paul Trusler) – up to 3m tall and weighing half a ton, Stirton's thunderbird is the largest bird that ever lived. It is one of eight species of flightless birds that have been grouped together in the uniquely Australian family Dromornithidae. For millions of years, from the Tertiary until between 50,000 and 35,000 years ago, these birds, called mihirungs by the Aborigines, were the dominant frugivores and browsers on the Australian continent. Australia lost 94 per cent of its megafauna, including the mihirungs, sometime during the Pleistocene

1600 AD. The compact divaricate growth pattern of many New Zealand plants is considered an anachronism that once evolved as a defence against browsing moas.

When Paul Martin analysed the dates of numerous megafaunal extinction events that took place worldwide he found that they roughly coincided with the first arrival of humans in the respective areas. Ever since humans radiated out of Africa and Eurasia, their appearance on new continents and islands seems to have been inevitably followed by a mass exodus of large-bodied animal species. The most recent major event of this kind took place 13,000 years ago in North America, shortly after melting ice sheets allowed the first humans to cross the land bridge between Siberia and Alaska. Martin's overkill hypothesis, nicknamed Blitzkrieg theory, proposes that humans blindly hunted the big beasts out of existence. For hunting stone-age cultures, large animals were the easiest to track. Although attacking them was risky, they provided the most food for the least effort and brought prestige to whoever managed to slay them. Little has changed today as the trophy collections of passionate hunters confirm.

The only places on our planet where the Age of Great Mammals has not yet come to an end are Africa and parts of tropical Asia. Famous for its wildlife, Africa boasts an astonishingly diverse megafauna that still includes five species of megaherbivores with a body weight of more than a ton: elephant, giraffe, hippopotamus and two species of rhinoceros. Their survival is a paradox. Africa is the cradle of mankind; people have existed there longer than anywhere else in the world. It should therefore have been the first continent to lose its megafauna. However, it is precisely the long period during which Africa's big mammals lived alongside humans that allowed them to adapt to our increasingly sophisticated hunting skills. For African wildlife, hominids have always been just another carnivorous predator. In principle, the same applies to megafauna in Asia where hominids are known to have been present for nearly two million years. Anywhere else in the world where humans turned up suddenly, their technically advanced predatory lifestyle took the unsuspecting animals by surprise, leaving them no time to adapt. This is especially true of island faunas, as the fate of the famously fearless dodo tragically demonstrates.

Statistically, it is almost impossible that the simultaneous appearance of humans and widespread extinction of large animals in an area could be coincidental in all the cases listed above (and many others not mentioned). Multiple Ice Ages came and went but only the end of the last one was marked by a mass extinction. If climate change rather than humans were to blame, the Wrangel Island mammoths would not have survived for thousands of years after the last Ice Age ended. Moreover, other megafaunal extinctions such as those of Australia, Madagascar, the Mascarene Islands and New Zealand, cannot be linked to drastic climate changes. The giant moa and some other species were undoubtedly hunted to extinction. However, in Pleistocene North America, millions of giant animals disappeared so rapidly that human hunting alone could not have been the only cause. Humans and their domestic animals such as dogs could have introduced diseases to which the New World megafauna had no previous exposure and therefore no resistance. The same happened to native Americans when European explorers brought them smallpox, measles, whooping cough, cholera, typhoid, bubonic plague, hepatitis, and other new diseases; the population of Mexico, for example, shrank from an estimated 25 million to just one million within a century. Nevertheless, disease and the impact of an asteroid may merely have accelerated rather than caused the demise of mastodons, mammoths, ground sloths and the rest of the North American heavyweight menagerie. It might have taken a little longer, but eventually, hand axes, spear points and cleavers made of stone would undoubtedly have achieved the same result.

As a logical consequence, the extinction of the Pleistocene megafauna must have left many plants without their most effective dispersal agents. Unfortunately, because of the permanent disappearance of the animals in question, co-adaptation of extant plants to Pleistocene dispersers remains even harder to prove than the causes of current climate change. Little research into anachronistic fruits has been done in the New World. In temperate Eurasia very few of the original mammalian dispersers survive. Many edible fruits have been introduced and cultivated for millennia, which makes it difficult to reconstruct their original co-evolutionary ties. Apples, quinces, medlars, peaches and pears probably evolved as mammal fruits. Even less is known about the natural dispersers of oversized fruits in tropical Asia such as breadfruit, jakfruit and the bowling-ball-sized amphisarca of the enigmatic parasite *Rafflesia*. With their hard outer rind, oily pulp and smell of rotten coconut, *Rafflesia* fruits seem perfectly adapted to the Asian elephant (*Elephas maximus*) that once lived in the same rainforests.

How much more evidence is needed to prove the reality of anachronistic fruits? Today more than ever, we are aware of the unprecedented speed at which an exponentially growing human population drives both animals and plants into extinction. Taking the slow pace of evolution into account, the existence of plants waiting for a disperser that will never come is merely an inevitable consequence of our disrespect for all other living beings on Earth.

opposite: *Kyllinga squamulata* (Cyperaceae) – Asian spikesedge; native to tropical Africa, Madagascar, India and Indochina – fruit (pseudanthecium); inside the loose bag formed of fused bracts lies a flat, discoid achene. The flat, wing-like shape is clearly an adaptation to wind dispersal whereas the sharply lateral lobes could be interpreted as an additional adaptation to animal dispersal (epizoochory); 3.7mm long

Rafflesia keithii (Rafflesiaceae) – rafflesia; endemic to Borneo – there are about 25 species of rafflesias, all of which are native to south-east Asia. They are chlorophyll-less endo-parasites, that live inside the tissue of their host plants, lianas of the genus *Tetrastigma* spp. (Vitaceae). The only visible part of rafflesias are the huge flowers that range from 20cm (*R. manillana*) to more than a metre in diameter (*R. arnoldii*), the largest single flower of any plant species. Little is known about the dispersal ecology of *Rafflesia*. The fruit is a large ball-shaped amphisarcum with an oily pulp that smells of rotten coconut. Size, texture and odour indicate that *Rafflesia* fruits are co-adapted to dispersal by the Asian elephant (*Elephas maximus*)

THE MILLENNIUM SEED BANK PROJECT

A fruitful partnership for survival

The first animals to disappear, as we continue to decimate natural habitats on a catastrophically short timescale, are mammals and birds. With the extinction of these animals, many plants lose their dispersers with serious consequences for their survival. This problem is exacerbated by a fast changing climate. Global warming causes the preferred climatic range of plants and animals to shrink or shift. These changes happen at a pace and to a degree to which many species find it impossible to adjust by migration or evolutionary adaptation, either because of human-caused habitat fragmentation or simply for geographical reasons (for example, natural boundaries set by mountain ranges and, on islands, by the surrounding water). Unlike animals, many seed-plants can survive long periods of unfavourable conditions in the form of their seeds. Most spermatophytes produce desiccation-tolerant (*orthodox*) seeds that keep their viability for many years if kept in a dry state. Therefore, with their seeds, fruits literally hold the key to the survival of a species, whether they give rise to a gigantic tree or a tiny herb. The astonishing ability of orthodox seeds to survive for a long time in a dry state is their most significant quality. Because of their small size and longevity, they provide an extremely efficient means of preserving plant germplasm. Experiments have shown that the longevity of seeds increases as their moisture content decreases and the ambient temperature drops. This quantifiable relationship between water content and ambient temperature is summarized in Harrington's 1973 rule of thumb, which predicts a doubling of storage life for every one per cent reduction in moisture content (based on the fresh weight), and for every 5° Celsius (10° Fahrenheit) reduction in storage temperature. This simple rule forms the theoretical basis for institutions devoted to preserving plant germplasm in the form of seeds, so-called *seed banks*. In seed banks, seeds are stored in air-tight containers at low temperatures (for example minus 20° C). Under these conditions, seed longevity varies from a few decades for the shortest-lived species to over a thousand years for the longest lived. Throughout the world, there are many seed banking enterprises that focus on the preservation of the genetic diversity of thousands of human-bred varieties of major crops, especially cereals, and their wild ancestors.

Until now, only a few seed banks have applied this technology to save wild plant species from the growing risk of extinction. One such exception is the Millennium Seed Bank Project (MSBP) of the Royal Botanic Gardens, Kew. The MSBP comprises 115 partner institutions in 53 countries, which all have the common goal of collecting and storing the seeds of thousands of wild species. This international conservation project established to mark the new millennium was founded in 2000 with funding from the UK Millennium Commission, The Wellcome Trust, Orange plc, and other corporate and private sponsors. The initial goal is to conserve the seeds of 10 per cent of the world's wild seed plant species by 2010 – approximately 24,000 species (based on Mabberley's 1987 conservative estimate of 242,000 species of spermtophytes worldwide). During its second phase, it aims to conserve 25 per cent of all wild seed plant species by 2020, while promoting the conservation and sustainable use of plant diversity on our planet. Although the seeds collected are to support a wide variety of projects in agriculture, forestry, horticulture and habitat repair now and in the near future, the majority are stored long term with the option of use in the distant future.

In view of the current massive destruction of natural habitats and extinction of species everywhere on Earth, collecting seeds for an uncertain distant future may seem a futile mission. The pace of environmental destruction leaves little hope that many plants will ever return to their unique natural environments that sustained them so perfectly for millions of years. Despite the bleak prospects, we owe it to future generations to fight the catastrophic loss of biodiversity. Even if today habitats are destroyed to an extent that seems irreparable with current means, one day mankind may have the knowledge and technology to restore what it thought it had lost. After all, under seed banking conditions, many seeds remain viable for hundreds of years. Would anybody in the Middle Ages have believed that people would one day walk on the Moon? This may seem a far-fetched analogy, but too much that is irreplaceable is at stake to limit our hopes and visions for the future.

To put our situation into perspective, the fossil record tells us that life on our planet has already experienced five major global mass extinctions. After each disaster the recovery of global biodiversity took millions or even tens of millions of years. As a comparison, the first upright-walking hominids appeared some 4-6 million years ago and modern humans like us have existed for no longer than about 200,000 years. Therefore, we have to act now and take precautionary measures, such as seed banking, if there is to be a chance of short-term recovery for our environment in the future.

Thankfully, we are becoming increasingly aware of our catastrophic impact on the environment. Perverse as it may seem, the emerging panic over the threats of overpopulation and climate change sends out a ray of hope. As yet another step forward in the evolution of *Homo sapiens sapiens,* ("knowledgeable man") the human race may find a way to honour its hitherto arrogant name and let reason prevail over selfish instincts. Perhaps, one day, if humans manage to survive, a new subspecies will evolve that may proudly call itself *Homo sapiens illuminens* ("enlightened man").

Eucalyptus macrocarpa (Myrtaceae) – mottlecah; native to Western Australia – a distinct species with striking silvery-grey foliage and spectacular red flowers up to 10cm in diameter. The fruit is a 6-7cm wide capsule shedding many small, angular, brown seeds. below: **immature fruit**; bottom: **flower**; opposite: **seed, 3.6mm long**

below: *Krameria erecta* (Krameriaceae) – Pima rhatany; native to the southern USA and northern Mexico – fruits (achenes) as they arrived at the Millennium Seed Bank after they were collected in Arizona. Adapted to attach themselves to the fur of animals, the fruits are covered by long barbed spines; the fruit (without spines) is 8mm long

bottom: a fruit of the Pima rhatany mounted on an aluminium stub and sputter-coated with platinum. The ultra-thin platinum layer covering the specimen improves the emission of secondary electrons and, very importantly, increases conduction thereby reducing electrostatic charging of the object

below: The Hitachi S-4700 Scanning Electron Microscope (SEM) at the Jodrell Laboratory of the Royal Botanic Gardens, Kew, that was used to produce the microscopic images for this book

bottom: the aluminium stub with the fruit inside the SEM's specimen chamber where, in a high vacuum, the object is scanned with an electron beam

below: a scanning electron micrograph shot at the lowest possible magnification. For the SEM, which is designed to provide magnified images of extremely small objects, the fruit of the Pima rhatany is huge

bottom: after assembling the partial images into a single composite image of the entire fruit, the artist's work can begin. The original black-and-white image is digitally transformed, carefully worked over, using a graphic tablet that allows the same sensitivity as a brush or finger. In this way each image becomes hand crafted, artistically unique, and not solely the product of digital technology

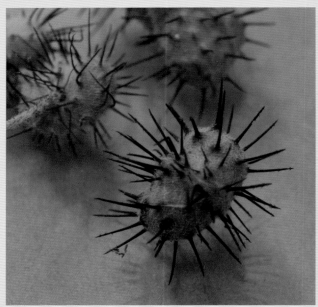

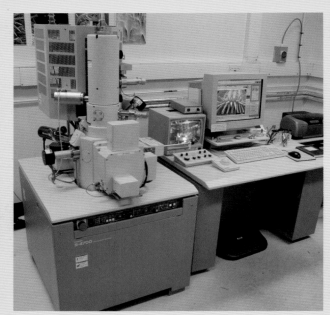

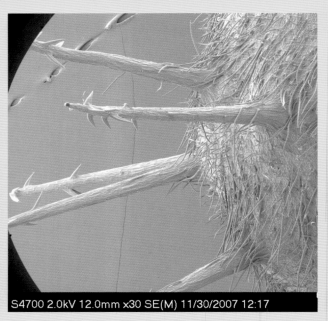

S4700 2.0kV 12.0mm x30 SE(M) 11/30/2007 12:17

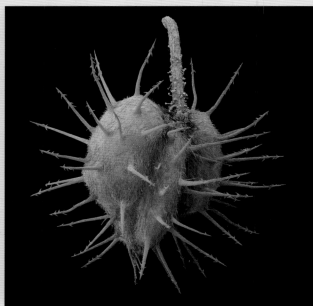

LUSCIOUSNESS

The crafted image in a digital environment

ROB KESSELER

It was my intention when developing the images for this book that they should not only reflect their subject but they should extend the creative evolution and interpretation of microscopic plant material developed from my previous two books. In *Pollen – The Hidden Sexuality of Flowers* (Kesseler & Harley), the colouring of the pollen specimens was intended to reflect the soft, ethereal, numinous quality of pollen grains and the subtle chromatic variations that occur naturally creating a sense of wonderment that something so small could be so vital for the procreation of plant life. Being a relatively new field of study I was concerned to not take too many creative liberties, using colour for emphasis and clarity, a strict fusion of artistic sensibility and scientific rigour.

In *Seeds – Time Capsules of Life* (Kesseler & Stuppy), I switched from the older analogue scanning electron microscope (SEM), from which images are transferred onto high resolution Polaroid negatives, to a newer digital model. This enabled the production of much higher resolution images of remarkable clarity, revealing an unimaginable diversity of surface topographies and forms. Unlike pollen, seeds in their naturally desiccated state appear in an endless variety of browns and blacks; to reveal their architectonic complexities it became necessary to use a more vibrant colour palette.

The development of digital imaging has been as swift as it has been impressive, to which the images of outer space developed from data sent back from the Hubble Telescope bear witness. However, to retain the trace of the artist's hand within the field of science imaging is a challenging task in a climate where programmes are constantly being developed to facilitate the production of visual spectacle. With the images in this book I wanted to move the artistic nature of depicting microscopic plant imagery to a more complex level. Since the Renaissance, artists have extended their illusory powers in depicting cornucopian tables overflowing with almost orgiastic displays of prime specimens. In the seventeenth century, artist Bartolomeo Bimbi, under the patronage of the Medicis created a series of canvases devoted to single fruits; in one painting alone he depicted no fewer than 115 different varieties of pear and in another 34 different lemons. It was my intention from the start to try and match this luscious outpouring.

Scanning electron microscopes were developed to enable higher resolution images than conventional microscopy allowed. Individual specimens are cleaned, dried, and attached to a small aluminium stub before being coated with an infinitesimally fine layer of gold or platinum. In this state they are like small, exquisitely made, precious metal jewels. They are then placed inside a vacuum chamber and bombarded with electron particles; the resulting data provide the digital image. Pollen grains are small, many hundreds can be captured in one frame; seeds are bigger, one seed fills one frame. Fruit, even small fruit, by its very nature tends to be much larger, most of it too large to fit in the SEM. The image of the young strawberry fruit on the cover of this book is made up of over forty separate frames carefully stitched together, cleaned up, tonally readjusted, and finally coloured, a process taking many long, intensive hours.

Whereas working in watercolour or pastel is an additive process – building up washes and layers of colour, covering what lies beneath – these images evolve from flat, grey micrographs, subsequently translated into many colour layers, carefully worked over or eroded away using a graphic tablet with the same sensitivity as a brush or finger. In this way each image becomes *hand crafted,* artistically unique, and not solely the product of digital technology.

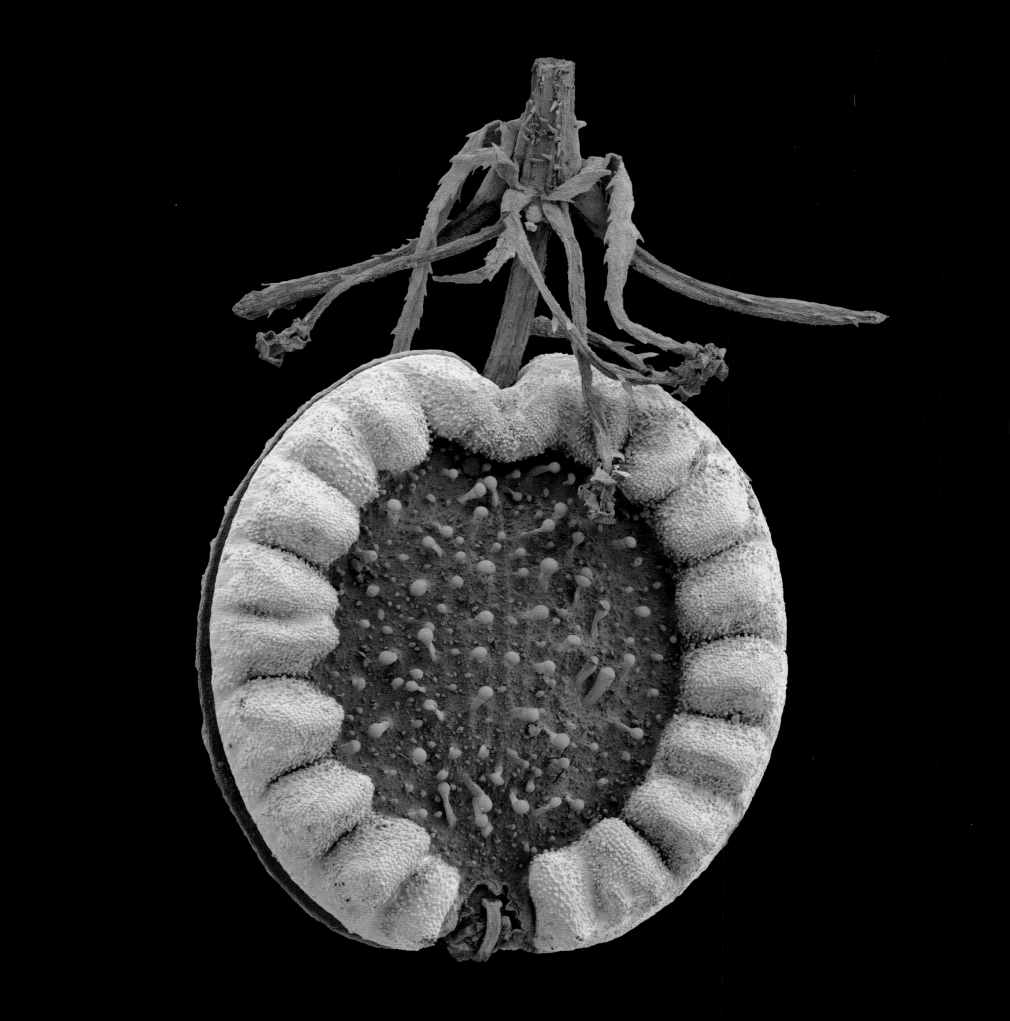

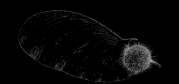

APPENDICES

Tordylium apulum (Apiaceae) – Roman pimpernel; native to Europe
and western Asia – fruit (polachenarium); as typical of the members
of the carrot family, the fruit eventually splits into two fruitlets. The
flat shape and swollen margin whose very light, spongy tissue is
subdivided into a ring of air bags indicate adaptation to wind
dispersal; 5mm long

GLOSSARY

EXPLANATORY NOTES

Like other scientific disciplines, botany in general and carpology in particular have their own vocabulary. When used for the first time in the text, technical terms are italicised. To avoid spoiling the flow of the text a glossary is included for those readers not yet familiar with these terms. The common names of plants are used whenever possible, but their Latin names are indispensable because they are unique and understood by naturalists worldwide, whatever their mother language. Common names are different in every language and some species may have several common names, and one common name may refer to several different species. Latin names generally consist of two parts, the genus name and the species name (e.g. *Liriodendron tulipifera*). A full Latin name includes its formal describer(s) at the end as in the *Index of Plants Illustrated* that follows. A group of closely-related species forms a genus and a group of closely-related genera forms a family (e.g. Magnoliaceae). With the application of molecular techniques to elucidate the natural relationships of plants, their classification – especially in the case of the flowering plants – has undergone profound changes. The boundaries of many long-established families have changed and some have been split (most recently, for example, the Scrophulariaceae). For the familial classification we have adopted Peter Stevens's system (Stevens, P.F. [2001 onwards]. Angiosperm Phylogeny Website. Version 8, June 2007), which largely follows the latest classification of the Angiosperm Phylogeny Group, an international team of scientists jointly researching the natural relationships of the angiosperms. The website of IPNI (International Plant Names Index; www.ipni.org), the International Legume Database & Information Service (www.ildis.org), the World Checklist of Monocotyledons (apps.kew.org/wcsp/home.do) and the W3TROPICOS website of the Missouri Botanical Garden (http://mobot.mobot.org/W3T/Search/vast.html) proved very helpful in verifying Latin names.

With the listed exceptions, the majority of photographs included here are the original work of the authors. Photographs were taken with Nikon digital cameras (model D100 and D200) using 60mm micro nikkor and 35–105 macro nikkor lenses. Digital scanning electron micrographs were produced with a Hitachi S-4700 Scanning Electron Microscope (SEM). Rob Kesseler subsequently coloured the original black-and-white SEM-images but left them unaltered in any other way. The choice of colours was inspired by the natural colours of the plant or its flowers, the structure and function of the seed coat, or simply by the intuition of the artist.

The fruits illustrated in this book have been sourced mainly from the collections of the Royal Botanic Gardens, Kew, including the living collections of Kew Gardens and Wakehurst Place, the Millennium Seed Bank, Kew's Herbarium Seed Collection (now housed at the Millennium Seed Bank) and the carpological and other collections of the Kew Herbarium.

Frequently used abbreviations: sp. = species (singular); spp. = species (plural); definitions of fruit types modified after Spjut (1994)

achene (Greek: *a* + *khainein* = to yawn): a small indehiscent, usually single-seeded fruit with a dry pericarp that is contiguous to the seed but distinguishable from the seed coat, e.g. sunflower (*Helianthus annuus*, Asteraceae).

achenetum: a multiple fruit of indehiscent carpels with the pericarp of each (achene-type) fruitlet contiguous to the seed.

aerial seed bank: see serotiny.

aggregate fruit = see multiple fruit.

amphisarcum (Greek: *amphi* = on both sides, around + *sarx* = flesh): a simple indehiscent fruit characterized by a pericarp differentiated externally into a dry crust and internally into one or more fleshy layers, e.g. baobab (*Adansonia* spp., Malvaceae).

anemoballism (Greek: *anemos* = wind + *ballistes*, from *ballein* = to throw): a form of indirect wind dispersal, i.e. wind does not transport the diaspore directly but exerts its influence on the fruit. The fruit (mostly a capsule) is usually exposed on a long, flexible stalk that swings in the wind, thereby flinging out the diaspores, e.g. poppy (*Papaver rhoeas*, Papaveraceae).

anemoballist: a plant dispersing its diaspores by anemoballism; see anemoballism.

anemochorous: wind dispersed; see anemochory.

anemochory (Greek: *anemos* = wind + *chorein* = to disperse): dispersal of diaspores by wind.

angiosperms (Greek: *angeion* = vessel, small container + *sperma* = seed): division of the seed plants (spermatophytes) that bear ovules and seeds in closed megasporophylls (carpels), in contrast to gymnosperms, which have exposed ovules and seeds, borne "naked" on the megasporophylls. Angiosperms are distinguished by a unique process of sexual reproduction called "double fertilization." According to the number of leaves (cotyledons) present in the embryo

two major groups are distinguished, the Monocotyledons and the Dicotyledons. Angiosperms are commonly referred to as "flowering plants" even though the reproductive organs of some gymnosperms are also borne in structures that fulfil the definition of a flower.

anthecosum (Greek: *anthos* = flower+ *oikos* = house + *osum*): a compound fruit of the Poaceae consisting of fused parts of branches, leaves, or glumes that form a burr or involucre around the florets; e.g. *Cenchrus spinifex*.

anther (Medieval Latin: *anthera* = pollen, derived from Greek: *antheros* = flowery, from *anthos* = flower): the pollen-bearing part of a microsporophyll (stamen) of the angiosperms. An anther consists of two fertile halves called "thecae," each bearing two pollen sacs (= microsporangia), which usually dehisce with longitudinal slits, valves or pores. The two thecae are connected by a sterile part called "connective" which is also the point where the anther is fixed to the filament.

anthocarp: see anthocarpous fruit.

anthocarpous fruit (= anthocarp; Greek: *anthos* = flower + *karpos* = fruit): a fruit in which not only the gynoecium but also other floral parts have undergone a marked development during post-fertilization to aid in the dissemination of the seed.

anthophyta/anthophytes (Greek: *anthos* = flower + *phyton* = plant): literally "flowering plants," a term often used synonymously with angiosperms. However, anthophytes also include some gymnosperms, the extinct cycad-like Bennettitales, the closely related *Pentoxylon* and the present-day Gnetales order (comprising the three genera *Ephedra*, *Gnetum*, and *Welwitschia*).

aperture: in pollen grains a preformed opening in the pollen wall through which the pollen tube penetrates.

apocarpous gynoecium (Greek: *apo* = being apart from + *karpos* = fruit): a gynoecium consisting of two or more separate carpels, each carpel forming an individual pistil.

archegonium (New Latin, from Greek: *arkhegonos* = offspring; from *arkhein* = to begin + *gonos* = seed, procreation): often flask-shaped, multi-cellular female sexual organ of a female or bisexual gametophyte producing and containing the female egg cell(s). Archegonia are fully developed in mosses, ferns and fern allies in the broadest sense, but only rudimentary in gymnosperms. In angiosperms true archegonia are absent (with the three-celled egg apparatus as the homolog).

aril (Latin: *arillus* = grape seed): edible seed appendages of various origin in gymnosperms and angiosperms. Arils usually developed as a reward for animal dispersers.

arillocarpium (Latin: *arillus* = grape seed, here referring to aril, a fleshy outgrowth around the seed, + Greek: *karpos* = fruit): a fruit of the conifers consisting of a seed covered by a fleshy appendage ("aril"), e.g. Taxaceae.

-arium: ending that in carpological terminology indicates a schizocarpic fruit.

autochory (Greek: *autos* = self + *chorein* = to disperse): self-dispersal.

bacca (Latin: berry): see berry.

baccarium (Latin: *bacca* = berry + *-arium*): a schizocarpic fruit whose indehiscent fleshy fruitlets resemble berries.

baccetum: a multiple fruit consisting of indehiscent berry-like fruitlets (carpels).

ballistic dispersal: dispersal of diaspores through direct or indirect catapult mechanisms, i.e. explosively dehiscent fruits or movement of plant parts by wind (anemoballism) and passing animals, respectively.

berry: a simple fruit whose pericarp (fruit wall) is entirely fleshy.

bibacca (Latin: double berry): a compound fruit composed of two mature ovaries that are partially fused, e.g. *Lonicera xylosteum* (Caprifoliaceae).

drupa (Latin *drupa*: overripe olive, from Greek: *dryppa* = olive) = drupe.

drupe: an indehiscent simple fruit with a fleshy mesocarp and a hardened endocarp that produces one or more stones.

drupetum: a multiple fruit of indehiscent carpels with the pericarp of each drupe-like fruitlet (drupelet) differentiated into a thin skin-like epicarp, a fleshy mesocarp and a hardened endocarp.

ballistic dispersal: mode of dispersal by which the diaspores are actively or passively catapulted away from the plant. This can happen either suddenly or by an external trigger.

Bennettitales: extinct order of gymnospermous seed plants that first appeared in the Triassic period (248-206 million years ago) and became extinct towards the end of the Cretaceous (142-65 million years ago). Because of their superficial resemblance to cycads they are also called cycadeoids ("cycad-like").

bract: a reduced or rudimentary leaf in the region of the flower or inflorescence. Bracts can either be small, green and inconspicuous or large and brightly coloured.

bracteole (New Latin: *bracteole*, diminutive of *bractea* = gold leaf): small specialized bracts immediately subtending a flower. Contrary to bracts, bracteoles are only ever associated with a single flower.

Caenozoic era (Greek: *kainos* = new + *zoion* = living being or animal, meaning "new animals"): time period from 65 million years ago until today. The Caenozoic is divided into two periods, which themselves are further divided into epochs, named the Tertiary (Paleocene, Eocene, Oligocene, Miocene and Pliocene epochs) and Quaternary (Pleistocene

and Holocene epochs.). An alternative classification divides the Caenozoic era into the Paleogene (Paleocene to Oligocene) and the Neogene (Miocene to Holocene) periods.

caltrop: a structure that consists of four spines that are arranged to point to the four corners of a tetrahedron so that, however it falls, it will sit on three of the spines with the fourth one pointing up in the air. Caltrops were first used as a means to slow down pursuers on horseback but later proved equally effective on car tyres.

calyx (Greek: *kalyx* = cup): the summary of the sepals of a flower, i.e. the outer whorl of floral leaves in a perianth.

camara (Greek: *kamara* = vault): an indehiscent or tardily dehiscent fruit, formed by a single carpel. Camaras may be internally dry (e.g. peanut, *Arachis hypogaea*, Leguminosae) or fleshy (e.g. tamarind, *Tamarindus indica*, Leguminosae).

capitulum (Latin: small head; diminutive of *caput* =head): an inflorescence with a condensed main axis bearing sessile flowers in a dense cluster, usually surrounded by an involucre of bracts, e.g. Asteraceae, Dipsacaceae.

capsiconum (Latin: *capsa* = box + *conum*, from Latin: *conus* = cone): a compound fruit composed of capsular fruitlets; e.g. *Liquidambar styraciflua* (Hamamelidaceae).

capsule (Latin: *capsula*, diminutive of *capsa* = box, capsule): a dehiscent fruit developing from a syncarpous gynoecium (i.e. composed of more than one carpel) and dispersing the seeds by opening the pericarp.

carcerulus (diminutive of Latin: *carcer* = prison): a simple indehiscent fruit formed by a syncarpous gynoecium, bearing one or more seeds surrounded by an air space within a firm pericarp.

Carboniferous: geological time period 354-290 million years before present.

carpel (New Latin: *carpellum* = little fruit; originally from Greek: *karpos* = fruit): in angiosperms a fertile leaf that encloses one or more ovules (megasporophyll). Carpels are usually differentiated into an ovule-bearing part (ovary), a style and a stigma. The carpels of a flower can either be separate from each other to form an apocarpous gynoecium or joined to form a syncarpous gynoecium.

carpophore (Greek: *karpos* = fruit + *phorein* = to bear; literally "fruit bearer"): the central column (axis) of a schizocarpous fruit from which the fruitlets separate at maturity except at one point as in the typical fruit of the carrot family (Apiaceae).

caryopsis (Greek: *karyon* = walnut or any nut, kernel + *-opsis* = resemblance): the traditional name of the fruit (nut) of the members of the grass family (Poaceae). The caryopsis is very similar to the achene. The only difference is that in a caryopsis the pericarp is not distinguishable from the seed coat except under high magnification.

caulicarpy (Latin: *caulis* = stem + Greek: *carpos* = fruit): the bearing of fruits directly on the bark the main trunk or the major branches of a tree.

Cell theory: Theory published in 1839 by Matthias Jakob Schleiden and Theodor Schwann, claiming that all organisms are composed of similar basic units of organization, called cells. In 1858, Rudolf Virchow completed the classical cell theory by adding his conclusion that all cells come from pre-existing cells.

central placentation: type of placentation in which the ovules are borne on a free-standing placenta in the centre of the ovary, as in Primulaceae, for example.

chiropterochorous: bat dispersed; see also chiropterochory.

chiropterochory (New Latin: *chiroptera* = bats, from Greek: *kheir* = hand + *pteron* = feather, wing + *chorein* = to disperse): dispersal of plant diaspores by bats.

circumscissile capsule: see pyxidium.

climacteric fruits: fruits that enter a sudden and irreversible burst of ripening marked by a distinct increase in respiration (CO_2-production) and the production of the volatile phytohormone ethylene. In contrast, non-climacteric fruits undergo a gradual and continuous ripening process.

clubmoss: common name for a member of the Lycophyta, a group of seedless, spore-producing vascular plants. In the Carboniferous time period (354-290 million years ago) tree-like clubmoss-relatives such a *Lepidodendron* and *Sigillaria* together with horsetail-relatives were the major components of the giant forests that thrived in the extensive swamps occupying large parts of our planet. Today, Lycophytes are represented by about 1280 species of herbaceous plants such as clubmosses (*Lycopodium* spp.), selaginellas (*Selaginella* spp.) and the aquatic quillworts (*Isoetes* spp.).

coccarium (Greek: *kokkos* = grain, seed + arium): a schizocarpic fruit characterized by fruitlets opening along their ventral and dorsal suture; capsular fruits, which do not break up entirely into their carpels but display loculicidal, septicidal and septifragal dehiscence simultaneously, are also included here (e.g. the fruits of Euphorbiaceae, Hamamelidaceae and some Rutaceae).

coccum (Greek: *kokkos* = grain, seed): a simple dehiscent fruit consisting of one carpel that opens along two sutures (the coccum of the Leguminosae is traditionally termed a legume).

columella (diminutive of Latin: *columna* = column): the persistent central axis of a capsule or schizocarpic fruit.

compound fruit: a fruit derived from more than one flower. Most modern textbooks apply this meaning to multiple fruit, a term which should, however, correctly be used for fruits developing from flowers with an apocarpous gynoecium; see also explanation under multiple fruit.

compound pistil: a pistil formed by two or more joined carpels.

conifers (Latin: *conus* = cone + *ferre* = to carry, to bear): group of the gymnosperms generally distinguished by needle- or scale-like leaves and unisexual flowers borne in cones. Well-known examples of conifers are pines, spruces and firs.

Cordatiales: extinct order of paleozoic gymnosperms considered to be directly related to modern conifers. The Cordaitales were trees up to 30m tall with strap-like leaves. They were abundant during the Cretaceous period (142-65 million years ago) and became extinct in the early Permian (290-248 million years ago).

corolla (Latin *corolla* = small garland or crown): the summary of the petals of a flower, i.e. the inner whorl of floral leaves in a perianth.

cotyledon (Greek: *kotyle* = hollow object; alluding to the often spoon- or bowl-shape of the seed leaves): the first leaf (in Monocotyledons) or pair of leaves (in Dicotyledons) of the embryo.

craspedium (Greek: *kraspedon* = border): a fruit consisting of a single carpel that disarticulates into one-seeded segments, the seed-bearing segments separating transversely from each other and separating longitudinally from a marginal frame.

Cretaceous: geological time period 65-142 million years ago.

cryptogams (Greek: *kryptos* = hidden + *gamein* = to marry, to copulate): old collective term referring to all plants without recognizable flowers. Cryptogams include algae, fungi (not really plants), mosses, ferns and fern allies. The Greek meaning – "those who copulate in secret" – refers to the absence of flowers as obvious indicators of sexual propagation.

cultigen: a species that is known only in cultivation.

cupule (Latin: *cupula* = little cask): cup-like organ surrounding the female flowers and fruits of oaks (*Quercus* spp.), beeches (*Fagus* spp.), chestnuts (*Castanea* spp.) and other Fagaceae.

cycadeoids: see Bennettitales.

cycads (Greek: *kykas* = palm, alluding to their palm-like look): ancient gymnosperms superficially resembling palms. Cycads are woody plants generally distinguished by thick, unbranched trunks, large, palm-like pinnate leaves and big cones. Cycads are living fossils that were an important source of food for the dinosaurs.

cypsela (Greek: *kypselé* = box, hollow vessel): a single-seeded fruit with longitudinally oriented awns, bristles, feathers or similar structures derived from accessory parts of the flower or inflorescence. Cypselas are typically found in the Asteraceae and Dipsacaceae but also in some Cyperaceae, Proteaceae and other families.

denticidal capsule: a capsular fruit dehiscing regularly along sutures but incompletely – not more than one-fifth of the length of the capsule.

Devonian: geological time period 417-354 million years ago.

diaspore (Greek: *diaspora* = dispersion, dissemination): the smallest unit of seed dispersal in plants. Diaspores can be seeds, fruitlets of compound or schizocarpic fruits, entire fruits or seedlings (e.g. mangroves).

diclesium (New Latin from Greek: *di-* = two, double + *klesis* = a higher calling + *-ium*): a simple anthocarpous fruit consisting of a mature ovary covered partly or entirely by a loose or tightly adhering fruiting-perianth, e.g. tomatillo (*Physalis philadelphica*, Solanaceae), Cape gooseberry (*Physalis peruviana*), bird catcher (*Pisonia brunoniana*, Nyctaginaceae).

Dicots = Dicotyledons.

Dicotyledons (Greek: *di* = two + cotyledon): one of the two major groups of the angiosperms distinguished by the presence of two opposite leaves (cotyledons) in the embryo. Other typical characters of the Dicotyledons are reticulate leaf venation, floral organs usually in fours or fives, vascular bundles arranged in a circle, a persistent primary root system developing from the radicle, and secondary thickening (present in trees and shrubs, usually absent in herbaceous plants). The Dicotyledons where long considered a homogenous entity. Only recently have they been split into two groups (magnoliids) and Rosidae (eudicots).

dioecious: see dioecy.

dioecy (Greek: *di* = double + *oikos* = house): (1) the formation of male and female sexual organs on separate gametophytes (e.g. in some mosses and ferns); (2) in seed plants the formation of male and female flowers on separate individuals.

diplochory (from Greek, *diplous* = double + *chorein* = to disperse; literally meaning "two-phase dispersal"): seed dispersal by a sequence of two or more steps or phases, each involving a different dispersal agent.

dormancy (Latin: *dormire* = to sleep): generally referring to a quiescent period in the life of a plant. When referring to seeds, dormancy summarizes the various mechanisms that ensure the seeds do not germinate immediately, even under the most favourable conditions.

drupetum: a multiple fruit of indehiscent carpels with the pericarp of each drupe-like fruitlet (drupelet) differentiated into a thin skin-like epicarp, a fleshy mesocarp and a hardened endocarp.

elaiosome (Greek: *elaion* = oil + *soma* = body): literally meaning "oil body;" a general ecological term referring to fleshy and edible appendages of diaspores, usually in the context of ant dispersal.

embryo (Latin: *embryo* = unborn foetus, germ, originally from Greek: *embryon*: *en-* = in + *bryein* = to be full to bursting): in plants the young sporophyte developing from the egg cell after fertilization.

endocarp (Greek: *endon* = inside + *karpos* = fruit): the innermost layer of the fruit wall (pericarp) forming the hard stone in drupes.

endosperm (Greek: *endon* = inside + *sperma* = seed): nutritive tissue in seeds. Generally, the term endosperm refers only to the nutritive tissue in the seeds of the angiosperms where it represents a (usually) triploid tissue as a result of the double fertilization. The food storage of gymnosperm seeds consists of the haploid tissue of the megagametophyte. To distinguish the two different types, the nutritive tissues of gymnosperms and angiosperms are called "primary endosperm" and "secondary endosperm," respectively.

endozoochorous: dispersal through transport in an animal's gut. See also endozoochory.

endozoochory (Greek: *endon* = inside + *zoon* = animal + *chorein* = to disperse): dispersal of the diaspores of a plant by being eaten and carried inside the gut of animals (and humans); the usually hard seeds or endocarps pass the intestines undamaged and are deposited with the faeces.

epicarp (Greek: *epi* = on, upon + *karpos* = fruit): the outermost layer of the fruit wall (pericarp), mostly a soft skin or leathery peel.

epigynous flower (from Greek: *epi* = above + *gyne* = female): a flower in which sepals, petals and stamens are inserted above the ovary. As a result, the ovary is inferior and no longer visible.

epispermatium (Greek: *epi* = above + *spermatos* = seed): a gymnospermous seed of the Podocarpaceae subtended or enclosed by a swollen appendage.

epizoochorous: dispersal by adhesion to the outside of an animal; see also epizoochory.

epizoochory (Greek: *epi* = on, upon + *zoon* = animal + *chorein* = to disperse): dispersal of diaspores on the surface of a body of an animal. Epizoochorous diaspores adhere to the fleece, coat or feathers of animals or the clothes of humans by hooks or sticky substances.

-etum: ending that in carpological terminology indicates a multiple fruit.

family: one of the main units in the hierarchical system of the taxonomic classification of living organisms. The major classification units are (in descending order) class, order, family, genus and species.

filament (Latin: *filum* = thread, string): the stalk of a stamen.

fissuricidal capsule (Latin: *fissura* = fissure): a capsular fruit opening irregularly by one or more parallel slits, or regularly along sutures between a closed apex and base.

flower: a reproductive short shoot with determinate growth bearing at least one of the sexual reproductive organs (male or female). This definition of a flower applies to the reproductive structures of both angiosperms (flowering plants) and gymnosperms.

flowering plants: meaning regionally different depending on the definition of flower. In continental Europe considered to comprise both gymnosperms and angiosperms, in Anglo-America and the UK only applied to angiosperms. In a strict scientific sense, "flowering plants" are circumscribed as defined under "anthophyta."

follicarium: a schizocarpic fruit in which the carpels are distinctly separate from one another before maturing and dehiscing along their ventral sutures; e.g. Apocynaceae.

follicetum: a multiple fruit of dehiscent fruitlets that open only along one suture (dorsally or ventrally).

follicle (Latin: *folliculus* = little bag, diminutive of *follis* = bellows): a fruit or fruitlet derived from a single carpel dehiscing along one (usually the ventral) suture, e.g. the fruitlets of the marsh marigold (*Caltha palustris*, Ranunculaceae).

folliconum (Latin: *folliculus* = little bag + *conum*, from Latin: *conus* = cone): a compound fruit composed of united follicular fruitlets; e.g. *Banksia menziesii* (Proteaceae).

foraminicidal capsule: a capsular fruit opening by irregular diverging cracks or slits, e.g. snapdragons (*Antirrhinum* spp., Plantaginaceae).

frugivore (Latin: *frug-* = fruit + *vorare* = to swallow, devour): a fruit-eating animal. Facultative frugivores, such as many birds in temperate regions, eat mainly fruit when available but also other plant and animal matter, whereas obligate frugivores are specialized animals, predominantly in the tropics, who feed exclusively on fruits.

fruit: any coherent seed-bearing structure, including domesticated fruits bred to be seedless.

fruitlet: a separate dispersal unit of a fruit that may be (1) a carpel or half-carpel of a mature schizocarpic fruit, (2) a single carpel of a mature multiple fruit, or (3) a mature (mono- or multicarpellate) ovary of a compound fruit.

funiculus/funicle (Latin: *funiculus* = slender rope): the stalk by which an ovule or seed is connected to the placenta in the ovary. The funiculus acts like an "umbilical cord," supplying the developing ovule and seed with water and nutrients from the parent plant.

gametophyte (Greek: *gametes* = spouse + *phyton* = plant): the generally haploid generation in a plant's life cycle producing gametes. Examples are the prothallus of the ferns or the megaprothallium (= female gametophyte) and germinated pollen grain (= microprothallium = male gametophyte) of the seed plants.

geocarpy (Greek: *ge* = earth + *karpos* = fruit): a condition in plants whose fruits ripen underground, e.g. peanut (*Arachis hypogaea*, Fabaceae), aardvark cucumber (*Citrullus humifructus*, Cucurbitaceae).

glandarium: a fruit formed by a schizocarpous gynoecium on an enlarged, fleshy receptacle.

glandetum: a multiple fruit of indehiscent carpels that mature on an enlarged receptacle, the fruitlets embedded in the receptacle; e.g. *Fragaria* spp. (Rosaceae).

glans (Latin for acorn): an indehiscent fruit composed of a mature ovary subtended or enclosed by an aril-like structure that may be derived from united bracts (e.g. in Fagaceae) or a swollen pedicel (e.g. in *Anacardium occidentale*), receptacle or perianth.

Gnetales: heterogeneous group of gymnosperms comprising just three families with three genera (*Gnetum, Ephedra, Welwitschia*) and a total of 95 species.

granivores (Latin: *grani-*, from *granum* = grain + *-vorus*, from *vorare* = to swallow, to devour): seed-eating animals.

gymnosperms (Greek: *gymnos* = naked + *sperma* = seed): inhomogeneous group of seed plants bearing their ovules on open megasporophylls (or ovuliferous scales in conifers) and not in closed megasporophylls (= carpels) like angiosperms. Gymnosperms comprise three distantly related groups: conifers (8 families, 69 genera, 630 species), cycads (3 families, 11 genera, 292 species) and Gnetales (3 families, 3 genera, 95 species).

gynoecium (Greek: *gyne* = woman + *oikos* = house): the summary of all carpels in a flower, irrespective of whether they are joined or separate.

gynodioecy (Greek: *gyne* = female, woman + *di* = double + *oikos* = house): condition in which bisexual and female flowers are borne on different plants of the same species. In certain *Ficus* spp. (including *Ficus carica*, the edible fig) the syconia of so-called "male" trees bear male flowers as well as both kinds of female flowers (short- and long-style) whilst female trees only bear syconia with purely (long-styled) female flowers.

herbivore (Latin: *herba* = vegetation + *vorare* = to swallow, devour): a plant-eating animal.

hermaphrodite: an individual or structure bearing both male and female reproductive organs, e.g. a bisexual flower has both fertile stamens and carpels. In Greek mythology *Hermaphroditos* was the name of the handsome son of Hermes and Aphrodite, who became united with the nymph Salmacis in one body – half man, half woman.

horsetail: common name for a member of the Sphenopyhta, a group of seedless, spore-producing vascular plants. Three hundred million years ago in the Carboniferous time period, lowland forest and swamps consisted of a great variety of spore-producing trees, the most prominent being relatives of clubmosses and horsetails. Today, the sphenophytes have only one surviving genus, *Equisetum*, with about 15 species worldwide.

hydrochorous: water-dispersed. See also hydrochory.

hydrochory (Greek: *hydor* = water + *chorein* = to disperse): dispersal of plant diaspores by water. Hydrochory can further be subdivided into nautohydrochory (dispersal by water currents) and ombrohydrochory (dispersal by rain and/or dew).

hygrochastic: see hygrochasy.

hygrochasy (Greek: *hygros* = wet, moist + *chasis* = crack, gullet): production of hygroscopically dehiscing capsules that open only when wet (and often close again upon desiccation), e.g. the capsules of many succulent Aizoaceae.

hypanthium (Greek: *hypo* = under, beneath + *anthos* = flower): a cup-shaped to tubular organ in the flower derived from the receptacle carrying the sepals and petals or tepals, and the stamens. In perigynous flowers the hypanthium surrounds the gynoecium but stays separate from it. In epigynous flowers the the dorsal parts of the gynoecium are included in the formation of the hypanthium, which results in an inferior ovary.

hypogynous flower (from Greek: *hypo* = under, beneath + *gyne* = female): a flower in which sepals, petals and stamens are inserted below the exposed and clearly visible (superior) ovary.

inferior ovary: see epigynous flower.

inflorescence: part of a plant which bears a group of flowers; inflorescences can be a loose group of flowers (as in lilies) or highly condensed and differentiated structures resembling an individual flower as in the sunflower family (Asteraceae).

infructescence: the flowers of an inflorescence at the fruiting stage.

involucre: one or more whorls of bracts at the base of an inflorescence, e.g. in a capitulum of the Asteraceae and Dipsacaceae.

Jurassic: geological time period 206-142 million years ago.

legume (Latin: *legumen* = bean): the typical dehiscent fruit of the Leguminosae, developed from a single carpel that opens along two sutures (dorsally and ventrally) with the seeds attached to the ventral suture.

loculicidal capsule: a capsular fruit opening completely along dorsal sutures, the valves consisting of the two halves of adjoining carpels.

loculus/locule (Latin: *loculus* = little place, diminutive of *locus* = place): one of the seed-bearing cavities of a gynoecium; a gynoecium may have only one loculus if monocarpellate or if pluricarpellate and aseptate (i.e lacking septae between the carpels).

mammaliochorous: mammal-dispersed; see also mammaliochory.

mammaliochory (Latin: *mamma* = breast + *chorein* = to disperse): dispersal of plant diaspores by mammals.

megagametophyte (Greek: *megas* = big + *gametes* = spouse + *phyton* = plant): female gametophyte developed from the megaspore, eventually producing the female gametes (egg cells). The megagametophyte of the gymnosperms gives rise to the archegonia and persists as the seed's nutritious tissue; in angiosperms the homolog of the megagametophyte is the embryo sac.

megasporangium (Greek: *megas* = big + *sporos* = germ, spore + *angeion* = vessel, small container) the organ of a *sporophyte* that produces the female megaspores. The term is usually applied to cryptogams, whereas the homologous structure in seed plants is called nucellus.

megaspore (Greek: *megas* = big + *sporos* = germ, spore): the larger spore formed by heterosporous plants that gives rise to a female gametophyte.

megasporophyll (Greek: *megas* = big + *sporos* = germ, spore + *phyllon* = leaf): specialized fertile leaf producing megasporangia with female spores, e.g. the carpel of the angiosperms.

mericarp: a fruitlet representing half a carpel, as in the schizocarpic fruits of the Lamiaceae.

mesocarp (Greek: *mesos* = middle + *karpos* = fruit): the fleshy middle layer of the pericarp (fruit wall) in a drupe.

Mesozoic era (Greek: *mesos* = middle + *zoion* = animal, meaning "middle animals"): time from 248-65 million years ago, spanning the Triassic, Jurassic and Cretaceous periods.

microbasarium (*micros* = small + *basis* = base + *arium*): a fruit derived from a schizocarpous gynoecium which at maturity disarticulates into discrete, seed-containing, half-carpels (mericarps); e.g. Boraginaceae, Lamiaceae.

micrometer (Greek: *mikros*): one millionth of a metre (a thousandth of a millimetre), abbreviated µm.

micropyle (Greek: *mikros* = small + *pyle* = gate): the opening of the integument(s) at the apex of the ovule or seed, usually acting as a passage for the pollen tube on its way to the embryo sac. The micropyle is formed by one or both integuments, the outer one producing the exostome, the inner one the endostome.

microsporangium (Greek: *mikros* = small + *sporos* = germ, spore + *angeion* = vessel, small container): the organ of a sporophyte that produces the male microspores. The term is usually applied to cryptogams, whereas the homologous structure in seed plants is called pollen sac.

microspore (Greek: *mikros* = small + *sporos* = germ, spore): the smaller spore formed by heterosporous plants that gives rise to a male gametophyte.

microsporophyll (Greek: *mikros* = small + *sporos* = germ, spore + *phyllon* = leaf): specialized fertile leaf producing microsporangia with male spores, e.g. the stamen of the angiosperms.

monocarp: a whole carpel of a schizocarpous fruit that functions as a diaspore; see also fruitlet.

monocarpellate: consisting of a single carpel only.

Monocots = Monocotyledons.

Monocotyledons (Greek: *monos* = one + *cotyledon*): one of the two major groups of the angiosperms distinguished by the presence of only one leaf (cotyledon) in the embryo. Other typical characters of the Monocotyledons are parallel leaf venation, floral organs usually arranged in whorls of three, scattered vascular bundles, a rudimentary primary root, which is soon replaced by lateral adventitious roots (i.e. roots formed by the stem), and the lack of secondary thickening, which is why most Monocotyledons are herbaceous plants (if secondary thickening is present, as in *Agave*, *Aloe*, *Dracaena*, *Xanthorrhoea* and others, then it is different from Dicots). Monocotyledons include grasses, sedges, rushes, lilies, orchids, bananas, aroids, palms and their relatives.

monoecious: see monoecy.

monoecy (Greek: *monos* = one + *oikos* = house): (1) the formation of male and female sexual organs on the same gametophyte (e.g. in many mosses and ferns), (2) in seed plants the formation of male and female flowers on the same individual.

morphology (Greek: *morphe* = shape + *logos* = word, speech): the study of form in the widest sense but mostly restricted to the external structure of an organism as opposed to anatomy, which refers to the internal structure of an organism.

multiple fruit: a fruit that develops from an apocarpous gynoecium. Most modern authors apply this term to a fruit that is derived from more than one flower (compound fruit) and call fruits developing from apocarpous gynoecia "aggregate fruits." However, "aggregate fruit" is historically synonymous with "compound fruit," both defined as being composed of more than one flower (Spjut and Thieret 1989). Spjut and Thieret (1989) traced the confusion to Lindley (1832) who had reversed meanings for aggregate and multiple as defined by de Candolle (1813). English text books have generally adopted Lindley's errors, while non-English text books have followed de Candolle's (1813) definitions, or have employed other related terms. To avoid further confusion between aggregate and multiple, Spjut and Thieret (1989) recommended the term compound fruit be adopted instead of aggregate fruit for fruits composed of more than one flower, and that the original and correct meaning for multiple fruit be maintained. The distinction between multiple and compound fruits was first made by Gaertner (1788), but more clearly made by Link (1798).

myrmecochorous: ant-dispersed; see also myrmecochory.

myrmecochory (Greek: *myrmex* = ant + *chorein* = to disperse): dispersal of plant diaspores by ants.

nucellus (New Latin: *nucellus* = little nut): the megasporangium of the seed plants.

nuculanium: a simple fruit with a dry pericarp, which is differentiated into a hard endocarp and an outer fibrous or coriaceous layer that may or may not be dehiscent; e.g. *Cocos nucifera* (Arecaceae), *Prunus dulcis* (Rosaceae).

nut: a dry, indehiscent, usually single-seeded fruit where the pericarp is contiguous to the seed.

nutlet: diminutive of nut, referring to an individual nut-like carpel or half-carpel (mericarp) of a fruit derived from an apocarpous or schizocarpous gynoecium.

ombrohydrochory (Greek: *ombros* = a shower of rain + *hydor* = water + *chorein* = to disperse): dispersal of diaspores by rain or dew, either directly by flushing the seeds out of their fruits (splash-rain dispersal), or indirectly by triggering a springboard mechanism (rain ballism).

ornithochorous: bird-dispersed; see also ornithochory.

ornithochory (Greek: *ornis* = bird + *chorein* = to disperse): dispersal of plant diaspores by birds.

-osum: ending that in carpological terminology indicates a compound fruit.

ovary (New Latin: *ovarium* = a place or device containing eggs, from Latin: *ovum* = egg): the enlarged, usually lower portion of a pistil containing the ovules.

ovule (New Latin: *ovulum* = small egg): the integumented megasporangium of the seed plants which, after fertilization of its egg cell, develops into the seed.

Paleozoic era (Greek: *palaios* = ancient + *zoion* = living being or animal, meaning "ancient animals"): time period from 540-248 million years before present, spanning the Cambrian, Ordovician, Silurian, Devonian, Carboniferous and Permian periods.

Pangaea: the ancient supercontinent in which all continents of the earth were once joined before they were separated by continental drift.

pappus (Latin: *pappus* = old man, from Greek: *pappos* = old man, old man's beard): bristles, awns, hairs or scales that develop at the upper margin of the fruit of the Asteraceae, possibly representing a reduced calyx. A pappus is often an adaptation for wind dispersal of the fruit (cypsela), e.g. in dandelion (*Taraxacum officinale*), meadow salsify (*Tragopogon pratensis*).

parietal placentation: type of placentation in which the ovules are attached to placentas on the walls of the ovary.

parthenogenesis (Greek: *parthenos* = virgin + Greek: *genesis* = birth, beginning) form of asexual reproduction whereby an egg cell develops into an embryo without prior fertilization by a male gamete. Parthenogenesis is usually the result of an abnormal meiosis resulting in an egg nucleus with an unreduced number of chromosomes (e.g. in dandelion, *Taraxacum officinale*, Asteraceae).

passerines: birds belonging to the order Passeriformes, better known as "perching birds" or "song birds." Over five thousand species, more than half of all known species of birds, belong to the passerines. Familiar examples are sparrows, finches and thrushes.

perianth (Greek: *peri* = around + *anthos* = flower): the floral envelope that is clearly differentiated into calyx (outer perianth whorl) and corolla (inner perianth whorl).

pericarp (New Latin: *pericarpum*, from Greek: *peri* = around + *karpos* = fruit): the wall of the ovary at the fruiting stage. The pericarp can be homogenous (as in berries) or differentiated into three layers (as in drupes) called epicarp, mesocarp, and endocarp.

perigone: floral envelope composed of uniform floral leaves, i.e. without differentiation into sepals (calyx) and petals (corolla).

perigynous flower (from Greek: *peri* = around + *gyne* = female): a flower in which the gynoecium is surrounded by a hypanthium but stays separate from it.

Permian: geological time period 290-248 million years ago.

petal (New Latin: *petalum*, from Greek *petalon* = leaf): in flowers where the outer whorl of the perianth is different from the inner whorl the elements of the inner whorl of the floral envelope are addressed as petals. The summary of the petals forms the often brightly coloured, showy corolla of a flower.

pistil (Latin: *pistillum* = pestle; alluding to the shape): an individual ovary with one or more styles and stigmas, composed of one or more carpels; introduced in 1700 by Tournefort. Nowadays, many authors use the term pistil only to refer to syncarpous ovaries or omit its usage and replace it with gynoecium.

placenta (New Latin: *placenta* = flat cake, originally from Greek *plakoenta*, accusative of *plakoeis* = flat, related to *plax* = anything flat): a region within the ovary where the ovules are formed and remain attached (usually via a funiculus) to the parent plant until the seeds are mature. In botany the term was adopted from the similar structure to which the embryo is attached in animals and humans.

Pleistocene epoch (Greek: *pleistos* = most + *kainos* = new): geological time period from 1.8 million to 11,550 years ago.

pluricarpellate: consisting of two or more carpels.

pod: colloquially used as a general term for any dry fruit composed of one or more carpels with a firm pericarp surrounding a cavity and containing one or more seeds. Some botanists restricted the usage of the term pod to certain fruits of the legume family (Leguminosae).

polachenarium: a schizocarpic fruit in which the fruitlets at maturity separate longitudinally from one another and remain attached to a central column (carpophore) formed by the ventral vascular bundles of the carpels, e.g. Apiaceae.

pollen (Latin for *fine flour*): the microspores of the seed plants, able to germinate on or near the megasporangium to produce a very small and strongly simplified microgametophyte.

pollen sac: the microsporangium of the angiosperms; one anther typically bears four pollen sacs.

pollen tube: tube-like structure formed by the germinating pollen grain. In cycads and *Ginkgo* the pollen tube releases the motile sperm directly into the pollen chamber from where they swim to the archegonia. In conifers and angiosperms the pollen tube delivers the sperm nuclei straight to the egg cells.

pollination drop: drop of liquid secreted by the micropyle of many gymnosperms as a means to collect pollen. The pollination drop is finally reabsorbed and the captured pollen sucked into the pollen chamber.

pome (Latin: *pomum* = fruit): an indehiscent simple anthocarpous fruit composed of a thick fleshy hypanthium and a pericarp differentiated into a thin fleshy outer layer (which is fused with the hypanthium) and a crustaceous or stony endocarp, e.g. Rosaceae-Maloideae such as apple (*Malus pumila*, Rosaceae), pear (*Pyrus communis*), quince (*Cydonia oblonga*).

pometum: a multiple fruit of carpels embedded in a hypanthium or receptacle that is not divided into more than one cavity, e.g. rose hips (*Rosa* spp., Rosaceae).

poricidal capsule: a capsular fruit that opens with a localised pore in each loculus.

pseudanthecium (*pseudos* + *anthos* + *oikos* + *ium*): a fruit of the Cyperaceae in which the mature achene-like ovary is enclosed by a loose or inflated sac of modified connate bracts (e.g. *Kyllinga squamulata*).

pseudocarp (Greek: *pseudos* = false, the lie, from *pseudein* = to lie + *karpos* = fruit): meaning "false fruit;" a term used in modern textbooks to denote a fruit in which not only the gynoecium but also other floral parts participate. The correct term for such a fruit is anthocarp.

pseudosamara (Greek: *pseudos* = false + Latin: *samara* = elm fruit) an anthocarpous fruit bearing distal wings longer than the mature ovary; e.g. accrescent sepals in Dipterocarpaceae.

pteridosperms (Greek: *pteris* = fern + *spermatos* = seed): fossil group of gymnosperms superficially resembling ferns; therefore also called "seed ferns."

pyrene (Greek: *pyren* = the stone of a fruit): the hard, bony endocarp of a drupe, usually simply referred to as a "stone." The stones of drupes are usually single-seeded but there are also multi-seeded stones (e.g. *Pleiogynium timoriense*, Anacardiaceae). The term pyrene is mostly used when a drupe contains more than one stone (e.g. *Ilex* spp., Aquifoliaceae, *Uapaca* spp., Phyllanthaceae); it is also used to refer to the entire fruit, if a multi-stoned drupe.

pyxidium (New Latin from Greek: *pyxidion*, diminutive of *pyxis* = box): a capsular fruit opening with a lid that is created by a transverse suture cutting across all loculi of the fruit.

rain ballism: see ombrohydrochory.

raphe (Greek: *raphe* = seam, suture, from *rhaptein* = to sew): area of the seed coat in which the continuation of the funicular vascular bundle runs from the hilum to the chalaza. The raphe is longest in anatropous seeds, shorter in moderately campylotropous seeds, and entirely absent in strongly campylotropous and atropous seeds.

Restinga: a distinct type of tropical and subtropical forest found on acidic, nutrient-poor soils on the the Atlantic coast of Brazil.

sarcotesta (Greek: *sarko* = flesh + Latin: *testa* = shell): a fleshy seed coat.

samara (Latin name for the fruit of the elm): a winged nut or achene in which the wing(s) is (are) longer than the seeded portion.

samarium: a schizocarpic fruit breaking up into indehiscent fruitlets that bear wings longer than the seeded portion, e.g. *Acer* spp., *Dipteronia* spp. (both Sapindaceae).

scanning electron microscope: a scientific instrument that produces highly-magnified images at extremely high resolution by using an electron beam to scan the specimen.

schizocarpic fruit (New Latin, from Greek: *skhizo-*, from *skhizein* = to split + *karpos* = fruit): one in which the carpels are partially or completely joined at the time of pollination but separate at maturity into their carpellary constituents, sometimes further dividing into mericarps, each part functioning as a seed dispersal unit.

seed: the organ of the seed plants (spermatophyta) that encloses the embryo together with a nutritious tissue inside a protective seed coat. Seeds develop from integumented megasporangia (ovules), the defining organ of the seed plants.

seed plants: plants that produce seeds, see spermatophyta.

sepal (New Latin: *sepalum*, an invented word, perhaps a combination of Latin: *petalum* and Greek: *skepe* = cover, blanket): in flowers where the outer whorl of the perianth is different from the inner whorl, the elements of the outer whorl are addressed as sepals. The summary of the sepals forms the generally inconspicuous green calyx of a flower.

septae: plural of septum.

septicidal capsule: a capsular fruit opening completely along ventral sutures, each valve consisting of the whole carpel with the placenta attached.

septifragal capsule: a capsular fruit that incompletely opens along the dorsal or ventral sutures by a break in the partitions (septae) nearer the central axis, leaving a persistent columella after the valves have separated.

septum (plural septae; Latin: *dissepimentum* = wall, division): partition within an ovary.

serotiny (Latin: *serotinus* = coming late, from *sero* = at a late hour, from *serus* = late): late in developing or blooming, in context with seeds referring to the condition in plants that maintain an aerial seed bank by holding on to their fruits and seeds for a long time after they have matured. Serotiny is an adaptation to fire-prone habitats; the fruits of serotinous plants release their seeds after exposure to high temperatures.

simple fruit: a fruit that develops from *one flower* with only *one pistil*, in which a pistil can be either a single carpel or several joined carpels.

single nut: see camara.

soil seed bank: the summary of all viable seeds present on and in the ground.

sorosus (Latin from Greek: *soros* = heap): a compound fruit composed of many succulent fruitlets that develop on a peduncle where the fruitlets may be either free from each other (e.g. *Broussonetia papyrifera*, *Morus nigra*; both Moraceae) or fused (e.g. *Ananas comosus*, Bromeliaceae, *Cornus kousa* subsp. *chinensis*, Cornaceae).

sperm nucleus: the extremely reduced, non-motile male gamete of conifers and angiosperms.

spermatophyta/spermatophytes (Greek: *spermatos* = seed + *phyton* = plant): seed-producing plants. Group of plants characterized by the female gametophyte being developed and retained within an integumented megasporangium (ovule), which after fertilization of the egg cell develops into a seed. The spermatophytes comprise two major groups, the gymnosperms and the angiosperms.

splash-rain dispersal: see ombrohydrochory.

sporangium (Greek: *sporos* = germ, spore + *angeion* = vessel, container): container with an outer cellular wall and a core of cells which give rise to spores.

spore: a cell serving asexual reproduction.

sporophyll (Greek: *sporos* = germ, spore + *phyllon* = leaf): fertile leaf carrying one or more sporangia. Heterosporous plants usually have specialized microsporophylls producing male (micro-)spores and megasporophylls producing female (mega-)spores.

sporophyte (*sporos* = germ, spore + *phyton* = plant): literally "the plant which produces the spores;" the diploid generation in the life cycle of plants that produces asexual, haploid spores giving rise to haploid gametophytes.

stamen (Latin: *stamen* = thread): the microsporophyll of the angiosperms, consisting of the sterile filament that carries the fertile anther at the apex; each anther bears four pollen sacs (microsporangia) containing the pollen grains (microspores).

stigma (Greek = spot, scar): the upper end of a carpel able to receive pollen grains; the stigma is usually elevated above the ovary by a style.
style (Greek: *stylos* = column, pillar): in angiosperms the narrow, elongated part of a carpel or pistil connecting stigma and ovary through which the pollen tubes grow down into the ovary.
superior ovary: see hypogynous flower.
suture (Latin: *sutura* = seam): a line marking a junction or seam of union of organs, sometimes representing preformed lines of dehiscence along which, for example, a carpel of a dehiscent fruit opens; the dorsal suture of a carpel usually coincides with the central vascular bundle ("midrib") of the carpel, the ventral suture usually coincides with the line of fusion of the carpellary margins.
syconium (Greek: *sykon* = a fig): a fleshy compound fruit whose fruitlets are enclosed in an infolded peduncle (inflorescence axis).
syncarpium (Greek: *syn* = *together* + *karpos* = fruit): a multiple fruit derived from a flower with distinct carpels at the flowering stage that become fused at maturity.
syncarpous gynoecium (Greek: *syn* = together + *karpos* = fruit and *gyne* = woman + *oikos* = house; literally "a joint women's house"): a gynoecium consisting of two or more joint carpels.
tepal (= *tepalum*, made-up Latin word (anagram of *petalum*) invented in analogy to petal and sepal): a member of a perianth that is not differentiated into a calyx and corolla, i.e. a floral leaf of a perigone.
Tertiary: geological time period 65-2million years ago.
Triassic: geological time period 248-206 million years ago.
trymetum: a multiple fruit characterized by mature ovaries that develop within a hypanthium or united bracts, and upon maturity are dispersed by the unfolding or splitting of the hypanthium or bracts; e.g. Monimiaceae (*Palmeria scandens*).
trymosum: a compound fruit consisting of mature ovaries that develop within united bracts or a receptacle, and at maturity are released by splitting or other movement of the bracts or receptacle; e.g. Fagaceae (e.g. *Fagus sylvatica, Castanea sativa*), Moraceae (e.g. *Dorstenia* spp.).
trymoconum: a compound fruit composed of fruitlets that are arranged in a conelike structure, where each fruitlet disperses its mature ovary by dehiscent bracts; e.g. Casuarinaceae *Casuarina* spp.; *Allocasuarina* spp.).
zoochorous: animal dispersed; see zoochory.
zoochory (Greek: *zoon* = animal + *chorein* = to disperse): dispersal of plant diaspores by animals.
zygote (Greek: *zygotos* = joined together): fertilized (diploid) egg cell.

BIBLIOGRAPHY

FRUIT & BOTANY

Amico, G. & Aizen, M.A. (2000) Mistletoe seed dispersal by a marsupial. *Nature* 408: 929-930.
Armstrong, W.P. A nonprofit natural history textbook dedicated to little-known facts and trivia about natural history subjects. www.waynes-word.com
Babweteera, F., Savill, P. & Brown, N.N. (2007) *Balanites wilsoniana*: regeneration with and without elephants. *Biological Conservation* 134: 40-47.
Barnea, A., Yom-Tov, Y. & Friedman J. (1991) Does ingestion by birds affect seed germination? *Functional Ecology* 5: 394-402.
Beattie, A.J. (1985) *The evolutionary ecology of ant-plant mutualism*. Cambridge University Press, Cambridge, UK
Beattie, A.J. & Culver, D.C. (1982) Inhumation: how ants and other invertebrates help seeds. *Nature* 297: 627.
Bell, A.D. (1991) *Plant form – An illustrated guide to flowering plant morphology*. Oxford University Press, Oxford, UK.
Bischoff, G. W. (1830) *Handbuch der botanischen Terminologie und Systemkunde*, Vol. 1. Johann Leonhard Schrag, Nürnberg.
Bollen A.; van Elsacker, L.; Ganzhorn, J.U. (2004) Tree dispersal strategies in the littoral forest of Sainte Luce (SE-Madagascar). *Oecologia* 139(4): 604-616.
Bond, W. & Sillingsby, P. (1984) Collapse of an ant-plant mutualism: the argentine ant (*Iridomyrmex humilis*) and myrmecochorous Proteaceae. *Ecology* 65: 1031-1037.
Bouman, F., Boesewinkel, D., Bregman, R., Devente, N. & Oostermeijer, G. (2000) *Verspreiding van zaden*. KNNV Uitgeverij, Utrecht.
Boutin, S., Wauters, L., McAdam, A., Humphries, M. Tosi, G. & Dhondt, A. (2006) Anticipatory reproduction and population growth in seed predators. *Science* 314: 1928-1930.
Bresinsky, A. (1963) Bau, Entwicklungsgeschichte und Inhaltsstoffe der Elaiosomen. Studien zur myrmekochoren Verbreitung von Samen und Früchten. *Bibliotheca Botanica* 126: 1-54.
Brodie, H.J. (1955) Springboard plant dispersal mechanisms operated by rain. *Canadian Journal of Botany* 33: 15-167.
Brown, R. (1827) On the structure of the inimpregnated ovulum in phaenogamous plants. In: King, P.P.: *Narrative of a Survey of the Intertropical and Western Coasts of Australia* 2: 539-565.
Burgt, X.M. van der (1997) Explosive seed dispersal of the rainforest tree *Tetraberlinia morelina* (Leguminosae – Caesalpiniodeae) in Gabon. *Journal of Ecology* 13: 145-151.
Burtt, B.D. (1929) A record of fruits and seeds dispersed by mammals and birds from the Singida District of Tanganyika Territory. *The Journal of Ecology* 17(2): 351-355.
Candolle, A.P. de (1813) *Théorie élémentaire de la botanique ou exposition de la classification naturelle et de l'art de décrire et d'étudier les végétaux*, second edition. Chez Déterville, Paris.
Cochrane, E.P. (2003) The need to be eaten: *Balanites wilsoniana* with and without elephant seed-dispersal. *Journal of Tropical Ecology* 19: 579-589.
Culver, D.C. & Beattie, A.J. (1980) The fate of *Viola* seeds dispersed by ants. *American Journal of Botany* 67: 710-714.
Dalton, R. (2007) Blast in the past? *Nature* 447: 256-257.
Desvaux, N.A. (1813) Essai sur les différents genres des fruits des plantes phanérogames. *Journal de Botanique, appliquée à l'Agriculture, à la Pharmacie, à la Médecine et aux Arts* 2: 161-181.
Dinerstein, E. & Wemmer C.M. (1988) Fruits *Rhinoceros* eat: Dispersal of *Trewia nudiflora* (Euphorbiaceae) in Lowland Nepal. *Ecology* 69: 1768-1774.
Dumortier, B.C. (1835) *Essai carpographique présentant une nouvelle classification des fruits*. M. Hayez, Imprimeur de l'Académie Royale, Brussels, Belgium
Fenner, M. & Thompson, K. (2005) *The ecology of seeds*. Cambridge University Press, Cambridge, UK.
Gaertner, J. (1788, 1790, 1791, 1792) *De fructibus et seminibus plantarum*. 4 Vols. Academiae Carolinae, Stuttgart.
Galetii, M. (2002) Seed dispersal of mimetic fruits: parasitism, mutualism, aposematism or exaptation? *In*: D.J. Levey, W.R. Silva & M. Galetti (eds.): *Seed dispersal and frugivory: ecology, evolution and conservation*. CABI Publishing, UK.
Govaerts, R. (2001) How many species of seed plants are there? *Taxon* 50, 1085-1090.
Gunn, C.R. & Dennis, J.V. (1999) *World guide to tropical drift seeds and fruits* (reprint of the 1976 edition). Krieger Publishing Company, Malabar, Florida, USA.
Harrington, J.F. (1973) Biochemical basis of seed longevity. *Seed Science and Technology* 1: 453-461.
Heywood, V.H., Brummit, R.K., Culham, A. & Seberg, O. (2007) *Flowering Plant Families of the World*. Royal Botanic Gardens, Kew, London, UK.
Howe, H.F. (1985) Gomphothere fruits: a critique. *The American Naturalist* 125(6): 853-865.
Howe, H.F. & Smallwood, J. (1982) Ecology of seed dispersal. *Annual Review of Ecology and Systematics* 13: 201-228.
Howe, H.F. & Vande Kerckhove, G.A. (1980) Nutmeg Dispersal by Tropical Birds. *Science* 210(4472): 925-927.
Jackson, B.D. (1928) *A Glossary of Botanic Terms*. Hafner Publishing Company, New York, USA.
Janick, J. & Paull, R.E. (eds.) 2008 *The encyclopedia of fruit and nuts*. CABI Publishing, UK.
Janzen, D.H. (1977) Why fruits rot, seeds mold, and meat spoils. *The American Naturalist* 111(980): 691-713.
Janzen, D.H. (1984) Dispersal of small seeds by big herbivores: foliage is the fruit. *The American Naturalist* 123: 338-353.
Janzen, D.H. & Martin, P.S. (1982) Neotropical anachronisms: the fruits the gomphotheres ate. *Science* 215: 19-27.
Jones, D.L. (2002) *Cycads of the world*, 2nd edition. Reed New Holland Publishers, Sydney, Auckland, London, Cape Town.
Jordano, P. (1995) Angiosperm fleshy fruits and seed dispersers – a comparative analysis of adaptation and constraints in plant-animal interactions. *American Naturalist* 145: 163-191
Judd, W.S., Campbell, S., Kellogg, E.A., Stevens, P.F. & M.J. Donoghue (2002) *Plant Systematics – a phylogenetic approach*. Sinauer Associates, Inc., Sunderland, MA, USA.
Kelly, D. (1994) The evolutionary ecology of mast seeding. *Trends in Ecology and Evolution* 9: 465-470.
Kesseler, R. & Stuppy, W. (2006) *Seeds – Time Capsules of Life*. Papadakis, London, UK.
Kislev, M., Hartmann, A. & Bar-Yosef, O. (2006) Early domesticated fig in the Jordan Valley. *Science* 312: 1372-1374.
Levey, D.J., Silva, W.R. & Galeti, M. (eds.) (2002) *Seed Dispersal and Frugivory: Ecology, Evolution and Conservation*. CABI Publishing, UK.
Leins, P. (2000) *Blüte und Frucht*. Schweizerbart'sche Verlagsbuchhandlung. Stuttgart, Berlin, 390 pp.
Lindley, J. (1832) *An introduction to botany*. Longman, Rees, Orme, Brown, Green & Longmans, London.
Lindley, J. (1848) *An introduction to botany*, 4th edition. Longman, Brown, Green and Longmans, London.
Loewer, P. (2005) *Seeds – the definitive guide to growing, history and lore*. Timber Press, Portland, Cambridge, USA.
Mabberley, D.J. (1997) *The plant-book*, 2nd edition. Cambridge University Press, Cambridge
Mack, A.L. (2000) Did fleshy fruit pulp evolve as a defence against seed loss rather than as a dispersal mechanism? *Journal of Bioscience* 25 (1): 93-97
Mauseth, J.D. (2003) *Botany – an introduction to plant biology*, 3rd edition. Jones and Bartlett Publishers Inc., Boston, USA.
Mirbel, C.F. (1813) Nouvelle classification des fruits. *Nouveau Bulletin des Sciences, publié par la Société Philomatique de Paris* 3: 313-319.
Morton, J. (1987) Breadfruit. p. 50–58. In: *Fruits of warm climates*. Julia F. Morton, Miami, FL.
Murray, P.R. & Vickers-Rich (2004) *Magnificent Mihirungs: the colossal flightless birds of the Australian dreamtime*. Indiana University Press, USA.
Noble, J.C. (1975) The effects of emus (*Dromais novohollandiae* Latham) on the distribution of the nitre bush (*Nitraria billardierei* DC.). *The Journal of Ecology* 63(3): 979-984
Parolin, P. (2005) Ombrohydrochory: rain-operated seed dispersal in plants – with special regard to jet-action dispersal in Aizoaceae. *Flora – Morphology, Distribution, Functional Ecology of Plants* 201(7): 511-518.
Phillips, H. (1820) *Pomarium Britannicum: an historical and botanical account of fruits, known in Great Britain*. T. & J. Allman, London, UK.
Pijl, L. van der (1982) *Principles of dispersal in higher plants*, 3rd edition. Springer, Berlin, Heidelberg, New York.
Popenoe, W. (1920) *Manual of tropical and subtropical fruits*. The Macmillan Company, New York, USA
Prance, G.T. & Mori, S.A. (1978) Observations on the fruits and seeds of neotropical Lecythidaceae. *Brittonia* 30: 21-33
Raven, P.H., Evert, R.F. & Eichhorn, S.E. (1999) *Biology of plants*. W.H. Freeman, New York.
Rheede van Oudtshoorn, K. van & Rooyen, M.W. van (1999) *Dispersal biology of desert plants. Adaptations of Desert Organisms*. Springer-Verlag, Berlin.
Rick, C.M. & Bowman, R.I. (1961) Galapagos tomatoes and tortoises. *Evolution* 15(4): 407-417.
Ridley, H.N. (1930) *Dispersal of plants throughout the world*. L. Reeve & Co., Ashford, UK.
Roeper, J. (1826) Observationes aliquot in florum inflorescentariumque naturam. *Linnaea* 1: 433-466
Roth, I. (1977) *Fruits of angiosperms*. Gebrüder Borntraeger, Berlin & Stuttgart. 675 pp.
Sachs, J. von (1868) *Lehrbuch der Botanik nach dem gegenwärtigen Stand der Wissenschaft*, 1st edition. Leipzig
Schleiden, J. M. (1849) *Principles of scientific botany*. Translated by E. Lankester (of the German 2nd edition, published in 2 volumes, 1845-1846; the 1st German edition was published 1842-1843). Longman, Brown, Green & Longmans, London.
Sernander, R. (1906) Entwurf einer Monographie der europäischen Myrmekochoren. *Kungliga Svenska Vetenskapsakademiens Handlingar* 41: 1-410.
Sorensen, A.E. (1986) Seed dispersal by adhesion. *Annual Review of Ecology and Systematics* 17: 443-463.
Spjut, R.W. (1994) A systematic treatment of fruit types. *Memoirs of the New York Botanical Garden* 70: 1-182.
Spjut, R. W. and J. Thieret (1989) Confusion between multiple and aggregate fruits. *The Botanical Review* 55: 53-72.
Temple, S.A. (1977) Plant-animal mutualism: co-evolution with Dodo leads to near extinction of plant. *Science* 197: 885-886.
Tiffney, B.H. (2004) Vertebrate dispersal of seed plants through time. *Annual Review of Ecology and Systematics* 35: 1-29.
Tournefort, Joseph Pitton de (1694) *Eléments de botanique, ou méthode pour connaître les plantes* (Elements of botany, or method for getting to know the plants).
Ulbrich, E. (1928) *Biologie der Früchte und Samen (Karpobiologie)*. Springer, Berlin, Heidelberg, New York.
Vander Wall, S.B. & Longland, W.S. (2004) Diplochory: are two seed dispersers better than one? *Trends in Ecology and Evolution* 19(3): 155-161.
Willdenow, C.L. (1802) Grundriß der Kräuterkunde zu Vorlesungen entworfen, 3rd ed, Berlin
Willdenow, C.L. (1811) The principles of botany and of vegetable physiology. Trans. from German. University Press, Edinburgh.
Willson, M.F. (1993) Mammals as seed-dispersal mutualists in North America. *Oikos* 67: 159-176.
Witmer, M.C. & Cheke, A.S. (1991) The dodo and the tambalocoque tree: an obligate mutualism reconsidered. *Oikos* 61(1): 133-137.
Zona, S. & Henderson, A. (1989) A review of animal-mediated seed-dispersal of palms. *Selbyana* 11: 6-21.

ART

Adam, H.C. (1999) *Karl Blossfeldt*. Prestel, Munich
Diffey, T.J. (1993) Natural Beauty without Metaphysics. Published in, *Landscape, natural beauty and the arts*. Cambridge University Press, Cambridge, UK
Ede, S (2000) *Strange and Charmed*. Calouste Gulbenkian Foundation, London
Frankel, F. (2002) *Envisioning Science, The Design and Craft of the Science Image*. MIT Press, Cambridge MA, USA
Gamwell, L. (2002) *Exploring the Invisible, Art Science and the Spiritual*. Princeton University Press, Princeton NJ, USA
Haeckel, E. (1904) *Art Forms in Nature*. Reprinted 1998. Prestel, Munich
Kesseler, R. (2001) *Pollinate*. Grizedale Arts and The Wordsworth Trust, Cumbria.
Stafford, B.M. (1994) *Artful Science, Enlightenment, Entertainment and the Eclipse of the Visual Image*. MIT Press, Cambridge MA.
Stafford, B.M. (1996) *Good Looking, Essays on the Virtue of Images*. MIT Press, Cambridge MA.
Thomas, A. (1997) *The Beauty of Another Order, Photography in Science*. Yale University Press, New Haven
Tongiorgi Tomasi, L. (2002) *The Flowering of Florence, Botanical Art for the Medici*, Lund Humphries

INDEX OF PLANTS ILLUSTRATED

Plant family	Latin name, and authority for name	Page	Common British name	Country
Leguminosae	*Abrus precatorius* L.	194	crab's eye, jequirity bean, paternoster pea, rosary pea	pantropical
Leguminosae	*Acacia cyclops* A. Cunn. & G. Don	213	coastal wattle	south-western Australia
Leguminosae	*Acacia vittata* R.S.Cowan & Maslin	92, 93	Lake Logue wattle	Western Australia
Ranunculaceae	*Actaea pachypoda* Elliot	207	white baneberry, doll's eyes	eastern North America
Actinidiaceae	*Actinidia deliciosa* (A. Chev.) C.F. Liang & A.R. Ferguson	58, 59	kiwi, Chinese gooseberry	southern China
Malvaceae	*Adansonia rubrostipa* Jum. & H. Perrier	222	baobab	west coast of Madagascar
Rosaceae	*Agrimonia eupatoria* L.	164	common agrimony, church steeples, cockleburr	collected in the UK; native to Old World
Sapindaceae	*Alectryon excelsus* Gaertn.	215	titoki tree	New Zealand
Casuarinaceae	*Allocasuarina tesselata* (C.A. Gardner) L.A.S. Johnson	134, 135	she-oak	Western Australia
Cucurbitaceae	*Alsomitra macrocarpa* (Blume) Cogn.	146	no common name	Indomalesia
Malvaceae	*Alyogyne huegelii* (Endl.) Fryxell (syn. *Hibiscus huegelii* Endl.)	151	lilac hibiscus	south and south-west Australia
Anacardiaceae	*Anacardium occidentale* L.	74	cashew nut	north-eastern Brazil, widely cultivated throughout the tropics
Myrsinaceae	*Anagallis arvensis* L.	51	scarlet pimpernel	native to Europe but widely naturalized elsewhere
Bromeliaceae	*Ananas comosus* (L.) Merr.	116	pineapple	cultivated since antiquity, originally native to South America
Plantaginaceae	*Antirrhinum orontium* L.(syn. *Misopates orontium* (L.) Raf.)	154, 155	lesser snapdragon, weasel's snout	native to Europe, naturalized in North America
Asteraceae	*Arctium lappa* L.	170	greater burdock, edible burdock	temperate Eurasia
Apiaceae	*Artedia squamata* L.	147	crown flower	endemic to Cyprus and the Eastern Mediterranean (Israel, Lebanon, Jordan, Syria, Turkey)
Moraceae	*Artocarpus altilis* (Parkinson) Fosberg	122	breadfruit	Malay Peninsula and islands of the western Pacific
Moraceae	*Artocarpus heterophyllus* Lam.	123	jakfruit, jackfruit	probable origin India (western Ghats), widely cultivated throughout the tropics
Apocynaceae	*Asclepias physocarpa* (E. Mey.) Schltr. (syn. *Gomphocarpus physocarpus* E. Mey.)	153	balloon plant, balloon cotton-bush	native to south-east Africa
Annonaceae	*Asimina triloba* (L.) Dunal	239	pawpaw	eastern North America
Solanaceae	*Atropa belladonna* L.	196	deadly nightshade, naughty men's cherries, banewort	Europe, North Africa, western Asia, naturalized in North America
Oxalidaceae	*Averrhoa carambola* L.	62	star fruit, carambola	cultivated for centuries in south-east Asia, presumed origin India, Sri Lanka, Indonesia
Proteaceae	*Banksia candolleana* Meisn.	132	propeller banksia	Western Australia
Proteaceae	*Banksia menziesii* R. Br.	133	firewood banksia	Western Australia
Sapindaceae	*Blighia sapida* K.D. Koenig	198	ackee, akee, vegetable brain	native to West Africa, widely cultivated in the tropics, esp. in the Caribbean
Arecaceae	*Borassus aethiopum* Mart.	230	elephant palm, black rhun-palm	Africa
Moraceae	*Broussonetia papyrifera* (L.) Vent.	136	paper mulberry	native to east Asia (Japan, Taiwan), naturalized in North America and many Pacific islands incl. Hawai'i
Cyperaceae	*Bulbostylis hispidula* (Vahl) R.W. Haines subsp. *pyriformis* (Lye) R.W. Haines	12	no common name	East Africa
Arecaceae	*Calamus aruensis* Becc.	1	rattan palm	New Guinea to the Solomon Islands, Aru Islands and the tip of Cape York, Australia
Arecaceae	*Calamus longipinna* K. Schum. & Lauterb.	14	rattan palm	New Guinea & Solomon Islands
Arecaceae	*Calotis breviradiata* (Ising) G.L.R. Davis	18	short-rayed burr daisy	Australia
Sapindaceae	*Cardiospermum halicababum* L.	153	love-in-a-puff, balloon vine	tropical America, widely naturalized
Caricaceae	*Carica papaya* L.	226, 227	papaya, pawpaw, melon tree	native to tropical America, cultivated throughout the tropics
Leguminosae	*Carmichaelia aligera* G. Simpson	212	North Island broom, leafless broom	New Zealand
Fagaceae	*Castanea sativa* L.	73	sweet chestnut, Spanish chestnut, European chestnut	originally native to south-eastern Europe and the Mediterranean
Poaceae	*Cenchrus spinifex* Cav.	172	coastal sandbur, field sandbur	native to America
Leguminosae	*Centrolobium ochroxylum* Rose ex Rudd	78	amarillo de Guayaquil	Ecuador
Cupressaceae	*Chamaecyparis lawsoniana* (A. Murray) Parl.	43	Lawson's cypress, Port Orford cedar, Oregon cedar	north-western North America
Ranunculaceae	*Cimicifuga americana* Michaux	104, 105	American bugbane, mountain bugbane, summer cohosh	eastern North America
Rutaceae	*Citrus hystrix* DC.	46, 47, 56	kaffir lime, makrut	Indonesia
Rutaceae	*Citrus margarita* Lour. (syn. *Fortunella margarita* (Lour.) Swingle)	64, 65	oval kumquat	cultivated for centuries in Asia, probable origin southern China
Rutaceae	*Citrus medica* L. var. *sarcodactylis* (Hoola van Nooten) Swingle	66, 67	Buddha's Hand, fingered citron	ancient cultigen; originally from northern India
Rutaceae	*Citrus sinensis* (L.) Osbeck	57	sweet orange	cultivated since antiquity; presumed origin China or India
Verbenaceae	*Clerodendrum trichotomum* Thunb.	207	harlequin glory bower	Japan
Arecaceae	*Cocos nucifera* L.	102	coconut	found in all tropical regions
Cornaceae	*Cornus kousa* Hance subsp. *chinensis* (Osborn) Q.Y. Xiang	118, 119	Chinese dogwood	central and northern China
Hamamelidaceae	*Corylopsis sinensis* Hemsl. var. *calvescens* Rehder & E.H. Wilson	160	Chinese winter hazel	China
Betulaceae	*Corylus avellana* L.	49	hazel	Eurasia
Lecythidaceae	*Couroupita guianensis* Aubl.	99	cannonball tree	tropical America
Bignoniaceae	*Crescentia cujete* L.	239	calabash tree	tropical America
Cucurbitaceae	*Cucumis melo* L. subsp. *melo* var. *cantalupensis* Naudin 'Galia'	201	Galia muskmelon	cultivar, bred in Israel
Cycadaceae	*Cycas revoluta* Thunb.	25	sago palm	Japan
Cucurbitaceae	*Cyclanthera brachystachya* (Ser.) Cogn. (syn. *Cyclanthera explodens* Naudin)	162	seed-spitting gourd	Central and South America
Rosaceae	*Cydonia oblonga* Mill.	217	quince	cultivated since antiquity, possible origin Turkey and northern Iraq, naturalized in southern Europe
Boraginaceae	*Cynoglossum nervosum* Benth. ex C.B. Clarke	141	hairy hound's tongue, great hound's tongue	Pakistan, India
Podocarpaceae	*Dacrydium cupressinum* Soland. ex Forst. f.	176	rimu, red pine	New Zealand
Ruscaceae	*Dasylirion texanum* Scheele	145	Texas sotol, green sotol	Texas, northern Mexico (Coahuila, Chihuahua)
Solanaceae	*Datura ferox* L.	87	fierce thornapple, Chinese thornapple	south-western North America
Apiaceae	*Daucus carota* L.	108, 109	wild carrot, Queen Anne's lace	native to Europe and south-western Asia
Poaceae	*Deschampsia antarctica* Desv.	37	Antarctic hair grass	southern South America, maritime Antarctica
Dicksoniaceae	*Dicksonia antarctica* Labill.	24	soft tree-fern, man fern, Tasmanian tree fern	Australia (New South Wales, Victoria, Tasmania)
Ebenaceae	*Diospyros kaki* Thunb.	199	Japanese persimmon, kaki	east Asia
Dipterocarpaceae	*Dipterocarpus grandiflorus* (Blanco) Blanco	79	keruing belimbing (Malay)	south-east Asia
Sapindaceae	*Dipteronia sinensis* Oliv.	78	dipteronia, Jin Qian Feng	China
Moraceae	*Dorstenia contrajerva* L.	128	contrayerva, tusilla	from southern Mexico to northern South America
Salicaceae	*Dovyalis caffra* (Hook. f.) Warb.	60	kei apple	southern Africa
Winteraceae	*Drimys winteri* J.R. Forst. & G. Forst.	30, 31, 32, 33	Winter's bark tree	Mexico to Terra del Fuego
Malvaceae	*Durio zibethinus* Murray	224, 225	durian, civet fruit	south-east Asia
Cucurbitaceae	*Ecballium elaterium* (L.) A. Rich.	162	squirting cucumber	Mediterranean
Polygonaceae	*Emex australis* Steinh.	168	threecornerjack, three-cornered jack, doublegee (Australia)	native to southern Africa, weedy in Australia and elsewhere
Zamiaceae	*Encephalartos ferox* Bertol. f.	26	Zululand cycad, Tongaland cycad	southern Africa
Leguminosae	*Entada gigas* (L.) Fawc. & Rendle	95	sea heart	tropical America and Africa
Leguminosae	*Entada rheedii* Spreng. (syn. *Entada gigas* Gilbert & Boutique)	95	African dream herb, snuffbox sea bean	Africa, Asia, Australia, Indian Ocean, Pacific Ocean
Leguminosae	*Enterolobium cyclocarpum* (Jacq.) Griseb.	236	guanacaste, elephant-ear tree, earpod tree	tropical America
Apiaceae	*Eryngium creticum* Lam.	140	Crete eryngo	south-east Europe, western Asia and Egypt
Apiaceae	*Eryngium leavenworthii* Torr. & A. Gray	11	Leavenworth's eryngo	North America
Apiaceae	*Eryngium paniculatum* Cav. & Dombey ex F. Delaroche	10	cardoncillo	southern South America (Argentina, Chile)
Rutaceae	*Esenbeckia macrantha* Rose	86	no common name	Mexico
Myrtaceae	*Eucalyptus macrocarpa* Hook.	252, 253	mottlecah	Western Australia
Myrtaceae	*Eucalyptus regnans* F. Muell.	38	mountain ash, Tasmanian oak, swamp gum	southern Australia, Tasmania
Myrtaceae	*Eucalyptus virginea* Hopper & Wardell-Johnson	39	no common name	south-western Australia
Celastraceae	*Euonymus europaeus* L.	214	spindle tree, European spindle	from Europe to western Asia
Santalaceae	*Exocarpos sparteus* R.Br.	75	broom ballart	Australia
Moraceae	*Ficus carica* L.	124, 125, 128	common fig	ancient cultigen, esp. Mediterranean, originally probably from south-west Asia
Moraceae	*Ficus dammaropsis* Diels	129	dinner-plate fig, highland breadfruit	New Guinea
Moraceae	*Ficus sansibarica* Warb. subsp. *sansibarica*	129	knobbly fig	south-east Africa
Moraceae	*Ficus villosa* Blume	2, 3	villous fig, shaggy fig	tropical Asia
Rutaceae	*Flindersia australis* R. Br.	86	crow's ash, Australian teak	eastern Australia
Rosaceae	*Fragaria x ananassa* (Weston) Decne & Naudin	114, 115	garden strawberry	only in cultivation
Cyperaceae	*Fuirena mutali* Muasya & Nordal, ined.	145	no common name	Kenya
Amaryllidaceae	*Galanthus nivalis* L. subsp. *imperati* (Bertol.) Baker (syn. *Galanthus nivalis* L.)	45	snowdrop	garden variety, the wild form native to southern Europe

Asteraceae	*Galinsoga brachystephana* Regel	84	no common name	Central and South America
Rubiaceae	*Galium aparine* L.	166, 167	stickywilly, goosegrass, sticky bobs	Eurasia, America
Clusiaceae	*Garcinia mangostana* L.	222	mangosteen	south-east Asia
Ginkgoaceae	*Ginkgo biloba* L.	163	ginkgo, maidenhair tree	a relic in eastern China
Leguminosae	*Gleditsia triacanthos* L.	236	honey locust	eastern North America
Gnetaceae	*Gnetum* L. sp.	29	gnetum	photographed in New Guinea
Malvaceae	*Gossypium hirsutum* L. 'Bravo'	152	upland cotton, Mexican cotton	domesticated cultivar; the wild form native to Mexico
Leguminosae	*Gymnocladus dioica* (L.) K. Koch	237	Kentucky coffee	eastern USA
Poaceae	*Hackelochloa granularis* (L.) Kuntze	181	pitscale grass	pantropical
Proteaceae	*Hakea orthorhyncha* F. Muell.	91	bird beak hakea	Western Australia
Pedaliaceae	*Harpagophytum procumbens* DC. ex Meisn.	96	devil's claw, grapple plant	southern Africa, Madagascar
Sapindaceae	*Harpullia pendula* Planch. ex F. Muell.	210	tulipwood, Australian tulipwood	Australia (Queensland, New South Wales)
Zingiberaceae	*Hedychium horsfieldii* R.Br. ex Wall.	91	Java ginger	Java
Hernandiaceae	*Hernandia bivalvis* Benth.	206	grease nut	Queensland
Malvaceae	*Hibiscus mutabilis* L.	150	Confederate rose, cotton rosemallow, Dixie rosemallow	native to China and Japan, naturalized in the southern USA
Celastraceae	*Hippocratea parvifolia* Oliv.	146	no common name	southern Africa
Leguminosae	*Hippocrepis unisiliquosa* L.	97	horse-shoe vetch	native to Eurasia and Africa
Elaeagnaceae	*Hippophae rhamnoides* L.	70, 71	sea buckthorn, sallow thorn	Eurasia
Araliaceae	*Hydrocotyle coorowensis* H. Eichler ms	183	no common name	south-west Australia
Cactaceae	*Hylocereus undatus* (Haw.) Britton & Rose	63	dragon fruit, pitahaya	tropical America
Leguminosae	*Hymenaea courbaril* L.	238	stinking toe tree, jatobá, Brazilian copal	tropical America
Leguminosae	*Hymenaea martiana* Hayne	238	jatobá	South America
Solanaceae	*Hyoscyamus niger* L.	197	henbane, stinking nightshade	temperate Eurasia, naturalized in North America
Aquifoliaceae	*Ilex aquifolium* L.	199	holly, European holly	Europe, Mediterranean
Iliciaceae	*Illicium simonsii* Maxim.	103	no common name	Asia (India, China, Myanmar)
Leguminosae	*Inga feuillei* DC.	94	ice-cream bean, pacay	known only in cultivation, probably native to Bolivia and Peru
Iridaceae	*Iris foetidissima* L.	209	gladdon, stinking gladwyn, stinking iris	southern and western Europe to northern Africa
Juglandaceae	*Juglans regia* L.	72	common walnut, English walnut, Persian walnut	Eurasia (south-eastern Europe to western China)
Cupressaceae	*Juniperus flaccida* Schltdl.	62	weeping juniper, drooping juniper, Mexican juniper	southern Texas, Mexico
Bignoniaceae	*Kigelia africana* (Lam.) Benth.	229	sausage tree	tropical Africa
Krameriaceae	*Krameria erecta* Willd. ex Schult.	8, 254	littleleaf rhatany, Pima rhatany	southern USA and northern Mexico
Cyperaceae	*Kyllinga squamulata* Thonn. ex Vahl	249	Asian spikesedge	native to tropical Africa, Madagascar, India and Indochina; widely naturalized elsewhere
Lamiaceae	*Lamium album* L.	178, 179	white deadnettle	native to Eurasia, naturalized in eastern North America
Lamiaceae	*Lamium galeobdolon* (L.) L.	179	yellow archangel	Eurasia (western Europe to Iran)
Zamiaceae	*Lepidozamia peroffskyana* Regel	27, 42	pineapple zamia	eastern Australia
Altingiaceae	*Liquidambar styraciflua* L.	131	sweet gum, American red gum	North and Central America
Sapindaceae	*Litchi chinensis* Sonn. subsp. *chinensis*	20, 21	lychee	south China
Caprifoliaceae	*Lonicera xylosteum* L.	116	fly honeysuckle	Eurasia
Moraceae	*Maclura pomifera* (Raf.) CK Schneid.	240, 242, 243	osage orange, bow wood	North America
Zamiaceae	*Macrozamia lucida* L.A.S. Johnson	187	no common name	eastern Australia (south-eastern Queensland, north-eastern New South Wales)
Zamiaceae	*Macrozamia moorei* F. Muell.	245	no common name	Queensland
Zamiaceae	*Macrozamia riedlei* (Fisch. ex Gaudich.) C.A. Gardner	186	no common name	Western Australia
Magnoliaceae	*Magnolia* sp.	208	magnolia	Photographed in Yunnan, China
Euphorbiaceae	*Mallotus nudiflorus* (L.) Kulju & Welzen (syn. *Trewia nudiflora* L.)	232	no common name	From India, Nepal and Sri Lanka throughout south-east Asia.
Rosaceae	*Malus pumila* Mill.	188	apple	cultivated since antiquity, origin Asia
Anacardiaceae	*Mangifera indica* L.	200	mango	cultigen, probable origin somewhere between India and the Malay peninsula
Martyniaceae	*Martynia annua* L.	170	devil's claw, small-fruited devil's claw	America; widely naturalized in the tropics
Leguminosae	*Medicago orbicularis* (L.) Bartal.	61	blackdisk medick, button medick, button clover	Mediterranean
Leguminosae	*Medicago polymorpha* L.	165	burclover, toothed medick	Eurasia, North Africa
Myrtaceae	*Melaleuca araucarioides* Barlow	130	no common name	south-west Australia
Moraceae	*Morus nigra* L.	120, 121	black mulberry	cultivated since antiquity; originally probably from China
Myristicaceae	*Myristica fragrans* Houtt.	214	nutmeg	Indonesia (Moluccas)
Nitrariaceae	*Nitraria billardierei* DC.	232	Dillon bush, nitre bush	Australia
Arecaceae	*Nypa fruticans* Wurmb	156, 157	nipa palm, mangrove palm	southern Asia to northern Australia, naturalized in West Africa and Panamá.
Ochnaceae	*Ochna natalitia* (Meisn.) Walp.	206	coast boxwood, Natal plane	southern Africa
Lamiaceae	*Ocimum basilicum* L.	106, 107	sweet basil	cultivated for more than 5,000 years, originally from tropical Asia
Rhamnaceae	*Paliurus spina-christi* Mill.	149	Christ's thorn, Jerusalem thorn, Crown of thorns	Mediterranean and western Asia
Monimiaceae	*Palmeria scandens* F. Muell.	208	anchor vine, pomegranate vine	eastern Australia
Pandanaceae	*Pandanus odorifer* (Forssk.) Kuntze (syn. *Pandanus odoratissimus* L. f.)	117	fragrant screwpine, umbrella tree	tropical and subtropical Asia
Papaveraceae	*Papaver rhoeas* L.	88, 89	corn poppy, field poppy	Eurasia, North Africa
Leguminosae	*Pararchidendron pruinosum* (Benth.) I.C. Nielsen	211	snow wood, monkey's earrings	eastern Australia, New Guinea and Malesia
Passifloraceae	*Passiflora edulis* Sims	48	passionfruit, maracujá	South America
Lauraceae	*Persea americana* Mill.	53	avocado	Central America
Picrodendraceae	*Petalostigma pubescens* Domin	180	quinine bush	Papua New Guinea (Western Prov.) and north and east Australia
Phytolaccaceae	*Phytolacca acinosa* Roxb.	44, 110, 111	Indian pokeweed	east Asia (China to India)
Pinaceae	*Pinus coulteri* D. Don	42	Coulter pine, big-cone pine	south-western North America (California & Mexico)
Pinaceae	*Pinus sabiniana* Douglas ex D. Don	175	digger pine, gray pine, California foothill pine	California
Leguminosae	*Piscidia grandifolia* (Donn. Sm.) I.M. Johnst. var. *gentryi* Rudd	143	no common name	Mexico
Nyctaginaceae	*Pisonia brunoniana* Endl.	173	bird catcher tree, Australasian catchbird tree	Australia, New Zealand and other Pacific islands incl. Hawai'i
Leguminosae	*Pithecellobium excelsum* (Kunth) Mart.	216	chaquiro	South America (Ecuador, Peru)
Anacardiaceae	*Pleiogynium timoriense* (DC.) Leenh.	244	Burdekin plum, cocky apple	Central Malesia to Pacific
Martyniaceae	*Proboscidea altheifolia* (Benth.) Decne.	171	golden devil's claw, desert unicorn plant, yuca de caballo	southern USA and Mexico
Rosaceae	*Prunus armeniaca* L.	188	apricot	northern China
Rosaceae	*Prunus dulcis* (Mill.) D.A. Webb (syn. *Prunus amygdalus* Batsch)	72	almond	Western Asia (Levant)
Rosaceae	*Prunus persica* (L.) Batsch var. *persica*	68, 69	peach	originally from China, naturalized in North America, widely cultivated
Rosaceae	*Prunus spinosa* L.	189	sloe, blackthorn	Eurasia
Rosaceae	*Pyrus communis* L.	189	European pear	Eurasia
Rosaceae	*Pyrus pyrifolia* (Burm. f.) Nakai	112	Chinese pear, Japanese pear, Nashi pear, apple pear	east Asia (China, Laos, Vietnam)
Fagaceae	*Quercus robur* L.	73	common oak, pedunculate oak, English oak, European oak	Europe, Mediterranean
Rafflesiaceae	*Rafflesia keithii* Meifer	248	rafflesia	endemic to Borneo
Ranunculaceae	*Ranunculus parviflorus* L.	54	small-flowered buttercup	western Europe and Mediterranean, naturalized elsewhere in temperate areas
Ranunculaceae	*Ranunculus pygmaeus* Wahlenb.	55	pygmy buttercup, dwarf buttercup	northern Europe, eastern Alps, western Carpathians, North America
Strelitziaceae	*Ravenala madagascariensis* Sonn.	90	traveller's palm	Madagascar
Grossulariaceae	*Ribes rubrum* L.	184	redcurrant	Eurasia
Annonaceae	*Rollinia mucosa* (Jacq.) Baill.	102	biriba	South America
Rosaceae	*Rosa roxburghii* Tratt.	113	sweet chestnut rose, cili fruit	China
Rosaceae	*Rubus fruticosus* L.	100	blackberry, bramble	Europe, Mediterranean
Rosaceae	*Rubus idaeus* L.	103	raspberry	Eurasia, North America
Rosaceae	*Rubus laciniatus* Willd.	100	cutleaf blackberry, evergreen blackberry	only known in cultivation, origin unknown
Rosaceae	*Rubus phoenicolasius* Maxim.	100, 101	Japanese wineberry, wineberry	northern China, Korea, Japan; cultivated in Europe and North America
Apiaceae	*Rumia crithmifolia* (Willd.) Koso-Pol.	158, 159	no common name	Crimea
Amaranthaceae	*Salsola kali* L.	142	Russian thistle, blackbush, prickly saltwort, prickly glasswort	Europe; naturalized in North America, South Africa, Australia, New Zealand etc.
Rosaceae	*Sanguisorba minor* Scop. subsp. *muricata* (Spach) Briq.	148	salad burnet, small burnet	native to Eurasia and Africa
Dipsacaceae	*Scabiosa columbaria* L.	80, 81	small scabious	southern Europe, western Asia, northeast Africa
Dipsacaceae	*Scabiosa crenata* Cyr.	82, 83	no common name	central and eastern Mediterranean
Sapotaceae	*Sideroxylon grandiflorum* A. DC. (syn. *Calvaria major* (A. DC.) Dubard)	234	tambalocoque tree, dodo tree, Calvaria tree	endemic to Mauritius
Solanaceae	*Solanum betaceum* Cav. (syn. *Cyphomandra betacea* (Cav.) Sendtn.)	48	tree tomato, tamarillo	South America, widely cultivated in the tropics
Solanaceae	*Solanum luteum* Mill. subsp. *luteum*	198	yellow nightshade, red-fruited nightshade	Mediterranean
Solanaceae	*Solanum lycopersicum* L. var. *cerasiforme* (Dunal) Spooner et al.	23	tomato	first domesticated in Mexico but originally probably from South America (Andes)

Strelitziaceae	***Strelitzia reginae*** Aiton	*90*	bird-of-paradise flower	South Africa
Leguminosae	***Sutherlandia frutescens*** (L.) R. Br.	*152*	balloon pea	South Africa
Meliaceae	***Swietenia mahagoni*** (L.) Jacq.	*87*	Cuban mahogany, West Indian mahogany	tropical America
Myrtaceae	***Syzygium jambos*** (L.) Alston	*98*	rose apple	south-east Asia
Arecaceae	***Tahina spectabilis*** J. Dransf. & Rakotoarinivo	*52*	no common name	endemic to Madagascar
Leguminosae	***Tamarindus indica*** L.	*94, 241*	tamarind, Indian date	cultigen, probable origin tropical Africa
Taxaceae	***Taxus baccata*** L.	*163*	English yew	Europe, Mediterranean
Combretaceae	***Terminalia catappa*** L.	*156*	Indian almond, Malabar almond, tropical almond, sea almond	widely grown throughout the tropics as an ornamental tree; probable origin Indomalesia
Malvaceae	***Theobroma cacao*** L.	*223*	cacao	domesticated already in pre-Columbian times, originally from the Amazon rainforest
Leguminosae	***Tipuana tipu*** (Benth.) Kuntze	*143*	tipu tree	South America (Brazil, Bolivia, Argentina)
Apiaceae	***Tordylium apulum***	*256*	Mediterranean hartwort, Roman pimpernel	Europe to western Asia
Araliaceae	***Trachymene ceratocarpa*** (W. Fitzg.) Keighery & Rye	*182*	creeping carrot	Australia
Zygophyllaceae	***Tribulus terrestris*** L.	*169*	puncture vine, caltrop, devil's thorn	native to the Old World
Malpighiaceae	***Tristellateia africana*** S. Moore	*149*	helicopter fruit	native to Kenya, Tanzania
Ericaceae	***Vaccinium corymbosum*** L.	*204, 205*	blueberry, American blueberry, highbush blueberry	eastern North America
Valerianaceae	***Valerianella coronata*** (L.) DC.	*5*	no common name	central and southern Europe, North Africa, south-west and central Asia
Violaceae	***Viola sororia*** Willd.	*160*	common blue violet	eastern North America
Santalaceae	***Viscum album*** L.	*185, 190*	common mistletoe, European mistletoe	Eurasia
Vitaceae	***Vitis labrusca*** L. 'Isabella"	*16*	Isabella grape	cultivar, locally naturalized in Europe
Welwitschiaceae	***Welwitschia mirabilis*** Hook. f.	*28*	tree tumbo, tumboa	desert of South-West Africa (Angola, Namibia)
Araceae	***Wolffia columbiana*** H. Karst.	*40, 41*	Columbian water meal	North and South America
Asteraceae	***Xanthisma texanum*** DC.	*85*	Texas sleepy-daisy	south-eastern USA (Texas, Oklahoma, New Mexico)
Poaceae	***Zea mays*** L.	*76, 77*	maize	origin Central America, widely cultivated
Zosteraceae	***Zostera marina*** L.	*36*	eelgrass, grasswrack, ulva marina	Europe

FOOTNOTES

[1] In a scientific context, the usage of the terms "primitive" and "advanced" requires some explanation in order to avoid confusion. Referring to certain plants as "advanced" suggests that they possess some "improvements" in comparison to "primitive" plants. However, by definition, extant plants are equally evolved because they have all been around for the same length of time since life began and they are all well adapted to their specific environments. A modern gymnosperm is not less evolved than an angiosperm. Gymnosperms are more "primitive" than angiosperms only in the sense that they are more similar to the extinct ancestral forms from which they evolved.

[2] An alternative theory postulates that the primitive carpel was not conduplicative (i.e. formed through the folding of a megasporophyll) but ascidiate (i.e. formed from inception as a cylindrical outgrowth). Although formerly, the conduplicative carpel was considered the most basic type of carpel, the basal-most angiosperms (Amborellaceae, Nymphaeales and Austrobaileyales) have been found to predominantly possess ascidiate carpels which is why this type of carpel formation is currently widely considered the most primitive state of carpel evolution in the angiosperms. Nevertheless, conduplicative carpels are found in many angiosperms (e.g. Magnoliales) and provide a valid model to illustrate one of the possible pathways of carpel evolution. The conduplicative and ascidiate type of carpel development are so different from each other that the question as to how they are linked evolutionarily is still unanswered.

[3] An even more significant advantage angiosperms have over gymnosperms has to do with the way in which they produce their seeds. Whereas angiosperms put their energy into the formation of the expensive, energy-rich seed storage tissue (endosperm) only after successful fertilization of the ovule, gymnosperms such as ginkgos, cycads and conifers produce their storage tissue (the massive megagametophyte) in advance, before the egg cell is even fertilized. In the race of evolution conservation of resources is always a great advantage.

[4] In many basal angiosperms the carpel is closed only by mucilage, not by fusion of the carpel edges.

[5] Readers who would like to know more intimate details about the intriguing sex life of the angiosperms are referred to Seeds – Time Capsules of Life.

[6] The legume family can either be referred to as Leguminosae or Fabaceae. Since the use of Fabaceae is ambiguous because it may either refer to the legume family as a whole or only to subfamily Papilionioideae when considered as a separate family, we decided to use Leguminosae as recommended by Lewis, G. et al. (2005) Legumes of the World, Royal Botanic Gardens, Kew.

[7] Gwilym Lewis, personal observation.

PICTURE CREDITS

Page 23: © Mike Bailey & Steve Williams; page 28 & 29: Andrew McRobb © RBG Kew; page 36: © Vidar.a; page 38: © James Wood, Hobart, Tasmania; page 52: Hannelore Morales © RBG Kew; page 53: © Mike Bailey & Steve Williams; page 74: © Suzana Profeta; page 78 (bottom): © Gwilym Lewis; page 90 (bottom): © David Lee; page 94 (top): © Mike Bailey & Steve Williams; page 95: Legume section at Herbarium © RBG Kew; page 108: © Mike Bailey & Steve Williams; page 116 (bottom): Elly Vaes - shot on location at Fairchild Tropical Botanic Garden, courtesy Fairchild Tropical Botanic Garden; page 117: Elly Vaes - shot on location at Fairchild Tropical Botanic Garden, courtesy Fairchild Tropical Botanic Garden; page 122: © Tim Waters, www.flickr.com/photos/tim-waters; page 123 (top): © Alex V. Popovkin; page 123 (bottom): © Dinesh Valke; page 142: NHPA / Rich Kirchner; page 146 (top): © Stephen Lyle, BBC Bristol; page 176: NHPA / A.N.T. Photo Library; page 195: © Mike Bailey & Steve Williams; page 196 & 197: Paul Little © RBG Kew; page 198 (bottom): © Kolade Nurse; page 203: NHPA / Haroldo Palo Jr.; page 214: © Veronica Olivotto; page 215: © Trevor James; page 216: © Gwilym Lewis; page 219: Getty / Tim Laman; page 220: NHPA / Kevin Schafer; page 229: © Kim Wolhuter, www.wildcast.net; page 231: © Nigel Dennis / Africa Imagery; page 233: © Lucy Commander; page 233: © Filipe de Oliveira; page 235: © The Natural History Museum, London; page 239 (top): © Phillip Merritt; (bottom): Andrew McRobb © RBG Kew; page 246: *Dromornis stirtoni*, first published in Peter F. Murrey & Patricia Vickers-Rich (2004), *Magnificent Mihirungs*. Indiana University Press; page 248: © Jamili Nais

We gratefully acknowledge the granting of permission to use these images. Every possible attempt has been made to identify and contact copyright holders. Any errors or omissions are inadvertent and will be corrected in subsequent editions.

ACKNOWLEDGMENTS

Many people have in many ways, either directly or indirectly, contributed to the amazing wealth of material, knowledge and ideas that provided the material for this book. Although it is impossible to mention all the scientists whose painstaking observations and publications over decades have revealed so many fascinating facts about fruits, and the people who discovered, collected or grew the fruits shown here, we would like to give a special mention to the following.

We thank our publisher, Andreas Papadakis, for the support and freedom he has given us throughout the preparation of this book and his daughter Alexandra for using her outstanding taste and artistic skills to create the beautiful design that provides such a spectacular background for both text and images. We are deeply indebted to Richard Bateman, Paula Rudall and Richard Spjut for their very thorough reviews of the manuscript and to Sheila de Vallée for editing the text.

We wish to thank Ken Arnold, Head of Public Programmes, Wellcome Trust, for his perceptive preface and Professor Stephen D. Hopper, Director of the Royal Botanic Gardens, Kew, for his encouragement and helpful comments on the manuscript, and for accepting our invitation to write the foreword.

We are grateful to Paul Smith, Head of the Seed Conservation Department (SCD), and John Dickie, Head of Information Section (SCD), for allocating resources of the Millennium Seed Bank Project towards the preparation of this book. The Millennium Seed Bank Project is funded by the U.K. Millennium Commission and the Wellcome Trust. The Royal Botanic Gardens, Kew receives an annual grant in aid from the U.K. Department of Environment, Food and Rural Affairs.

We thank the staff of the Royal Botanic Gardens, Kew & Wakehurst Place, particularly the Micromorphology Section, all members of the Seed Conservation Department and the many partners of the Millennium Seed Bank Project all over the world who have contributed to the outstanding collection that provided us with many extraordinary and unusual examples of fruits. We especially acknowledge the use of material collected in the following partner countries of the Millennium Seed Bank Project (MSBP): Australia, Burkina Faso, Kenya, Lebanon, Mali, Mexico, Madagascar, South Africa, Ukraine and the United States. At the Kew Herbarium we would like to thank the members of the Legume Section, Palm Section, Malpighiales Section and the South-East Asian Regional Team for granting us access to their collections and the Micromorphology Section of the Jodrell Laboratory, especially Paula Rudall and Chrissie Prychid, for granting us access to their Scanning Electron Microscope and for technical support. At the SCD, we are grateful for the kind support of the members of the Curation Section, and especially Janet Terry, who was instrumental in sourcing material.

Furthermore, we both wish to thank colleagues and friends at Kew who kindly supported us in many different ways by offering their expertise, technical support and time to help us answer difficult questions, by providing us with helpful comments and ideas concerning the manuscript, by giving us access to important material and supporting us with photographs, in particular, at the **SCD**: John Adams, Matthew Daws, Ilse Kranner, Hannelore Morales, Emma York. At the **Herbarium**, WS would like to extend special thanks to Gwilym Lewis from the Legume Section, who never tired of answering questions and kindly allowed us to draw on his slide collection of Leguminosae. Also at the **Herbarium** our thanks to: Bill Baker, Gill Challen, Martin Cheek, Tom Cope, Aaron Davis, John Dransfield, David Goyder, Yvette Harvey, Petra Hoffmann, Terry Pennington, Brian Schrire, David Simpson, Tim Utteridge, Sue Zmarzty. Anne Griffin from the Kew **Library** deserves a very special thank you for the speed with which she tracked down the many books and scientific papers needed for our research. Also at the Library, our thanks to Julia Buckley and Anne Marshall. At **HPE**: David Cooke, Laura Giuffrida, Mike Marsh, Wesley Shaw. Also at Kew: Andrew McRobb and Paul Little (Media Resources), Mark Nesbitt (Centre for Economic Botany), Ian Parkinson (Wakehurst Place).

At the Natural History Museum, London, we thank Christopher Lyal for help with the identification of the Malagasy snout beetle.

Outside the United Kingdom we thank Reto Nyffeler, Institut für Systematische Botanik und Botanischer Garten der Universität Zürich, Switzerland, for providing us with material of *Wolffia columbiana* and Jean-Yves Rasplus, INRA – Centre de Biologie et de Gestion des Populations, Montferrier-sur-Lez, France, for providing us with specimens of *Blastophaga psenes*. For assistance in the field WS would like to thank Sarah Ashmore, Phillip Boyle, Richard Johnstone, Andrew Crawford, Andrew Orme, Andrew Pritchard and Tony Tyson-Donnelly in **Australia**; and Ismael Calzada and Ulises Guzmán in **Mexico**.

In **South Africa**, WS would like to thank Ernst van Jaarsveld and Anthony Hitchcock (Kirstenbosch Botanical Garden, Cape Town), and Johan Hurter (Lowveld National Botanical Garden, Nelspruit) for their time, hospitality, and permission to photograph plants in their collections; in **Australia**, the staff of Kings Park and Botanic Garden, Perth; Geelong Botanic Gardens; Brisbane Botanic Gardens at Mt Coot-tha; the Royal Botanic Gardens, Melbourne; the Royal Botanic Gardens, Sydney; and Mount Annan Botanic Garden, New South Wales for their hospitality and permission to photograph plants in their collections. In **New Zealand** WS would like to thank his friend and colleague Trevor James for his hospitality and company in the field while visiting his country and for photographing the fruits of the titoki tree (*Alectryon excelsus*, Sapindaceae) specifically for this book. Also in New Zealand, thanks to Jane Marshall and Phil Knightbridge (Department of Conservation, Hokitika) for sharing their ideas on dispersal ecology.

Other colleagues and friends who kindly supported us with images include Lucy Commander (Perth, Australia), Phil Knightbridge, Stephen Lyle (BBC Natural History Unit, Bristol), Andrew McRobb (Kew), Filipe de Oliveira (UK), Elly Vaes (Hawai'i) and James Wood (Hobart, Tasmania). We are particularly grateful to Peter Trusler (Australia) for permission to reproduce his wonderful painting of *Dromornis*.

At Central Saint Martins College of Art & Design we thank Jane Rapley OBE (Head of College), Jonathan Barratt (Dean of Graphic and Industrial Design), Kathryn Hearn (Course Director, Ceramic Design).

At Papadakis, our thanks to Hayley Williams for her painstaking work on the production of this book, and to Mike Bailey and Steve Williams for their photographs that were specially for this book.

As always looking after the rest of my life, Agalis Manessi. RK

Finally, my warmest thanks to my wife, Emma Lochner-Stuppy, for her love and support, always patient, always encouraging, never complaining, despite her husband's year-long mental and physical absence while working on this book. WS